U0315393

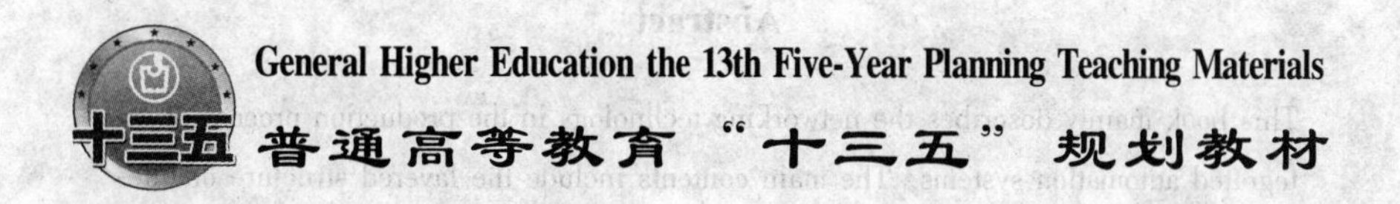

Control Network Technology

控制网络技术

李擎　刘艳　李江昀　张维存　编著

Beijing
Metallurgical Industry Press
2019

Abstract

This book mainly describes the networking technology in the production process of integrated automation systems. The main contents include the layered structure of integrated automation systems and an overview of control networks, the foundation of data communication and computer networks, concepts and functions of the OSI reference model, and information on a typical control network named Profinet. As an entity network standard, Profinet focuses on application and practice, and integrates network equipment and its use.

It is suitable as a textbook for relevant courses in colleges and universities and as a reference book for automation engineers and technicians in the field.

图书在版编目(CIP)数据

控制网络技术=Control Network Technology：英文/李擎等编著．—北京：冶金工业出版社，2019. 3

普通高等教育“十三五”规划教材

ISBN 978-7-5024-7957-2

Ⅰ.①控…　Ⅱ.①李…　Ⅲ.①工业控制系统—高等学校—教材—英文　Ⅳ.①TB4

中国版本图书馆 CIP 数据核字（2018）第 300868 号

出 版 人　谭学余

地　　址　北京市东城区嵩祝院北巷 39 号　邮编　100009　电话　(010)64027926

网　　址　www.cnmip.com.cn　电子信箱　yjcbs@ cnmip. com. cn

责任编辑　戈　兰　美术编辑　彭子赫　版式设计　孙跃红

责任校对　王永欣　责任印制　李玉山

ISBN 978-7-5024-7957-2

冶金工业出版社出版发行；各地新华书店经销；三河市双峰印刷装订有限公司印刷

2019 年 3 月第 1 版，2019 年 3 月第 1 次印刷

787mm×1092mm　1/16；18. 5 印张；470 千字；284 页

52. 00 元

冶金工业出版社　投稿电话　(010)64027932　投稿信箱　tougao@cnmip. com. cn

冶金工业出版社营销中心　电话　(010)64044283　传真　(010)64027893

冶金工业出版社天猫旗舰店　yjgycbs. tmall. com

Preface

The content of "Control Network Technology" can be summarized as "Computer Networks + Fieldbus". The background of writing this book is threefold: First, the background of the era in which the automation industry is located. Since human society entered into the twenty-first century, a new chapter has been launched in the information revolution based on computers and the Internet. The emergence of new technologies and concepts such as the Internet of things, cloud computing and Internet + has made the top-level design of the information revolution more clear and concrete. All aspects of human society are permeated by these elements, including the fields and industries radiated by the automation industry. The second is the technology reserve accumulated by advanced control systems. The four levels of integrated automation in modern industrial enterprises, namely, basic automation, process control system, manufacturing execution system and enterprise resource planning, have gradually changed from conception to reality. All of these four levels rely on network connection: at the bottom is fieldbus, and at the top is computer networks. Thirdly, the future expectation of upgrading the control field to intelligent control in an all-round way. With the introduction of German Industry 4.0 and Chinese Manufacturing 2025, the networking of control systems will be deeper and more popular. In such an era and professional background, automation engineering practice for professionals requires not only familiarity with fieldbus, but also sufficient knowledge of computer networks.

This book combines the technology of "computer network" and "fieldbus", deletes any content on computer networks which has little relevance to the practice of automation engineering, and integrates the knowledge of the two parts organically to make it more complete and systematic. It is suitable as a textbook for relevant courses in colleges and universities and as a reference book for automation engineers and technicians in the field.

The book is divided into eight chapters. Chapter 1 is an introduction which mainly describes the composition of enterprise integrated production systems, computer networks, and fieldbus. Chapter 2 covers data communication and introduces basic theories of data communication, such as signal transmission modes, channels, data coding, transmission media, etc. Chapter 3 introduces major concepts related to computer networks, including computer network topology, computer network hardware and software, the Internet composition and computer network performance indexes. Chapters 4, 5, 6 and 7 describe the data link layer, media access control sublayer, network layer and transport layer of OSI reference model respectively. Chapter 8 is about Profinet, a typical integrated network of field bus and computer networks, and introduces industrial ethernet, Profinet, network equipment, VLAN, VPN and so on. Chapter 1 is written by Li Jiangyun; Chapters 2 and 3 by Li Qing and Liu Yan; Chapters 4 and 5 by Li Qing and Li Jiangyun; Chapters 6 and 7 by Liu Yan and Li Jiangyun; Chapter 8 by Liu Yan and Zhang Weicun. The compilation process was also supported by postgraduate students from the School of Automation & Electrical Engineering, University of Science and Technology Beijing, such as Zuo Lei, Geng Jiahui, Su Zhenfeng, Wang Yaoping and Li Rui. The responsible editors of this book have also paid a lot of hard work for its publication. On the occasion of the official publication of this book, I would like to express my heartfelt thanks to them.

This textbook has been included in the school-level planning textbook of University of Science and Technology Beijing. The compilation and publication of the textbook have been funded by the teaching material development funds of University of Science and Technology Beijing.

Since my work background as well as the authors' level of expertise is limited, shortcomings and mistakes are inevitable. I hereby urge readers to criticize and correct.

Editor

Beijing, December 2018

Contents

1 Introduction

Goal:

(1) Understand the concept of control network.

(2) Master enterprise integrated automation system.

(3) Master the concept of computer networks and fieldbus.

1.1 Enterprise integrated automation system

In modern enterprises, computers have taken on more and more tasks in automatic control, office automation, business management, and marketing. Enterprise networks will be an important infrastructure for connecting the various workshops and departments, as well as exchanging information with the outside world. In the market economy and the information society, the networks play an important role in the comprehensive competitiveness of enterprises.

Fieldbus is a popular topic in the field of industrial control. The control network discussed in this book is a broader concept and technology comprising fieldbus. The control network generally refers to a computer network system that is characterized by the control of objects. Control network originates from computer network technology and has much in common with general information networks, but there are also differences. For instance, in an enterprise automation system, a single user is dispersed into a system by means of control network, usually with broadcast or multicast as the communication method. However, in information network, an autonomous system generally adopts a one-to-one communication with another autonomous system.

1.1.1 A layered architecture

With the development of enterprise integrated automation systems, it is necessary to closely link business decision-making, management, planning, scheduling, process optimization, fault diagnosis and field control. According to market demand, we should conduct comprehensive information processing to produce new products in the shortest time and with lower resource energy consumption. In addition, all levels of computers such as automatic control, office automation, business management, and marketing must be interconnected into network to achieve information communication and data sharing. Therefore, the information in the enterprise network is multi-layered.

Figure 1-1 is a schematic diagram of an enterprise network system based on a control network. The system is divided into the following four levels according to the functional structure: Enterprise resource planning (ERP), Manufacturing execution system (MES), Process control system (PCS) and Field control system (FCS). A complete enterprise network system is formed through network connection and information exchange between layers.

The early structure of the enterprise network system is complex and has many functional levels, including control, monitoring, scheduling, planning, management, and business decision-making. With the development and popularization of Internet technology, the structural level of enterprise network systems tends to be flat. At the same time, the division of functional levels is also more simplified. The lower layer is the FCS layer where the fieldbus is located. The top layer is the ERP layer. The information from the industrial site is controlled, optimally calculated and displayed in the PCS layer. The part of the multiple control management functions is interspersed in the middle of MES layer.

ERP and EMS functional layers mostly use Ethernet technology to form an information network. Therefore, the network integration between them and, the information interaction with the external Internet are better solved, and its information integration is relatively easy.

In Figure 1-1, the fieldbus network segments (H1, PROFIBUS, and LonWorks) are connected to the factory field devices. The foundation of an enterprise network is the FCS layer. At present, there are many types of control networks used in the FCS layer. The communication consistency within this layer network is very poor, except Distributed control system (DCS), Programmable logic controller (PLC), Supervisory control and data acquisition (SCADA), etc. The control network is quite different from the data network in several aspects, such as the communication protocol and the network node type. It is difficult to exchange information between control networks and between control networks and external internet networks. There are many obstacles to achieving interconnection and interoperability. Therefore, it is necessary to improve the data integration and exchange capabilities of the control network from the aspects of communication consistency and data exchange technology.

Figure 1-2 is the functional model and the hierarchical structure of an enterprise information system. The top layer is the decision-making layer; the bottom layer is the field control layer; and the bottom-up middle layers are the monitoring optimization layer, the scheduling layer, the planning layer, the management layer and the decision-making layer. Each functional module runs under the support of computer network and database. According to the internal and external information of the enterprise, the enterprise decision-making system provides decision support for the medium-term/

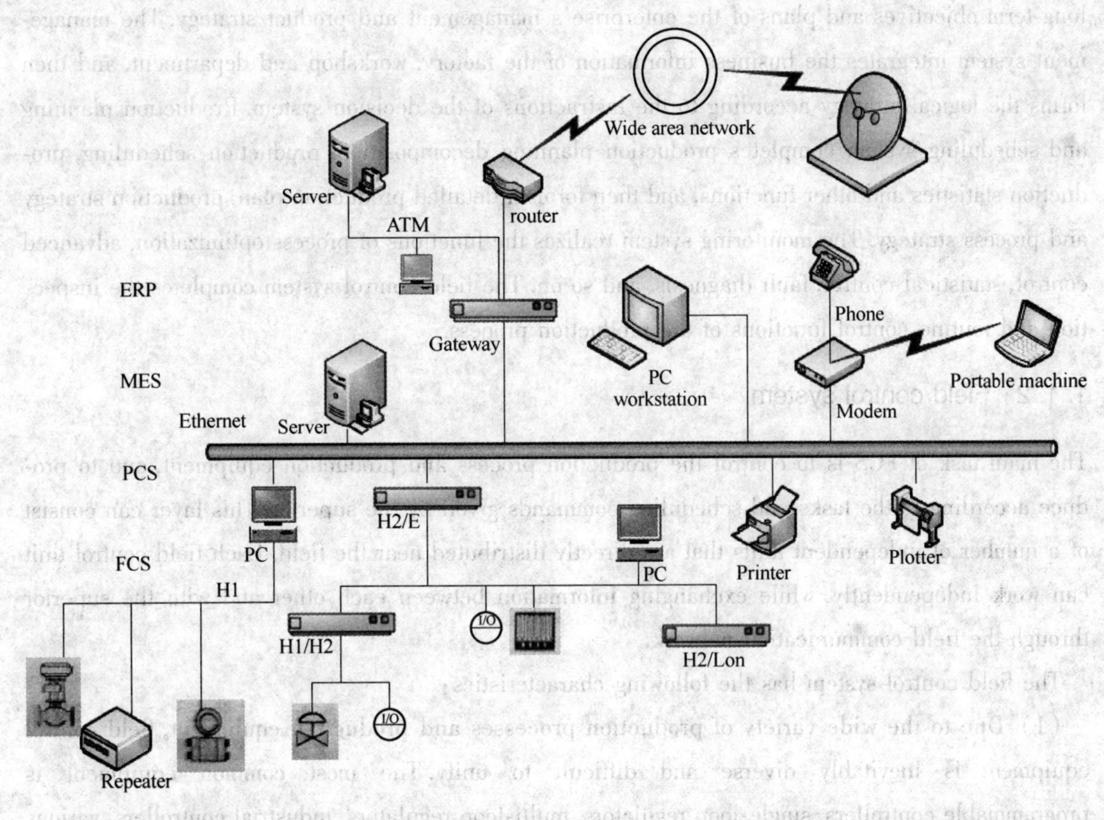

Figure 1-1 A schematic diagram of an enterprise network system based on a control network

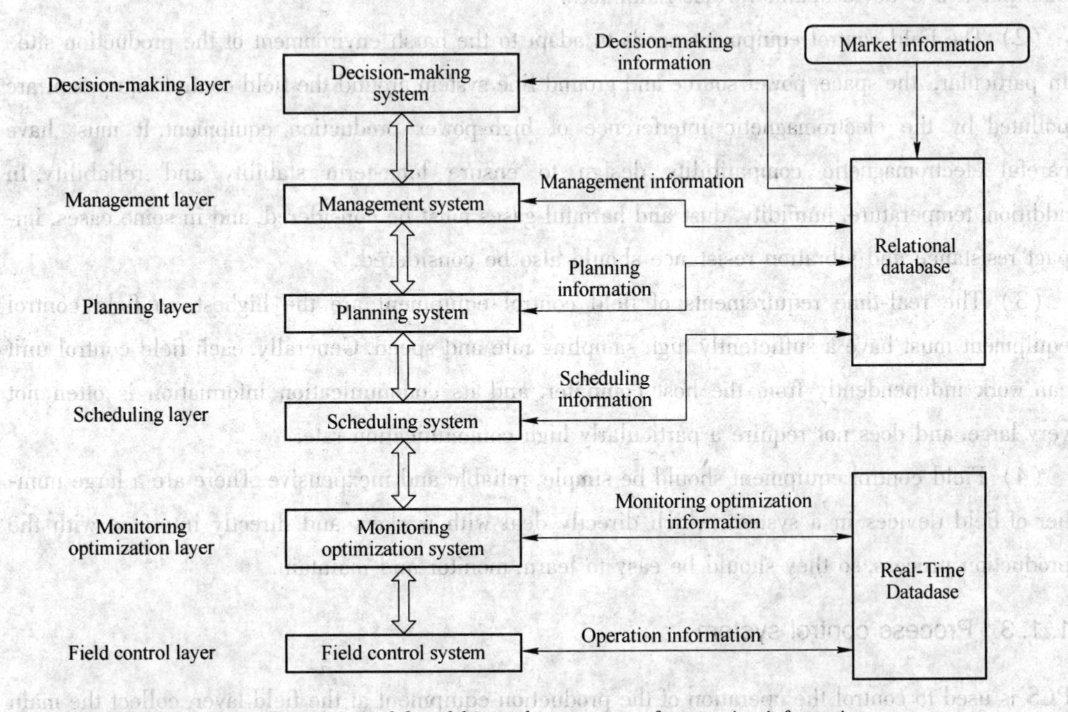

Figure 1-2 Function model and hierarchy structure of enterprise information system

long-term objectives and plans of the enterprise's management and product strategy. The management system integrates the business information of the factory, workshop and department, and then forms the logical strategy according to the instructions of the decision system. Production planning and scheduling system completes production planning decomposition, production scheduling, production statistics and other functions, and then forms a detailed production plan, production strategy and process strategy. The monitoring system realizes the functions of process optimization, advanced control, statistical control, fault diagnosis, and so on. The field control system completes the inspection and routine control functions of the production process.

1.1.2 Field control system

The main task of FCS is to control the production process and production equipment, and to produce according to the tasks and scheduling commands given by the superior. This layer can consist of a number of independent units that are directly distributed near the field. Each field control unit can work independently, while exchanging information between each other and with the superior through the field communication network.

The field control system has the following characteristics:

(1) Due to the wide variety of production processes and production equipment, field control equipment is inevitably diverse and difficult to unify. The most common equipment is programmable controllers, single-loop regulators, multi-loop regulators, industrial controllers, various distributed I/O devices, and various fieldbuses.

(2) The field control equipment needs to adapt to the harsh environment of the production site. In particular, the space, power source and ground line system around the field control equipment are polluted by the electromagnetic interference of high-power production equipment. It must have careful electromagnetic compatibility design to ensure long-term stability and reliability. In addition, temperature, humidity, dust and harmful gases must be considered, and in some cases, impact resistance and vibration resistance should also be considered.

(3) The real-time requirements of field control equipment are the highest, so field control equipment must have a sufficiently high sampling rate and speed. Generally, each field control unit can work independently from the host computer, and its communication information is often not very large, and does not require a particularly high communication rate.

(4) Field control equipment should be simple, reliable and inexpensive. There are a large number of field devices in a system, which directly deal with workers and directly interface with the production process, so they should be easy to learn, monitor and maintain.

1.1.3 Process control system

PCS is used to control the operation of the production equipment at the field layer, collect the main parameters and status of the production process, and report anomalies. At the same time, the necessary commands are issued to the field control unit, such as setting the set value, so that the production equipment can work in coordination and ensure the best control of the process. The process control system also has to undertake part of the management work, such as forming daily reports of each workshop, displaying trend curves and statistical charts of key parameters, and tracking product quality, so that the management of the workers on the workshop is scientific. This layer includes multiple monitoring hosts. These hosts communicate with superiors and subordinates, and also communicate with each other over the network.

PCS has the following characteristics:

(1) The process control system is composed of some process monitoring stations. Each station governs a portion of the field control unit equipment. There is also a requirement to establish certain communication links between stations and stations, especially where fault-tolerant support is required. If there is a management level above, the process control station must also be able to communicate with it.

(2) Process control system includes many functions such as system functions, operator functions and engineer functions. System functions are related to both operators and engineers. Operator functions are used while operating the plant. Engineer functions are used during the setup, commissioning, maintenance and management of the system. The process monitoring station displays the on-site process and field device information to the operator in real time. It displays this information intuitively and vividly by integrating graphics, text, data, tables and colors into one screen. The displayed content is not only the window for monitoring, but also the basis for guiding the operation.

(3) The process control station has a rich configuration function. This is necessary for building, adding, subtracting and changing systems, and makes it easy for engineers and operators to define and modify various parameters of the system.

1.1.4 Manufacturing execution system

MES is a production information management system for the manufacturing floor of the manufacturing enterprise. MES provides many management modules such as manufacturing data management, planned scheduling management, production scheduling management, inventory management, quality management, human resource management, work center/equipment management, tool tooling management, procurement management, cost management, project board management, production process control, underlying data integration analysis and upper layer data integration decomposition.

Today MES can offer real-time applications. They generate current and even historical maps for production equipment and can thus be used as a basis for optimization processes. As early as the beginning of the 1980s, work started on methods of this kind which were then known as production data acquisition or machine data collection. But while the main emphasis in the past was on achieving improvements in machine utilization, today the concern is predominantly to obtain real-time mapping of the value stream.

MES monitors the entire production process, starting with raw materials entering the factory and ending up with products entering storage. It records the materials and equipment used in the production process, the data results of the product inspection, and personnel in each production process. The collection of the information can be analyzed by the MES system, and the production schedule, target achievement status, product quality status, utilization status of the produced people, machines and materials can be presented in real time through the system report, so that the entire production site is completely transparent. Whenever you are in the office, the management of the company can see the status of the production site clearly through the Internet. The boss at the headquarters can also manage information through MES. Customers who are far away from abroad can also look over progress and product quality of their order.

The benefits of the MES system for the factory are as follows:

(1) Optimizing the production management mode of the enterprise, strengthening process management and control, and achieving refined management objectives.

(2) Strengthening the collaborative office capacity of each production department, improving work efficiency and reducing production costs.

(3) Improving the timeliness and accuracy of statistical analysis of production data, avoiding human interference and promoting the standardization of enterprise management.

(4) Providing effective and standardized management support for quality inspection of products, intermediate products and raw materials.

(5) Real-time control of information such as planning, scheduling, quality, process, and equipment operation, so that relevant departments can find problems and solve problems timely.

(6) The MES system can be used to establish a standardized production management information platform, so that the information between the internal control layer and the management layer can be interconnected, and then the core competitiveness of the enterprise is improved.

MES optimizes the management of the manufacturing process through feedback. It is mainly responsible for workshop production management and scheduling execution. A well-designed MES system can integrate management functions such as production scheduling, product tracking, quality control, equipment failure analysis, and network reporting on a unified platform. The system uses a unified database and network connection to provide workshop management information services for production departments, quality inspection departments, technology departments, logistics departments, etc. The system helps companies implement complete closed-loop production by emphasizing the overall optimization of the manufacturing process.

1.1.5 Enterprise resource planning

ERP is information systems that integrate processes in an organization using a common database and shared reporting tools. In other works, an ERP system helps the different parts of the organization share data and knowledge, reduce costs, and improve management of business processes. It integrates information technology and advanced management ideas, and provides decision-making means for employees and decision-making layers with systematic management ideas. It is a new generation of integrated management information system developed from Material Requirements Planning (MRP), which expands the functions of MRP, and its core idea is supply chain management. It goes beyond the traditional enterprise boundary, optimizes the resources of the enterprise from the scope of the supply chain and the operation mode of modern enterprises, and reflects the market requirements for the rational allocation of resources. It has a significant effect on improving business processes and the core competitiveness of enterprises.

ERP was introduced in the 1990s by an IT company in the United States based on the needs of computer information and enterprise supply chain management at the time, predicting the development trend of enterprise management information systems in the information era. ERP is an enterprise management software that integrates material resource management, human resource management, financial resource management, and information resource management integration. In addition to the existing standard features, it includes other features such as quality, process operation management, and adjustment reports.

ERP has the following characteristics:

(1) Practicality: It is more important to realize the "management tools" in the reality for ERP. ERP's main purpose is to balance and optimize the management of the comprehensive resources of human, finances, materials, information and time, so ERP is not only a software but also a management tool. It is a fusion of IT technology and management ideas.

(2) Integration: The most obvious feature is the integration of the entire enterprise information, which is more functional than the traditional single system.

(3) Flexibility: ERP adopts a modular design method so that it can be integrated according to the enterprise needs to increase the flexibility.

(4) Storage: Integrate data from all corners of the original enterprise to make data consistent and improve its accuracy.

(5) Convenience: In an integrated environment, the information generated within the enterprise can be obtained and applied anywhere in the enterprise.

(6) Real-time: ERP is the integrated management of the entire enterprise information, and the key to integrity is reflected in real-time and dynamic management.

(7) Timeliness: ERP management makes the actual work digital. Because workers' energy and ability are limited, it is necessary to have reliable information management tools to digitalize the work content and working methods.

1.2 Overview of computer networks

In each of the past three centuries there was a dominant technology. In the 18th century, with the advent of the industrial revolution, the era of large mechanical systems came. The 19th century is the era of steam engines. In the course of the 20th century, the key technology was the collection, processing and distribution of information. On the other side, we can also see that telephone networks around the world have been established; radio broadcasting and television have emerged; and the computer industry was born and is growing at an unimaginable speed.

Computer network is the combination of computer technology and communication technology. In the process of computer application, we need to collect, exchange, process, handle and transmit a lot of complex information, so the communication technology was introduced. Data communication is the basis of the realization of computer network. The application of computers is inseparable from the support of communication network environment. The widespread application of computer system has promoted the continuous development and renewal of the new technology of communication network. They complement each other, penetrate each other and promote each other.

1.2.1 Terminal-oriented computer network

In the early 50s, the way of batch processing had nothing to do with computers and communication. In 1946, the world's first computer named ENIAC(as shown in Figure 1-3 and Figure 1-4) was born at the University of Pennsylvania. In the next 10 years, computers had nothing to do with remote communication. Users must use computers in computer rooms to do scientific calculations at that time.

In the late 50s, the single machine system, which takes the host computer as the center, had the communication function. As shown in Figure 1-5, this is a remote centralized information processing system for computer hosts supporting multi-user terminals, so it reduced the time of remote users' journey. However, since data processing and communication control are simultaneous, the host was easily overloaded. Besides, the utilization rate of the line is relatively low, especially when the terminal rate is lower.

In the early 60s, there appeared communication system using communication control processor and concentrator, as shown in Figure 1-6. The front-end processor completed all communication tasks to let the host process data. The concentrator was connected to the front-end processor by the high speed communication line and was connected to the concentrated low speed terminal through the low-speed line.

Figure 1-3 The worlds first computer-ENIAC

Figure 1-4 Programmers are programming ENIAC

1.2.2 Packet switching network

In 1968, US Department of Defense Advanced Research Projects Agency developed ARPANET. At first, it aimed to connect computers of several universities, research institutes and companies to share resources. In the end, it became the core of the Internet. In November 1969, the experimental ARPANET was opened. In 1975, ARPANET had been connected to more than 100 hosts and finished the network experiment stage. It was transferred to the Defense Communica-

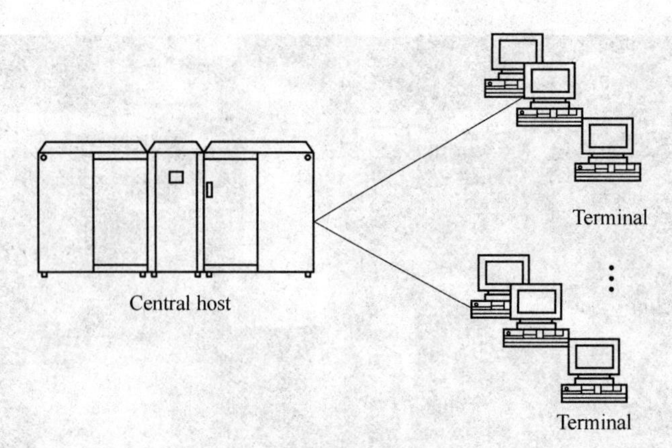

Figure 1-5 The single machine system

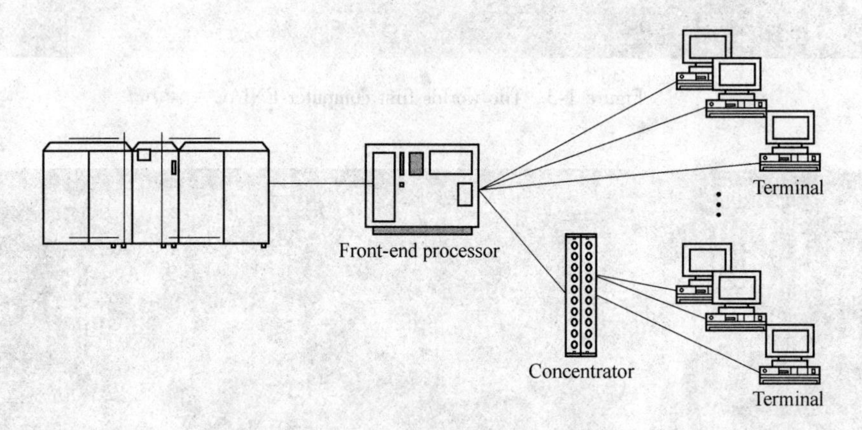

Figure 1-6 Communication system using communication control processor and concentrator

tions Bureau of the US Department of Defense. In January 1983, the conversion of ARPANET to TCP/IP ended. In the mid-1980s, ARPANET became the backbone of Internet. In 1990, ARPANET retired.

ARPANET is a milestone in the history of computer network development, marking the birth of modern computer network with the purpose of resource sharing. It proposed packet switching to implement data exchange, adopted the hierarchical network architecture model and put forward a two-level sub-network concept; involving communication subnet and resource subnet.

Packet switching is the technological foundation of modern computer networks. Circuit switching implemented the switching of lines among switches and established a dedicated communication line between two users. But there are some disadvantages of circuit switching. First, the computer's digital signal was burst and intermittent. Second, circuit switching has long access time, usually 10s to 20s. Third, it is difficult for circuit switching to adapt to communication between terminals and computers of different types, specifications and rates. Finally, the reliability is poor. Then, in August 1964, the concept of packet switching was put forward, which is more suitable for the exchange technology of computer communication. In December 1969, the ARPANET with packet switching technology (as shown in Figure 1-7) in the United States marked the beginning of the modern communication era.

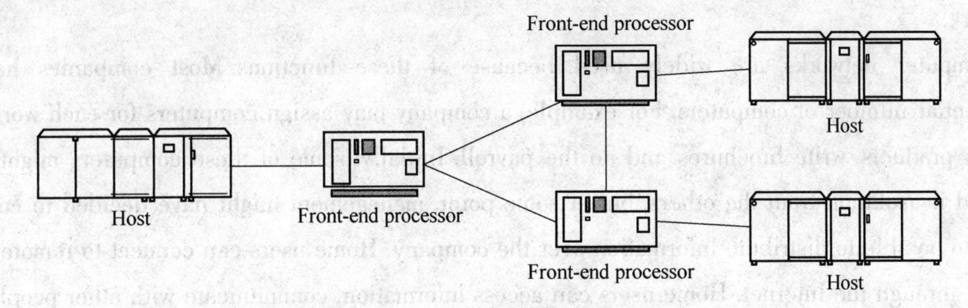

Figure 1-7 The packet switching network

As shown in Figure 1-8, host H_1 sends data to host H_6. Firstly, the data can be divided into a series of equal length packets, with additional information of destination and address. Then, the packet is sent to nodes connected to the H_1 in turn. Finally, when node A receives the packet, the received packet is stored in the buffer. Then according to the address information carried by the group, which node will be sent to the group is determined through a certain routing algorithm.

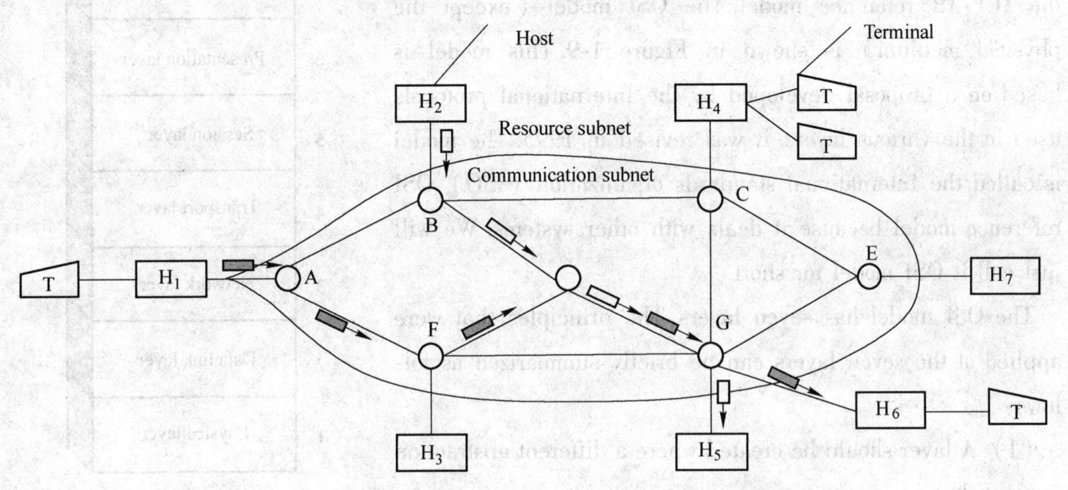

Figure 1-8 The host H_1 sends data to the host H_6

1.2.3 The function and application of computer network

The function of computer network mainly involves five aspects: data transmission, sharing of resources, the reliability and availability of computers, distributed processing and load balancing. Data transmission is the most basic function of computer networks. Sharing of resources is the most attractive function of computer networks. Computer networks have improved the reliability and availability of the computer. The reliability makes standby for each other possible. And the availability balances the burden of computers. When a node in a computer network is overloaded, it can assign a complex task to other node in the network, so as to utilize idle computer resources to improve the utilization of the whole system. Load balancing equates work tasks to all node in the network.

Computer networks are widely used because of these functions. Most companies have a substantial number of computers. For example, a company may assign computers for each worker to design products, write brochures and do the payroll. Initially, some of these computers might have worked in isolation from the others, but at some point, management might have decided to connect them to be able to distribute information over the company. Home users can connect to remote computers through the Internet. Home users can access information, communicate with other people and buy products and services with e-commerce as they are in company. There are many concrete application examples, such as online transaction processing, POS system, e-mail system and so on.

1.2.4 Open standardized network

We will discuss two important network architectures: the Open systems interconnection (OSI) reference model and the TCP/IP reference model. The OSI model (except the physical medium) is shown in Figure 1-9. This model is based on a proposal developed by the International protocols used in the various layers. It was revised in 1995. The model is called the International standards organization (ISO) OSI reference model because it deals with other systems. We will just call it OSI model for short.

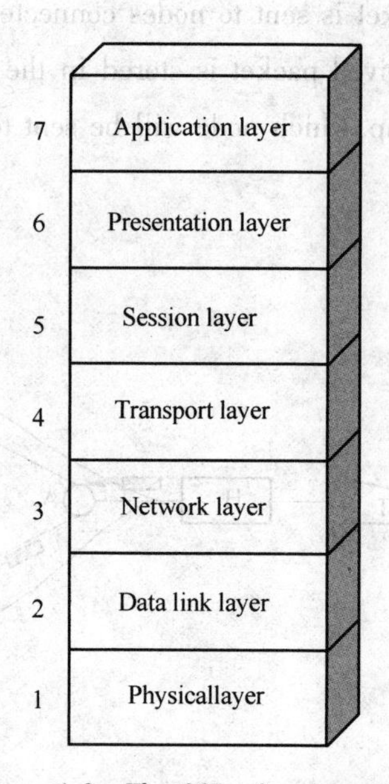

Figure 1-9 The OSI reference model

The OSI model has seven layers. The principles that were applied at the seven layers can be briefly summarized as follows:

(1) A layer should be created where a different abstraction is needed.

(2) Each layer should perform a well-defined function.

(3) The function of each layer should be chosen based on internationally standardized protocols.

(4) The layer boundaries should be chosen to minimize the information flow across interfaces.

(5) The number of layers should be large enough that distinct functions need not be thrown together in the same layer. It also should be small enough that architecture does not become unwieldy.

Let us now turn from the OSI reference model to the reference model used in the grandparent of all wide area computer networks, ARPANET, and its successor, the worldwide Internet. Although we will give a brief history of ARPANET later, it is useful to mention a few key aspects of it now. ARPANET was a research network sponsored by the U. S. Department of Defense. It eventually connected hundreds of universities and government installations, using leased telephone lines. After satellite and radio networks were added later, the existing protocols had trouble interworking with them, so a new reference architecture was needed. Thus, from nearly the beginning, one of the major design goals was to connect multiple networks in a seamless way. This architecture later became known as the TCP/IP reference model, after its two primary protocols. It was first described by Cerf and Kan, and later refined and defined as a standard in the Internet community. The design philosophy behind the model was discussed by Clark (1988).

The OSI and TCP/IP reference models have much in common. Both of them are based on the concept of a stack of independent protocols. Also, the function of the layers is roughly similar. For example, in both models the layers up through and including the transport layer are there to provide an end-to-end, network-independent transport service to processes wishing to communicate. These layers form the transport provider. Again in both models, the layers above transport are application-oriented users of the transport service. Despite these fundamental similarities, the two models also have many differences. Turning from philosophical matters to more specific ones, an obvious difference between the two models is the number of layers: the OSI model has seven layers and the TCP/IP model has four. Both have (inter) network, transport, and application layers, but the other layers are different. Another difference is in the area of connectionless/connection-oriented communication. The OSI model supports both connectionless and connection-oriented communication in network layers, but in the transport layer only connection-oriented communication is supported. Transport layer is important because the transport service is visible to the users. The TCP/IP model supports only one mode in the network layer (connectionless) but supports both in the transport layer. This gives the users a choice, which is especially important for simple request-response protocols.

1.3 Overview of industrial fieldbus

Fieldbus is one of the hotspots in the development of automation technology today, and is known as the computer local area network (LAN) in the field of automation.

Fieldbus is an industrial network system for real-time distributed control. Fieldbus is used to achieve the digital, serial communication in both directions between intelligent devices (such as transmitters, valve positioners and controllers) and high level devices (such as host computer, HMI device, or gateway). This means an intelligent field device can be connected to fieldbus to communicate with other devices or systems (Multi-point communication). FCS is a distributed control system which employs the fieldbus techniques. Figure 1-10 shows a fieldbus architecture.

The fieldbus control system consists of three parts: measurement system, control system and management system. The hardware and software of the communication part are its most distinctive parts. Figure 1-11 shows the fieldbus parts.

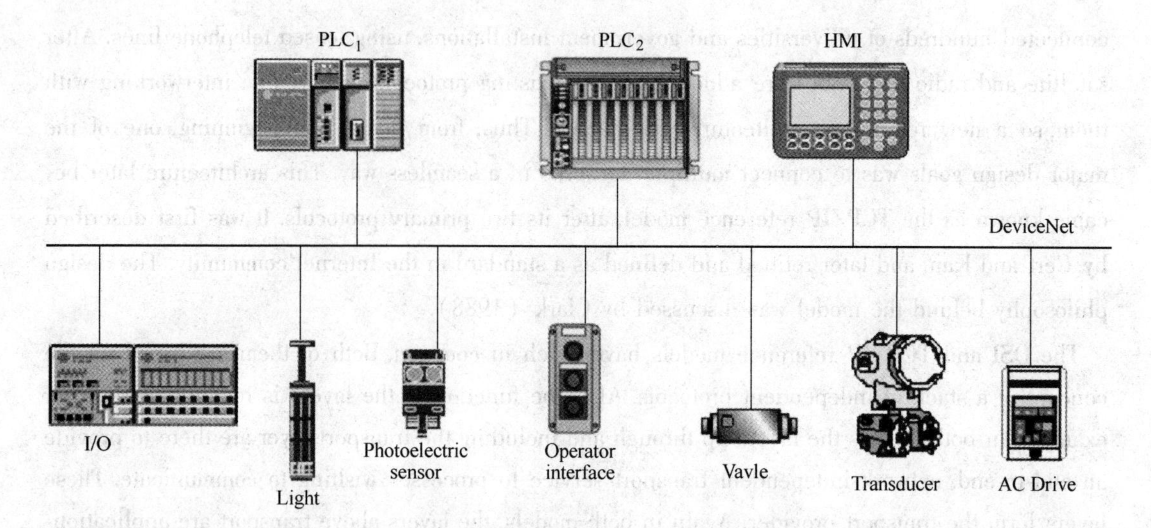

Figure 1-10 Fieldbus architecture

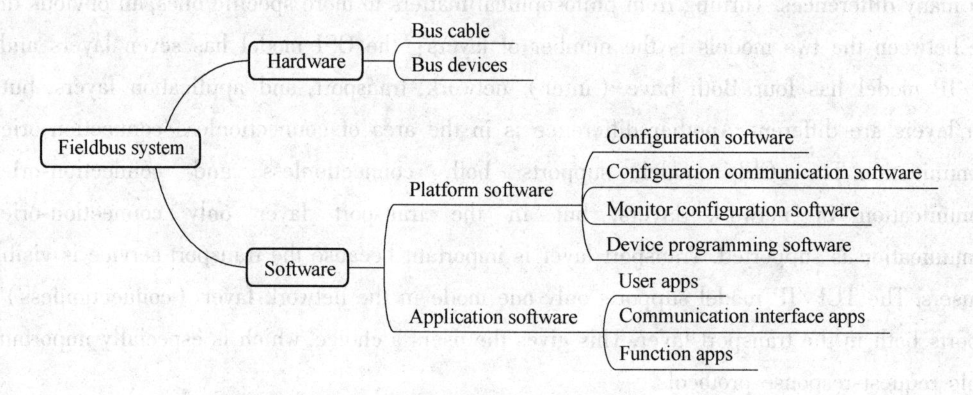

Figure 1-11 The fieldbus parts

Hardware includes bus cables and bus devices. Bus cables are the medium of information transmission, also known as communication media. And the device connected to the communication line is called bus device, also known as bus equipment, node (main node, slave node), site (main station, slave station).

Software is an important part of the fieldbus system. It includes system platform software and system application software. System platform software provides the environment, conditions or tools for the system construction, operation and application programming. It includes configuration tool software, configuration communication software, monitoring configuration software and device programming software. System application software is designed to implement the various functions of the system and equipment, including the system user program software, the device interface communication software and the device function software.

The platform software consists of configuration tool software, configuration communication software, monitoring configuration software and device programming software. Configuration tool software is used to exchange information with the configuration of a computer, network configuration and the field bus protocol/specification (Protocol/Specification) and configuration communication software, such as Siemens Step 7. Configuration communication software not only realizes the communication between the computer and the bus device, the bus device parameters, but also transfers the configuration of the bus device and the network configuration information to the bus device, such as SIEMENSSIMATIC. net. Monitoring configuration software, running on the monitoring computer, has the function of real-time display of the running state parameters of the field equipment, the fault alarm information, and the function of data recording, trend map analysis and report printing. Monitoring configuration software enables users to realize the monitoring function of the system through simple image configuration work. Monitoring configuration software is also called host monitoring configuration software, such as WinCC, Kingview. Device programming software is a platform that provides a programming environment of system application software. While the device is controller/PLC, the device programming software is the controller programming software, such as SIEMENS Step 7, GE ME.

Application software includes system user program software, device interface communication software and device function software. System user program software is a system-level user application written according to the system's process or function and other requirements. The program usually runs in a controller as a master station or in a computer. Device interface communication software, which is based on Fieldbus Protocol/specification, is used for communication between bus devices through bus cable. Device function software enables the bus device to realize its own function (excluding fieldbus communication part).

1. 3. 1 Control network and field bus

At present, the fieldbus control network has received widespread attention and developed rapidly. From a technical point of view, the access problems of the media access control layer and the device in the physical layer and the data link layer are better solved. Fieldbus technology is the combination of field control technology and modern electronic, computer and communication technologies. Influential fieldbuses include FF (foundation fieldbus), LonWorks, WorldFIP, PROFIBUS. CAN and HART.

The technical characteristics of the control network are as follows:

(1) High real-time and good time certainty are required.

(2) Most of the transmitted information is short frame information, and information exchange is frequent.

(3) Strong fault tolerance, reliability and security.

(4) Control network protocol is simple and practical, and works efficiently.

(5) Control network structure is highly dispersive.

(6) Control equipment is intelligence and autonomy of control functions.

(7) Efficient communication and easy integration with the information network.

Fieldbus is the combination of today's 3C technology, namely communication, computer and control technology. It is the intersection of three major technological developments: process control technology, automation instrumentation technology and computer network technology. Compared with DCS and other traditional systems, fieldbus (system) has the following features:

Field communication network: Fieldbus extends the communication line (bus cable) to the industrial field (manufacturing or process area), or the bus cable is directly installed on the industrial site. It is completely suitable for industrial field environment, because it is designed for this purpose.

Digital communication network: The signal transmission between the transmitter, sensor, actuator and controller at the bottom of the field is all digital. All of the data transfer between the upper/middle controller, monitor/monitor computer and other devices use a digital signal. The information exchange between devices is digital.

Open interconnection network: Fieldbus standards, protocols/specifications are open, and all manufacturers must comply with them. The fieldbus network is open, which can realize interconnection between the same layer network and network interconnection of different layers, regardless of which manufacturer it is. Users can share network resources.

Device interconnection network: Only one communication line is directly connected to N field devices, thus forming the interconnection network of field devices.

A highly decentralized system of structure and function: The system structure of fieldbus is highly distributed. The functions of the fieldbus system are highly dispersed, and the field devices are made up of distributed functional modules. The field bus discarded the control station of DCS and its input/output unit. It fundamentally changed the distributed control system of DCS centralization and dispersion, and realized the complete decentralized control by distributing the control function to the field equipment.

Interoperability and interchangeability networks: The field devices of different manufacturers can be interconnected, and they can exchange information with each other and configure them uniformly. Different manufacturers' performance similar to field devices can be replaced with each other.

1.3.2 Industrial data communication

The general system is used for information transmission several devices or between devices and humans. Early communication systems date back to the ancient times of using the fire and smoke of the Beacon Tower. The invention and use of electricity provides an effective tool for the development of communication systems. Modern communication systems typically use electronic or electrical equipment to communicate information between two or more points. Data communication refers to the process of exchanging information in binary form between two or more points. With the development of computer technology in recent years, especially the emergence of the Internet, computer data communication and network technology based on data exchange has developed rapidly. Therefore, data communication generally refers to communication in computer applications, and is used to transfer various files between one computer and another, and between one computer and peripherals such a printer. In addition to computers and their peripheral devices in the industrial process, there are a large number of transmitters. These transmitters detect the value and status of process parameters that control the production process. Industrial data communication refers to the process of transferring data. The process use data transmission techniques to transfer information between functional units of devices, between devices and devices, and between computers and computers. It has a big difference and a close relationship with ordinary computer communication, telegraph and traffic communication, so it can be considered a communication technology in the field of automation.

The direct data transfer between two points by means of a transmission medium is the simplest form of data communication. The traditional measurement control system, from input device to controller, from controller to output device, uses one-to-one parallel connection between devices, that is, point-to-point wiring information transmission. The signal is transmitted using analog signals such as voltage and current. If serial communication of digital signals is to be realized between multiple points, it is unnecessary to establish a direct line between each communication node, but a data channel is constructed by using a network connection form to generate a data communication network.

A large number of data communication systems for nodes generally use serial data communication. The biggest advantage of serial data communication is that it is economical. There are dozens, hundreds or even more sensors and actuators attached to the two wires, which have the advantages of simple installation and convenient communication. The two wires that implement serial data communication are called buses. In addition to transmitting the value of the measurement control, the bus can also transmit device status, parameter adjustment and fault diagnosis information.

The size of industrial data communication systems ranges from simple to complex, from two or three data nodes to thousands of devices. An automobile assembly line may have as many as 250000 I/O points, and there are thousands of measurement control devices in a common device in the petroleum refining process. The scale of communication and network systems made up of them is considerable.

In the field of industrial data communication, people usually divide the data transmission bus according to the length of the communication frame: sensor bus, device bus and field bus. The sensor bus belongs to the data bit level bus, and its communication frame has only a few or a dozen data bits, such as an ASI (actuator sensor interface) bus. The device bus belongs to a byte-level bus, and the communication length is generally several to tens of bytes, such as a CAN (controller area network) bus. Fieldbus is a block-level bus. It can transmit data blocks up to several hundred bytes in length. When the data block to be transmitted is longer, it can support packet transmission. However, the length of data frames directly transmitted in the fieldbus is generally only a few or tens of bytes. For example, Foundation Fieldbus, ControlNet, PROFIBUS, etc. are typical fieldbuses. However, in many applications, people are accustomed to referring to these buses of varying lengths as fieldbuses. This may be because they are all located at the production site.

It is worth noting that in the data frames directly related to the control role in industrial data communication, the real-time requirements for communication transmission cannot be ignored.

1. 3. 3 Early bus technology

Early fieldbuses, also known as industrial telephone lines, were used to transfer information between industrial measurement control devices. This pair of wires connects sensors, buttons and actuators between each other and connects them to some type of controller, such as a programmable controller (PLC) or an industrial computer. Then industrial control tasks are performed through them. The input of the system includes the position and value of the button, sensor, and contactor. The output includes the drive signal, contactor, switch, and so on.

Around the 1970s, at the same time as programmable controllers appeared, Culter-Hammer introduced a product called Directrol, the first device-level fieldbus system. It is technically ahead of its time, while programmable controllers and PCs did not have enough capabilities to match Directrol. In the early 1980s, it quietly disappeared.

Several device buses introduced in the mid-1980s are still in use today. They are General Electric's Genius I/O, Phoenix's Interbus-S, Turck's Sensoplex, and Process Data's P-Net, which were all influential bus technologies at the time.

Genius I/O is a generic type which is capable of handling longer data. It transfers data information from one PLC to another PLC and from PLC to human-machine interface, and manages I/O ports. Genius I/O belongs to the token bus, and the total number of network segment nodes can reach 32; the branch topology is limited; the transmission distance is 3500ft (1ft = 0. 3048m, transmission rate 153. 6 kbaud) to 7500 ft (transmission rate 38. 4 kbaud). The transmission medium is shielded twisted pair and optical cables are supported too; the transmission signal is frequency shift keying (FSK) modulated signal; the input of each node is 1024 bits and the output of the node is also 1024 bits; the transmission speed is 38. 4 kbaud to 153. 6 kbaud, with error detection and error correction.

Interbus-S is a unique fieldbus which is fast, accurate and efficient. It is designed for single-host systems. Interbus-S actually consists of two buses: the remote bus and the local bus. The remote node uses the same chip as the local node but performs different functions, so the two nodes are not interchangeable. Interbus-S is physically a straight line and is transmitted as a data loop. Despite its high implementation cost, it is a good choice when you need a large system with fast speed and long distances. Its remote bus can carry up to 64 nodes with a coverage distance of up to

12.8km. The input data for each node is 16 bits and the output data is 16 bits. For the programmer, it is like a 16-bit shift register. In the data loop, each node receives the message from the network to get the information it wants, adds its own information, and then sends the message back to the network. The signal follows the EIA-485 standard and has a transmission rate of 500 kbps. Due to its low overhead and high ratio of valid bits in the data frame, this message shifting is one of the most efficient device buses available today. Today Interbus-S has evolved into a subset of the IEC fieldbus standard.

As an early special bus, Sensoplex was originally developed for Ford Motor Company in Cologne, Germany. This project required that the nodes and data lines be connected directly to the welding arms of the machine. Due to the short distance to the electromagnetic field, the surrounding low electromagnetic field affects not only the data, but also the input power of the node. Sensoplex has a strong ability to withstand noise and electromagnetic interference. It is installed in the harsh environment mentioned above and works well. Sensoplex uses a master-slave architecture, connected with a 759 coaxial cable, and uses FSK to modulate the signal to support bus power and intrinsic safety. The total number of nodes is 32in the first generation and 64 to 120in the second generation. It uses a bus topology with unrestricted branches. The distance between the master node and the farthest slave node is 200mn, which can be increased to 400m using a repeater. The transmission rate is 187 kbaud, the address setting uses DIP switch, the data input bit of each node is 8bits, and the data output bit is also 8bits. The second-generation Sensoplex product, introduced in 1992, has intrinsic safety features and is widely used in flammable and explosive environments, in addition to the most explosive acetylene and hydrogen environments.

P-Net is the control network technology introduced by Process Data in Denmark in 1983. It is mainly used in animal support systems, dairy products, breweries, agricultural environment control and other applications, connecting I/O nodes into network systems. PNet's system scale can range from simple systems with several I/O nodes to complex systems with thousands of I/O nodes. So far, many communication reference models have only adopted the physical layer, data link layer and application layer in the ISO/OSI reference model, while PNet uses many network layers and transport layers which are not available in bus technology. Therefore, a control network with a relatively complicated network system structure can be formed. Its electrical specifications are based on the FIA-485 standard. It uses shielded twisted pair and the transmission rate is 76.8 kbps. It has a maximum transmission distance of 1200m and can connect up to 125 devices per segment. In order to ensure the collection of real-time data, it is specified that the data is 56bits per frame. When the data is larger than 56bits, it will be automatically split into several consecutive frames for transmission. CRC cyclic redundancy detection and Hamming code correction are used for error checking. P-Net uses a multi-master network system with a segmented structure. Up to 32 master nodes can be connected in the network, and multiple slave nodes can be connected under the master node. Figure 1-12 shows the architecture of the P-Net multi-master network system.

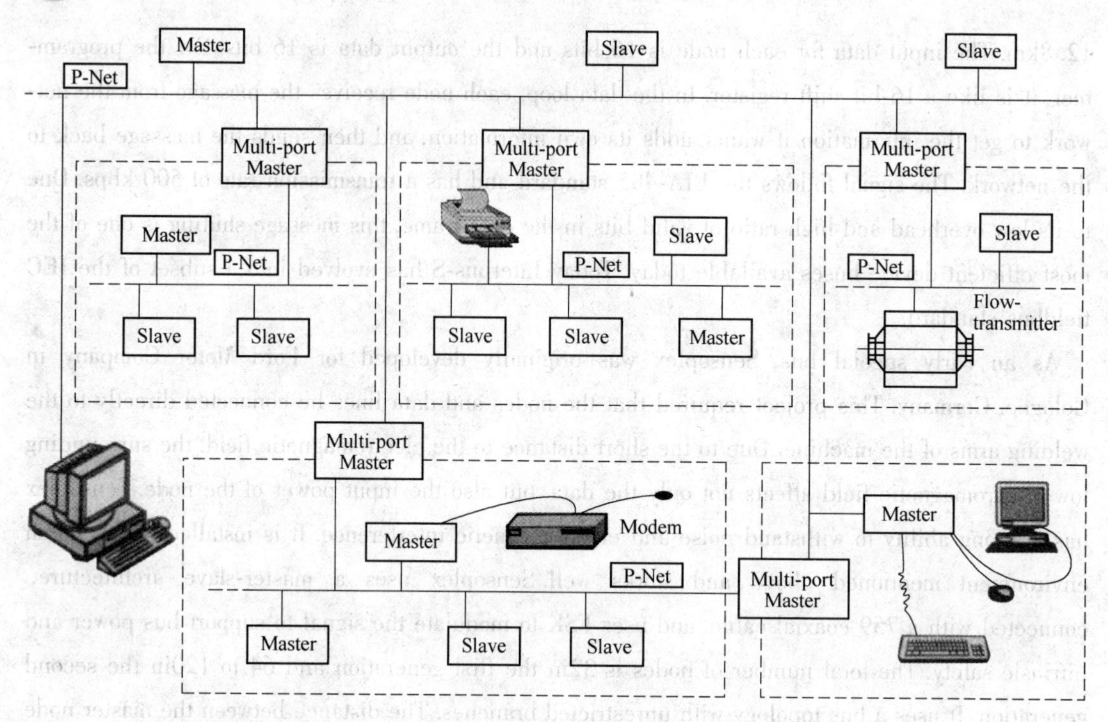

Figure 1-12 The architecture of the P-Net multi-master network system

The characteristics of industrial data communication and control networks can also be seen from these early bus technologies. The control network has distinctive features from voice communication and computer network in terms of network size, node type, length of communication frame, working environment, various communication technical parameters and problems to be solved. Most of the early buses used proprietary communication technology, which was self-contained and not open, and products from different manufacturers could not communicate with each other. The development of technology and user requirements has promoted the birth of open bus. "Open" means that communication protocols and specifications are open and unified, and products produced by different manufacturers can be interconnected and operated in the same communication network system.

The most typical open product paradigm in industrial data communication systems in the early days was the CAN bus chip. It has open and unified technical specifications. Many chip manufacturers, such as Motorola, Intel, Philips, Hitachi, are in production. The openness of CAN has indeed played a positive role in the development and promotion of CAN technology. Another feature of early bus technology is the short communication frame. It should be said that this is also one of the characteristics of industrial data communication. However, today's control network generally has longer communication frames and larger traffic than earlier bus communication. This also shows that the demand for information in the control network is increasing. The parameters of the production site are sent to the centralized control room through a unified analog signal and connected to the control panel.

1. 3. 4 New structure of control system

The emergence of the control network has led to the transformation of the traditional control system structure, and formed a new structure, the network integrated digging system, which is also called FCS. This is a new structural form of the control system after the base-type pneumatic instrument control system, the electric unit combined analog instrument control system, the centralized digital control system and the DCS.

Before the 1950s, due to the small scale of production at that time, the inspection and control instruments were still in the initial stage of development. Only pneumatic instruments of base-type were used. These instruments were installed at the production site and had simple measurement and control functions. These instruments generally cannot be transmitted to other instruments or systems while working, that is, each measurement and control point can only become a closed state, and were unable to communicate with the outside world. The operator can only know the status of the production process by visiting the production site.

With the expansion of the production scale, the operators need to grasp the operating parameters and information of multiple points comprehensively. It is necessary to carry out operation control according to the information of multiple points at the same time. Therefore, the unit combination meter of pneumatic and electric series appeared, so that the centralized control room appeared. Operators can sit in the control room and look at the conditions throughout the production process, and can combine the signals of each unit meter into different types of control systems.

Since the transmission of analog signals requires one-to-one physical connection, the signal changes slowly, the overhead and difficulty of improving the calculation speed and accuracy are large, and the anti-interference ability of signal transmission is also poor. Therefore, people began to seek to replace analog signals with digital signals. Thus, direct digital control has emerged. Since the technology of digital computers was not developed at the time, the computer was expensive. People tried to replace as many control room meters as possible with one computer, so a centralized digital control system appeared. However, the reliability of the computer was still poor at that time. Once a certain fault occurred in the computer, it would cause all relevant control loops to be dysfunctional, and the production stoppage was serious. This dangerous centralized system structure was difficult for enterprises to accept.

With the improvement of computer reliability and the sharp drop in price, digital regulator, PLC, and DCS have emerged. This is the DCS which is adopted by many companies today. In the DCS system, the measurement transmission instrument is generally an analog instrument, so it is an analog digital hybrid system. This kind of system has made great progress in function and performance compared with analog instrument and centralized digital control system. On this basis, it can realize optimal control at the device level and workshop level. However, in the process of forming the DCS system, due to the defects of the system closure that existed in the early stage of the computer system, the products of the various manufacturers are self-contained systems, and the equipment of different manufacturers cannot be interconnected, so it is difficult to achieve interchange and interoperability. There are many difficulties in the network systems that make up a wider range of information sharing.

The control network breaks through the defects caused by the closure of the private network in the DCS system. It adopts an open and standardized solution to connect automation devices from different manufacturers, comply with the same protocol specification into a control network and combine them into various control systems. In this way, it achieves a variety of functions of integrated automation. This is the new network integrated control system. It changes the analog and digital signal mixing in the DCS system. The signal transmission of a simple control system needs to go through the process from the field to the control room, and then from the control room to the field. Since the field instrument as a node in the control network has communication and digital computing capabilities, it can support multiple control functions by relying on data communication between several control devices at the production site. This is completely produced under the support of the control network. The structure of the control system formed on site is also known as the fully distributed control system.

It should be pointed out that in the era of rapid development, the structural form of the control system is not static. The emergence and development of fieldbus technology has had a strong impact on DCS systems. Some DCS manufacturers are changing the structure of DCS to make it move in the direction of openness and compatibility.

1.3.5 The advantage of the control system to the network

The field adaptability, digitization, openness, dispersion, equipment connectivity, interoperability and interchangeability of the fieldbus determine and derive a series of advantages:

Large reduction in wire and attachment accessories: A bus cable is directly connected to equipment of N stations, so the number of cables is greatly reduced (a bus cable on the field bus replaces the original hundreds or even thousands of signals and control cables of DCS). The number of connection accessories, such as terminals, slot boxes, bridge frames, wiring boards and so on, is greatly reduced.

Instruments and input/output converters are reduced significantly: Using human-machine interface, display devices and monitoring computers instead of display instruments greatly reduced the number of instruments. The number of input/output converters (cards) is greatly reduced. The (4~20) mA line used in DCS can only obtain one measurement parameter, which is directly connected to the input/output unit in the control station, so the number of input/output units is higher. In a fieldbus system, a field device can measure a number of parameters and transmit them in the form of a required digital signal through a bus cable, so the need of single input/output converters (cards) is lower.

The cost of design, installation and commissioning is greatly reduced: Schematic design is simple and feasible in fieldbus, which is tedious in DCS. The workload of installation and proofreading is greatly reduced. In order to make debugging work flexible and convenient, the system is divided into several parts and debug separately. Powerful fault diagnosis function makes debugging easy and enjoyable.

The cost of maintenance is drastically reduced: The high reliability of the system greatly reduces the failure probability of the system . The powerful fault diagnosis function makes the early detection, location and elimination of the fault fast and effective. The system can run for a long time and the maintenance time is greatly reduced.

System reliability is improved: The high dispersion of system structure and function determines the high reliability of the system. The high reliability of communication is guaranteed by the strict regulations of fieldbus protocol/specification.

System measurement and control precision are improved: In field bus, all kinds of switch signal and analog signal are changed into digital signals. All bus devices communicate with digital signals to avoid the attenuation and deformation of the signal and reduce transmission error. In other words, fieldbus adopts the digitalized characteristics of digital signal communication, which fundamentally improves the measurement and control accuracy of the system.

The system has excellent remote monitoring functions: It can monitor the running status of field devices and systems remotely in the control room. Remote control of field equipment and system can be carried out in the control room.

The system has powerful (remote) fault diagnosis function: All kinds of faults can be diagnosed and displayed, such as open circuit and short circuit faults of bus devices and connectors, communication failures and power failures. It can transmit all kinds of states and fault information to the monitor/monitor computer in the control room, which greatly reduces the unnecessary on-site inspection for users and maintenance personnel. This is especially important when fieldbus is installed in a harsh environment.

Device configuration, network configuration and system integration are convenient and free: Users can set parameters on the field devices through the same layer network or the upper network, instead of configuring each device one by one at the scene. With the help of the network configuration tool software, the field bus network can be built quickly and conveniently, and the network parameters can be configured. Because of the openness, interoperability and interchangeability of fieldbus, users can freely integrate different manufacturers, different brands of products and networks, and then form the required system.

Site replacement and system expansion are more convenient: The field equipment has interoperability and interchangeability. The damaged equipment can be replaced by any device with similar functions, so as to achieve plug and play. Field devices can be added without new cables, so they can be connected to the original cables nearby. The configuration time required for system expansion is greatly reduced.

It creates important conditions for the construction of enterprise information system: The enterprise information system is composed of field control layer, process monitoring layer and enterprise management layer. Fieldbus is the field control layer of enterprise information system, and constitutes the basic framework of enterprise information system. The field bus creates conditions for the various operating states and fault information of the bus equipment and the system, and the access of various control information to the public data network, so that more decision basis and powerful support are available for the managers. Managers can use the information to make all kinds of correct decisions, and then the enterprise can obtain a higher overall profit.

Compared with the traditional DCS scheme, the West Sak ARCO oil field in Alaska, USA, where the natural environment is very bad, uses a field bus to save the situation as follows:

(1) Reduce the connection terminal by 84%;

(2) Reduce the number of I/O cards by 93%;

(3) Reduce control instrument panel space by 70%;

(4) Reduce indoor wiring by 98%;

(5) The maintenance workload (50~80)% is reduced by remote diagnosis;

(6) The required configuration time for the expanded oil field is reduced by 90%;

(7) The cost of saving cable is 69%.

Problems

1-1 The definition of fieldbus.

1-2 The composition of enterprise network system and the role of each part.

2 Data Communication Foundation

Goal:

(1) Grasp the Fourier Analysis and Nyquist's theorem.

(2) Master the basic communication concepts.

(3) Master the data coding methods.

The way of human information transmission has gone through five stages. The first stage is mainly language, which transmits information through primitive means such as manpower, horsepower and bonfire. The second stage is mainly carried out by means of words and postal means, which increases the means of information dissemination. The third stage is to spread and share data through printing, and expand the scope of information dissemination. The fourth stage, telegraph, telephone and radio become the main transmission channels, representing the entry of human communication into the electrical stage. In the fifth stage human communication entered the information age marked by the birth of electronic computers. The transmission methods include language information, data, and image history.

Modern data communication began in the early 1950s and evolved with the advancement of computer remote processing applications. Data communication is a new communication method that combines communication technology and computer technology. According to different transmission medium, data communication is divided into wired data communication and wireless data communication. However, they all connect the data terminal to the computer through the transmission channel, so that the data terminals in different locations can share the software, hardware and information resources.

Most of the early telematics systems were centered on one or several computers and connected to a large number of remote terminals by means of data communication to form a centralized processing system for terminals. In the late 1960s, starting from the birth of the ARPA computer network in the United States, a heterogeneous computer communication network for resource sharing began to emerge, opening up a new field of computer technology—networking and distributed processing technology. After the 1970s, computer networks and distributed processing technologies developed rapidly, which also promoted the development of data communications. In 1976, CCITT officially announced the X. 25 recommendation, an important standard for packet-switched data networks, and then refined and modified it several times, laying the foundation for the technical development of public and private data networks. In the late 1970s, the International Organization for Standardization (ISO) proposed an Open Systems Interconnection (OSI) reference model to promote the interconnection of different types of systems. It was officially adopted in 1984 and became an international standard. Since then, the development of computer network technology and applications has been carried out in accordance with this model.

2. 1 Basic data communication theory

Information can be transmitted on the line by changes in certain physical characteristics, such as voltage or current. If we use a single-valued function $f(t)$ with time t as the independent variable to represent the value of voltage or current, we can model the behavior of the signal and mathematically analyze the signal. The subject of this section is the mathematical analysis of the signal.

2. 1. 1 Basic concepts

Data communication systems refer to systems that connect data terminals distributed in different locations to computer systems through data circuits to implement data transmission, exchange, storage, and processing.

A data communication system is a combination of the computer and communication. It originated in the early 1950s when the United States established a semi-automatic ground-based air defense (SAGE) system that connected long-range radar and other equipment to computers to create an early data communication system. Subsequently, data communication systems have developed rapidly, and have evolved from centralized systems to distributed systems to computer networks. The centralized data communication system is composed of a central computer and multiple remote data terminal devices, which are centrally processed by the central computer. The system has simple management and control procedures, but the economy and reliability are poor. With the decentralized setup of central computers and the advent of pre-processors, concentrators, and smart terminals, distributed data communication systems characterized by decentralized communication processing functions and data processing functions have emerged. In the late 1960s, a number of independent computer systems were interconnected to achieve a resource-sharing computer network. This is the prototype of our modern data communication system.

The purpose of communication is to exchange information. Information is the content and interpretation of data. The information carrier can be numbers, text, sound, graphics, images, and video. Data is the physical symbol or meaningful entity that carries the message. A signal is an electrical or electromagnetic representation of data and is a specific representation of the data during transmission.

From the form of the signal, the signal can be divided into two types: analog signal and digital signal.

(1) Analog signal (continuous signal): A value in which data continuously changes within a certain interval. For a current, voltage, or electromagnetic wave that changes continuously over time, one of its parameters (such as amplitude, frequency, phase, etc.) can be used to represent the data to be transmitted. In addition, temperature, pressure, sound, video, etc. are analog signals.

(2) Digital signal (discrete signal): The data takes discrete values within a certain interval. Such as text information, numbers, integers, binary sequences, and so on. When a waveform using a time domain (or simply a time domain) represents a digital signal, the basic waveform representing different discrete values is called a symbol: when binary encoding is used, there are only two different symbols. One represents the 0 state and the other represents the 1 state.

From the specific representation of the signal in the transmission process, the signal can be divided into two types: baseband signal and broadband signal.

1) Baseband signal: "0" and "1" are represented by high and low voltages. This high and low level alternating matrix pulse signal is a baseband signal. In data communication, a digital data signal representing a binary bit sequence is a typical rectangular pulse signal. The inherent frequency band of a rectangular pulse signal is referred to as a base band, and the rectangular pulse signal is called the baseband signal.

2) Broadband signal: A combination of multiple baseband signals. In data communication, a channel is divided into a plurality of sub-channels, and audio, video, and digital signals are respectively transmitted, which is called broadband transmission. Broadband is a wider band than the audio bandwidth and includes most of the electromagnetic spectrum.

2.1.2 Fourier analysis

In the early 19th century, French mathematician Fourier found that any period of the T function $g(t)$ can be represented by an infinite series of sine and cosine functions, called the Fourier series. Fourier analysis mainly studies the Fourier transform of a function and its properties. It is a mathematical tool for analyzing signals, such as finding the bandwidth of a signal and analyzing complex signals. Its mathematical expression is as follows:

$$g(t) = \frac{1}{2}c + \sum_{n=1}^{\infty} a_n \sin(2\pi nft) + \sum_{n=1}^{\infty} b_n \cos(2\pi nft)$$

$$a_n = \frac{2}{T}\int_0^T g(t)\sin(2\pi nft)\,\mathrm{d}t \qquad (2\text{-}1)$$

$$b_n = \frac{2}{T}\int_0^T g(t)\cos(2\pi nft)\,\mathrm{d}t$$

$$c = \frac{2}{T}\int_0^T g(t)\,dt$$

where $f = 1/T$ is the fundamental frequency, a_n and b_n are the sine and cosine amplitudes of n times, and c is a constant. With the Fourier series, a function can be reconstructed; that is, if the period T is known and the amplitude is already given, the original function $g(t)$ can be obtained by summing equation (2-1).

2.1.3 Bandwidth-limited signals

According to the Fourier analysis, any periodic signal can be decomposed into a combination of multiple sine waves of different amplitude, frequency and phase, and the signal bandwidth refers to the highest frequency of these signals minus the lowest frequency.

Signal spectrum: Using Fourier series expansion, the frequency domain map represents the set of all frequency components contained in the signal, which is the combination of all the sine wave signals that make up the signal.

Signal bandwidth: The range of frequency components. Regardless of the analog signal or the digital signal, its main components occupy a certain frequency range, which is obtained by subtracting the lowest frequency from the highest frequency in the signal spectrum. For example, if a periodic signal is decomposed into five sine waves with frequencies of 100, 300, 500, 700 and 900 Hz, the bandwidth is 800Hz.

Channel Bandwidth: The range of frequencies that a channel can transmit without distortion of the signal. When the bandwidth of the channel is greater than the bandwidth of the transmitted signal, the signal can pass through the channel smoothly.

Here is a simple example to illustrate how the above mathematical principle applies to the transmission of the ASCII character "b" in data communication, and the character is encoded into 8-bit bytes. The bit pattern transmitted is 01100010. The left half of Figure 2-1 (a) shows the voltage output when the computer transmits the character. In the fourier analysis of the signal, you can get the following coefficients:

$$a_n = \frac{1}{\pi n}[\cos(\pi n/4) - \cos(3\pi n/4) + \cos(6\pi n/4) - \cos(7\pi n/4)]$$

$$b_n = \frac{1}{\pi n}[\sin(3\pi n/4) - \sin(\pi n/4) + \sin(7\pi n/4) - \sin(6\pi n/4)]$$

$$c = 3/4$$

The square root amplitudes of the first few terms $\sqrt{a^2 + b^2}$ are shown on the right side of Figure 2-1 (a). These values are of interest because their squares are proportional to the energy transmitted at the corresponding frequency.

All transmission facilities lose some energy during transmission. If all Fourier terms are equally attenuated, the resulting signal will decrease in amplitude, so will not deform, and it will have the

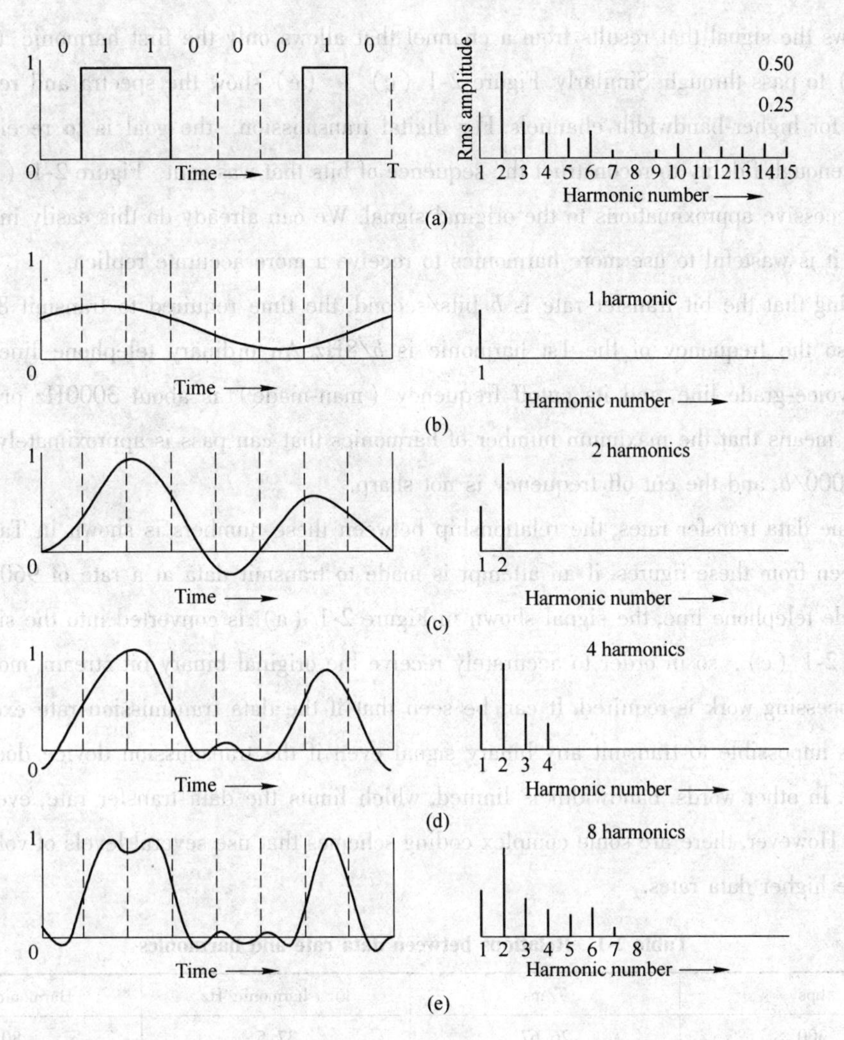

Figure 2-1 Fourier analysis

(a) A binary signal and its root-mean-square Fourier amplitudes; (b) ~ (e) Successive approximations to the original signal

same square wave shape as in Figure 2-1 (a). However, all transmission facilities are not the same for different Fourier attenuations and therefore cause signal distortion. In general, from 0 to a certain frequency f_c, the amplitude will not be attenuated during transmission, and the amplitude corresponding to the frequency above the cut off frequency f_c will be weakened to varying degrees. This frequency range in which the amplitude does not significantly decrease during transmission is called a bandwidth. The commonly quoted broadband refers to the frequency from 0 to a certain one that retains the general energy. Broadband is a physical property of a transmission medium and usually depends on the material composition, thickness and length of the medium. In some cases, a filter may be introduced into the circuit to limit the amount of bandwidth each customer can use.

Figure 2-1 (a) showes a binary signal and its root-mean-square Fourier amplitudes. Figure 2-1

(b) shows the signal that results from a channel that allows only the first harmonic (the fundamental, *f*) to pass through. Similarly, Figure 2-1 (c) ~ (e) show the spectra and reconstructed functions for higher-bandwidth channels. For digital transmission, the goal is to receive a signal with just enough fidelity to reconstruct the sequence of bits that was sent. Figure 2-1 (b) ~ (e) shows successive approximations to the original signal. We can already do this easily in Figure 2-1 (e), so it is wasteful to use more harmonics to receive a more accurate replica.

Assuming that the bit transfer rate is *b* bits/second, the time required to transmit 8bits is 8/*b* seconds, so the frequency of the 1st harmonic is *b*/8Hz. An ordinary telephone line is usually called a voice-grade line, and its cutoff frequency (man-made) is about 3000Hz or more. This limitation means that the maximum number of harmonics that can pass is approximately 3000 (*b*/8) or 24000/*b*, and the cut off frequency is not sharp.

For some data transfer rates, the relationship between these numbers is shown in Table 2-1. As can be seen from these figures, if an attempt is made to transmit data at a rate of 9600 bps on a voice-grade telephone line, the signal shown in Figure 2-1 (a) is converted into the signal shown in Figure 2-1 (c), so in order to accurately receive the original binary bit stream, more complex signal processing work is required. It can be seen that if the data transmission rate exceeds 38.4 kbps, it is impossible to transmit any binary signal even if the transmission device does not have any noise. In other words, bandwidth is limited, which limits the data transfer rate, even for ideal channels. However, there are some complex coding schemes that use several levels of voltage values to achieve higher data rates.

Table 2-1 Relations between data rate and harmonics

bps	*T*/ms	First harmonic/Hz	Harmonics sent
300	26.67	37.5	80
600	13.33	75	40
1200	6.67	150	20
2400	3.33	300	10
4800	1.67	600	5
9600	0.83	1200	2
19200	0.42	2400	1
38400	0.21	4800	0

2.1.4 The maximum data rate of a channel

As early as 1924, an AT&T engineer, Henry Nyquist, realized that even a perfect channel has a finite transmission capacity. He derived an equation expressing the maximum data rate for a finite bandwidth noiseless channel. In 1948, Claude Shannon carried Nyquist's work further and extended it to the case of a channel subject to random noise. We will just briefly summarize their classic results here.

Nyquist proved that if an arbitrary signal has been run through a low-pass filter of bandwidth *H*, the filterd signal can be completely reconstructed by marking only 2*H* (exact) samples per second. Sampling the line faster than 2*H* times per second is pointless because the higher frequency components that such sampling could recover have already been filtered out. If the signal consists of *V* discrete levels, Nyquist's theorem states:

$$\text{Maximum data rate} = 2H\log_2 V \text{ bits/sec}$$

For example, a noiseless 3kHz channel cannot transmit binary (i. e. , two-level) signals at a rate exceeding 6000bps.

So far we have considered only noiseless channels. If random noise is present, the situation deteriorates rapidly. And there is always random thermal noise present due to the motion of the molecules in the system. The amount of thermal noise present is measured by the ratio of the signal power to the noise power, known as the signal-to-noise ratio. If we denote the signal power by *S* and the noise power by *N*, the signal-to-noise ratio is *S/N*. Usually, the ratio itself is not quoted; instead, the quantity 10 lg(*S/N*) is given. These units are called decibels (dB). An *S/N* ratio of 10 is 10dB, a ratio of 100 is 20dB, a ratio of 1000 is 30dB, and so on. The manufacturers of stereo amplifiers often characterize the bandwidth (frequency range) over which their product is linear by giving the 3dB frequency on each end. These are the points at which the amplification factor has been approximately halved (because $\lg 3 \approx 0.5$).

Shnnon's major result is that the maximum data rate of a noisy channel whose bandwidth is *H* Hz, and whose signal-to-noise ratio is *S/N*, is given by

$$\text{Maximum number of bits/sec} = H\log_2(1 + S/N)$$

For example, a channel of 3000Hz bandwidth with a signal to thermal noise ratio of 30dB (typical parameters of the analog part of the telephone system) can never transmit much more than 30000 bps, no matter how many or how few signal levels are used and no matter how often or how infrequently samples are taken. Shannon's result was derived from information-theory arguments and applies to any channel subject to thermal noise. Counter examples should be treated in the same category as perpetual motion machines. It should be noted that this is only an upper bound and real systems rarely achieve it.

2.2 Signal transmission modes

Data transmission refers to the process of transferring data between a source and a sink over one or more links in accordance with appropriate procedures. It also refers to the operation of transferring data from one location to another by means of signals on the channel. The data transmission system usually consists of data channel equipment (DCE) at both ends of the transmission channel and the channel, and in some cases, a multiplexing device at both ends of the channel. The transport channel can be a dedicated communication channel or can be provided by a data switching network, a telephone switching network, or other types of switching network. The input and output devices of the data transmission system are terminals or computers, collectively referred to as data terminal equipment (DTE), and the data information sent by them is generally a combination of letters, numbers and symbols. Channel refers to the path through which information is transmitted. A channel is generally used to represent media that transmits information in a certain direction. A communication line typically includes a transmission channel and a reception channel.

The data transmission mode is the way that data is transmitted on the transmission channel. There are a variety of classification methods: according to the characteristics of the transmitted data signal, data transmission modes can be divided into baseband transmission, frequency band transmission and broadband transmission; alternatively according to be order of data transmission, there are parallel transmission and serial transmission; in terms of synchronization the modes include synchronous transmission and asynchronous transmission; if the flow direction and time are considered data transmission can be divided into simplex, half-duplex and full-duplex transmission.

2.2.1 Band transmission

(1) Baseband transmission: The baseband signal is directly sent to the line for transmission. The digital signal "1" or "0" is directly represented by two different voltages. The signal that is continuously changing at high and low level is the baseband signal. Baseband transmission is the direct transmission of digital signals on the channel, and the whole bandwidth of the transmission media is occupied by baseband signals. The baseband signal is unmodulated and can be directly transmitted after being driven by a code transform (or waveform transform). The baseband signal is characterized by a DC, low frequency, and high frequency component in the spectrum. As the frequency increases, its amplitude decreases correspondingly and eventually approaches zero. Baseband transmission is often used in short-distance data transmission, such as data communication between short-range computers or data transmission using twisted pair or coaxial cables in the local area network.

(2) Frequency-band transmission: The band signal is sent directly to the line for transmission. The frequency band transmission is to modulate the digital signal into an analog signal, and then sent and transmitted. When it reached the receier, the analog signal is demodulated into the original digital signal. Frequency-band transmission enables long-distance data communication, such as nationwide or global data communication using a telephone network.

(3) Broadband transmission: The spectrum of multiple baseband signals, audio signals and video signals are respectively transmitted to different frequency bands of one cable for transmission. The transmitted signals are modulated analog signals, and the signals do not interfere with

each other, and the utilization of the line is improved, the transmission distance is long, and multiple channels can be provided.

The main difference between broadband transmission and baseband transmission (Table 2-2): First, the data transmission rate is different. The baseband network has a data rate ranging from tens to hundreds of Mb/s, and the broadband network can reach Gb/s. Second, the broadband network can be divided into multiple baseband channels to provide multiple good communication paths.

Table 2-2 Difference between broadband transmission and baseband transmission

Baseband transmission	Digital signal	Occupied by a single signal	Two-way	Bus	Several kilometers
Broadband transmission	Analog signal	Using FDMA technology, multiple data channels	One-way	Bus or tree	Dozens of kilometers

2.2.2 Serial and parallel transmission

When communicating between a computer and a computer and between various components within the computer, the data transmission mode can be divided into parallel transmission and serial transmission according to the amount of data transmitted at a time (Figure 2-2).

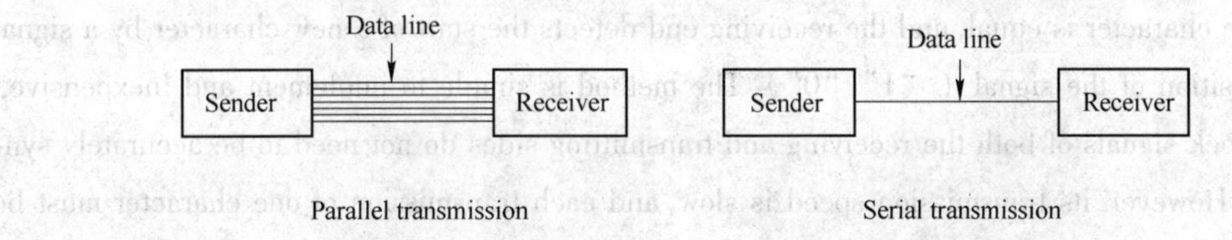

Figure 2-2 Parallel transmission and serial transmission

(1) Parallel transmission is a method in which binary codes are simultaneously transmitted on parallel channels. For example, 8 units of code characters should be transmitted simultaneously in parallel with 8 channels, one character at a time. There is no synchronization problem between the receiving and transmitting sides. So the transmission speed is fast, but when the distance is far away and the number of digits is more, the communication line is complicated and the cost is high; it is suitable for short distance and high speed.

(2) Serial transmission: is a method in which a binary code is transmitted bit by bit in chronological order on a channel in units of bits (symbols). Besides sending and receiving bit by bit, this mode confirms the characters and takes synchronisation measures. The communication line is simple, as long as a pair of transmission lines can be communicated, and the telephone line can be used, thus greatly reducing the cost, especially for long distance communication, but the speed of transmission is slow.

2.2.3 Asynchronous and synchronous transmission

During the serial transfer process, the data is transmitted one bit at a time, and the transmission and reception of each bit of data requires the control of a clock pulse. The sending end determines the start and end of the data bits by transmitting a clock signals. Before sending the data, a series of synchronous clocks are sent first, and the receiver adjusts the timing sequence to the synchronization frequency according to the frequency of the clock pulse to keep the synchronization. That is to say, the receiving. That is to say, the receiving end and the sending end must maintain the same pace, or the drift phenomenon will occur, and finally the data transmission will be wrong. However, it is not an easy task to keep two independent clocks in sync. Therefore, the following two methods are often used to solve this problem: asynchronous transmission and synchronous transmission.

(1) Asynchronous transmission is a method of synchronous character transmission, also known as start-stop synchronization. When sending a character code, the character is preceded by a "start" signal; the length is 1 symbol wide; the polarity is "0", that is, the space polarity; and a "stop" is added after the character is sent. The signal has a length of 1, 1.5 (used in International Code No. 2) or two symbols wide, and the polarity is "1", that is, the polarity of the mark. The receiving end can distinguish the transmitted characters by detecting the start and stop signals. Characters can be sent continuously or separately. When no characters are sent, the stop signal is sent continuously. The start time of each character can be arbitrary, the length of the symbols in one character is equal, and the receiving end detects the start of a new character by a signal to the transition of the signal ("1" "0"). The method is simple to implement and inexpensive, and the clock signals of both the receiving and transmitting sides do not need to be accurately synchronized. However, its transmission speed is slow, and each transmission of one character must be attached with 2~3 bits of transmission time, so it is more suitable for low-speed terminal or dialogue operation. The operating format of asynchronous transmission is shown in Figure 2-3.

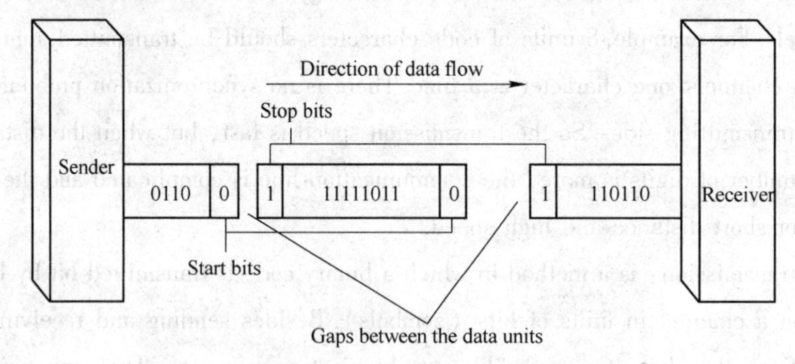

Figure 2-3 Asynchronous transmission

(2) Synchronous transmission is a bit (symbol) synchronous transmission method, as shown in Figure 2-4. This method must establish accurate bit timing signals on both the receiving and transmitting sides to correctly distinguish each data signal. In transmission, data is divided into groups (or frames), and one frame contains multiple character codes or multiple independent symbols. Before the data is transmitted, a predetermined sequence of frame synchronization symbols

must be added at the beginning of each frame. After the receiver detects the sequence flag, the start of the frame is determined, and synchronization between the two parties is established. The receiving end DCE extracts the bit timing signal from the received sequence to achieve bit (symbol) synchronization. Synchronous transmission does not add up and stop signals, and has high transmission efficiency. It can be used for data transmission above 2400bit/s, and allows users to transmit data composed of non-bits of 8 bits, and can communicate with mainframes that communicate in synchronous mode. However, synchronous transmission involves high technical and hardware requirements.

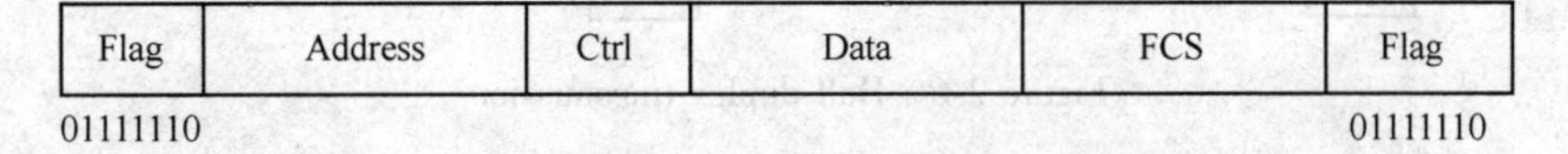

Figure 2-4 Synchronous transmission

2.2.4 Simplex, half-duplex and full-duplex transmission

According to the transmission directivity of data on the communication line and its relationship with time, data communication can be divided into three modes, namely simplex transmission, half-duplex transmission and full-duplex transmission.

2.2.4.1 Simplex transmission

Simplex transmission means that data can only be sent and received in a single direction. As shown in Figure 2-5, for stations A and B, only A can send data to the transmission line, and B can only receive data from the line. That is to say, in this way, the flow of data can only be from A to B, not from B to A.

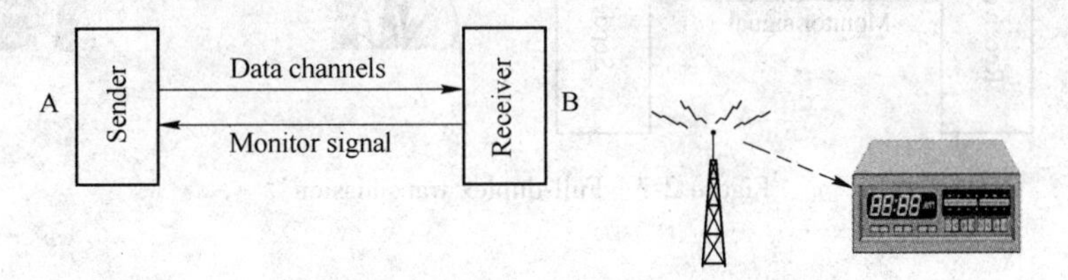

Figure 2-5 Simplex transmission

2.2.4.2 Half-duplex transmission

Half-duplex transmission means that data can be transmitted in two directions but not simultaneously, that is, alternately received and transmitted. As shown in the Figure 2-6, when using the half-duplex mode, data can be transmitted from A to B or from B to A. However, since there is only one transmission channel between A and B, the signal can only be transmitted in time. In this mode of operation, either A sends data to be received by B; or B sends data to A. While idle A and B would be in the receiving mode so as to respond to the other parties call at any time.

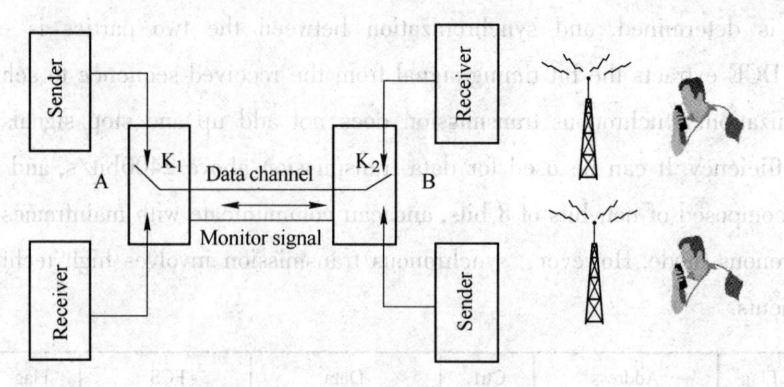

Figure 2-6 Half-duplex transmission

2.2.4.3 Full-duplex transmission

Full-duplex transmission means that data can be transmitted simultaneously in both directions, as shown in Figure 2-7. Generally, the four-wire line is full-duplex data transmission, and the second-line line can realize full-duplex data transmission. As shown in the figure, both stations A and B can receive data and send data at the same time. Because there are two transmission channels in the full-duplex mode, bidirectional transmission can be realized without switching the switches, thereby increasing the transmission rate.

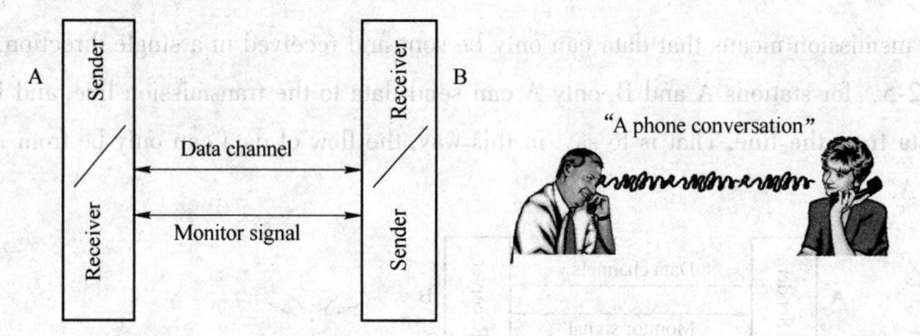

Figure 2-7 Full-duplex transmission

2.3 Channel

Information is abstract, but the transmitted information must pass through a specific medium. For example, the channel of the wireless telephone is the space through which the radio wave propagates, and the channel of the wired telephone is the cable. A channel is a communication line between a source and a sink during communication. A communication line often includes a transmission channel and a reception channel, and the channel is composed of a corresponding device for transmitting and receiving information and a transmission medium. A source is a device or computer that generates and transmits information during communication. A sink is a device or computer that receives and processes information during communication.

In theoretical research, the channel is often divided into a channel coder, a channel itself, and a channel decoder. One can change the encoder and decoder to get the best communication effect. Therefore, the encoder and decoder often refer to the parts that are easy to change and easy to design, and the channel refers to those parts that are relatively fixed. However, such divisions are stipulated in specific circumstances. For example, modems and error correction coding and decoding devices are generally considered to belong to channel coder and decoder, but sometimes the channel containing the modem is called a modulation channel; and an error correction coder is included. The channel of the decoder is called the code channel. All channels have an input set A, an output set B, and a link between them, such as the conditional probability $P(y \mid x)$, $x \in A$, $y \in B$. These parameters can be used to specify a channel.

2.3.1 Channel classification

According to different standards, the channel has the following types of division.

2.3.1.1 Physical channel and logical channel

A physical channel is a physical path used to carry signals or data. It consists of a transmission medium and associated communication equipment. The logical channel refers to adding some necessary control procedures on the basis of the physical channel to control the transmission of data, that is, the logical channel adds software or hardware procedures on the basis of the physical channel, thereby realizing reliable data transmission of the physical channel.

2.3.1.2 Cable channel and wireless channel

The physical channel can be divided into a wired channel and a wireless channel depending on the presence or absence of the transmission medium. The medium composition of the wired channel includes a cable, a twisted pair cable, a coaxial cable, an optical fiber, etc. , and the wireless channel includes radio, microwave communication, satellite communication, and the like.

2.3.1.3 Analog channel and digital channel

According to the type of signal transmitted in the channel, the physical channel can be further divided into an analog channel and a digital channel. An analog signal is transmitted on the analog channel, and a digital signal is transmitted on the digital channel. By adding a modem device on both sides of the channel, the transmission of the digital signal on the analog channel can be realized, and the transmission of the analog signal on the digital channel can also be realized.

2. 3. 1. 4 Dedicated channel and public channel

The physical channel can be divided into a dedicated channel and a common channel according to the manner in which the channel is used. The dedicated channel can be set up by users or leased from the communication department. It is mostly used for short distance and large transmission. The public channel can serve a large number of users.

2. 3. 2 Main technical indicators of the channel

Bit rate: The number of binary code bits transmitted per second (bit/s, b/s or bps). The data rate is also known as the bit rate or data transmission rate.

Baud Rate: The number of electrical signal units or symbols transmitted per second in Baud. The baud rate is also known as the modulation rate or symbol rate. In the field of electronic communication, the Baud rate, refers to the change in unit time after the signal is modulated, that is, the number of times the carrier parameter changes per unit time. It is a measure of the symbol transmission rate, and 1 baud means 1 symbol per second represents. The relationship between data rate and modulation rate can be illustrated by the Nyguist formula. If an N-ary signal, that is, the amount of information per symbol load, is used, the relationship between the data rate at the time of transmitting the signal and the baud rate is: $S = B \log_2 N$ (bps), where S is the bit rate and B is the baud rate. A signal can often carry multiple binary bits, so at a fixed signal transmission rate, the bit rate is often greater than the baud rate. In other words, multiple bits can be transmitted in one symbol.

Channel bandwidth: The range of frequencies occupied by a signal transmitted in a channel without distortion, usually referred to as the passband of the channel, in units of Hz.

Channel capacity: The maximum data rate that a channel can transmit; the maximum number of bits that a channel can transmit per unit of time. Generally, the channel capacity and channel bandwidth have a proportional relationship; the larger the bandwidth, the higher the capacity, and the higher the signal transmission rate.

Throughput: The total amount of information that the channel successfully transmits in a unit of time.

Propagation speed: The distance that the signal travels per unit of time on the communication line (meters per second).

Bit error rate: The bit error rate, the probability that a binary bit will be misdirected during transmission; the channel transmission reliability indicator. Bit error rate P = number of transmitted errors/total number of transmitted bits.

2. 3. 3 Modulation

Signals from sources are often referred to as baseband signals. Data signals representing various text or image files output by a computer belong to the baseband signal. The baseband signal often contains more low frequency components and even DC components, and many channels cannot transmit such low frequency components or DC components. In order to solve this problem, the baseband signal must be modulated.

Modulation can be divided into two broad categories. One is to simply transform the waveform of the baseband signal to match the channel characteristics. The transformed signal is still the baseband signal. This type of modulation is called baseband modulation. The other type requires modulation using a carrier to shift the frequency range of the baseband signal to a higher frequency band for transmission in the channel. The carrier-modulated signal is called a band-pass signal, and modulation using a carrier is called band-pass modulation. The basic bandpass modulation method is amplitude modulation (AM), frequency modulation (FM) and phase modulation (PM).

AM: the amplitude of the carrier varies with the baseband digital signal. For example, 0 or 1 corresponds to no carrier or carrier output, respectively.

FM: the frequency of the carrier varies with the baseband digital signal. For example, 0 or 1 corresponds to frequency f_1 or f_2, respectively.

PM: the initial phase of the carrier varies with the baseband digital signal. For example, 0 or 1 corresponds to a phase of 0 or 180 degrees, respectively.

In order to achieve a higher information transmission rate, a more technically complex multi-dimensional amplitude phase hybrid modulation method must be employed. For example, quadrature amplitude modulation (QAM), as shown in Figure 2-8.

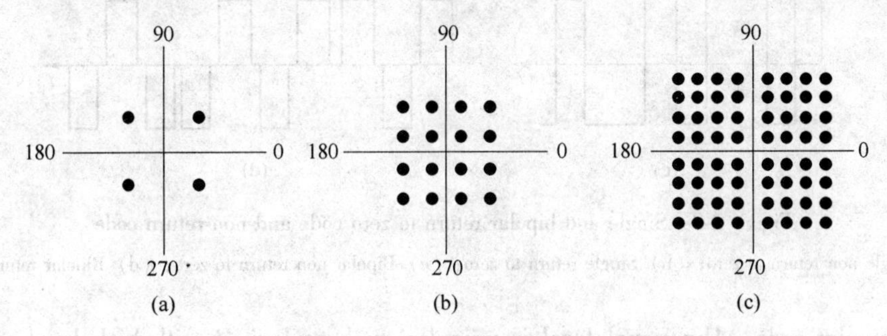

Figure 2-8 Quadrature amplitude modalation

(a) QPSK; (b) QAM-16; (c) QAM-64

2.4 Data coding

Digital signals cannot be transmitted directly in analog circuits, and analog signals cannot be transmitted in digital circuits. Therefore, the data is transformed from the original form to a representation suitable for processing, storage, and transmission. This process is called data encoding.

To pass the teaching data between the devices, the data must be encoded into a physical signal suitable for transmission to form an encoded waveform, and symbols 0, 1 are the basic units for transmitting data. Most of the signals transmitted in the industrial data network communication system are binary codes, and each bit can only take one of the two states of 1 or 0. Each of these bits is a symbol. The symbol is the basic unit of the data being transmitted. The use of different amplitudes, different frequencies, and different phases of the analog signal to express the 0, 1 state of the data is called analog data encoding. The high and low level rectangular pulse signals are used to express the 0, 1 state of the data, which is called digital data encoding.

2.4.1 Digital data coding

Digital data coding ore-encode the digital signal in the computer for baseband transmission when transmitting computer data in the digital channel. The digital signal modification methods mainly include non-return-to-zero coding, Manchester coding, and differential Manchester coding.

The following are some common types of digital data encodings, and Figure 2-9 (a) shows a typical waveform of a single, bipolar return to zero and non-return to zero code.

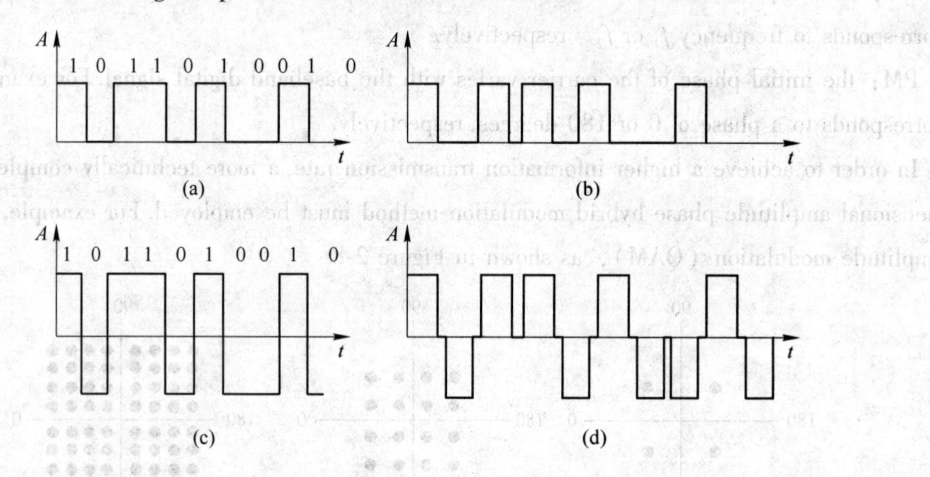

Figure 2-9 Single and bipolar return to zero code and non-return code

(a) Single non return to zero; (b) Single return to zero; (c) Bipolar non return to zero; (d) Bipolar return to zero

(1) Unipolar code: The signal level is unipolar, such as logic 1 with high level and logic 0 with low level signal encoding.

(2) Bipolar code: The signal level has positive and negative polarities. If logic 1 uses a positive level, logic 0 encodes with a negative level signal.

(3) The return-to-zero code (RZ): It after each bit of binary information transmission. For example, the logic 1 of the bipolar return-to-zero code returns to zero level only after a certain segment of the symbol time (such as half of the symbol time) and its logic 0 returns to zero level only after maintaining a negative level for half of the symbol time.

(4) Non-Return to zero (NRZ): A code that maintains the corresponding level of its logic state throughout the symbol time.

(5) Differential code. At the beginning of each clock cycle, the state of the data "1" and "0" is represented by the change in the signal level. For example, it is prescribed that at the start of the clock cycle, the signal level change represents "1", and the unchange represents "0". The differential code is either high level or low level according to the initial state signal, and there are two waveforms with opposite phases. Figure 2-10 shows the data waveform of an 8-bit data and its differential code waveform. When the initial signal is low, the code 1 is formed; oppositely, the code 2 is formed.

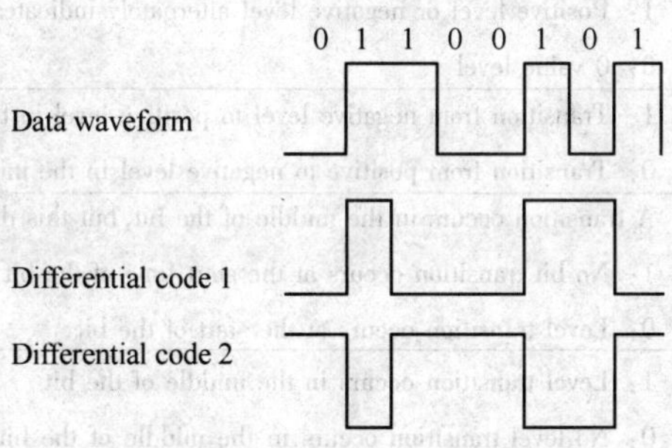

Figure 2-10 Differential code

(6) Manchester Encoding. This is the most commonly used baseband signal encoding in data communications. It has synchronization information that keeps each node on the network synchronized with the clock. In Manchester coding, time is divided into equally spaced segments by clock cycle, where each segment represents one bit. Each bit time is divided into two halves. The signal transmitted in the first half of the time period is the inverse of the transmitted bit value of the time period, and the bit value itself is transmitted in the second half of the time period. Therefore, jumping from a high level to a low level indicates 0, and a transition from a low level to a high level indicates 1. It can be seen that in one bit time, there is always a change in the signal level at the middle point, and this change in the signal level can be used as the synchronization information between the nodes. Figure 2-11 shows the process and waveform of Manchester encoding. Differential Manchester coding is a variant of Manchester encoding. The code must be hopped in the middle of each bit interval, and the differential code starts with the clock cycle start level, whether or not it represents the characteristics of the logic "1" or "0".

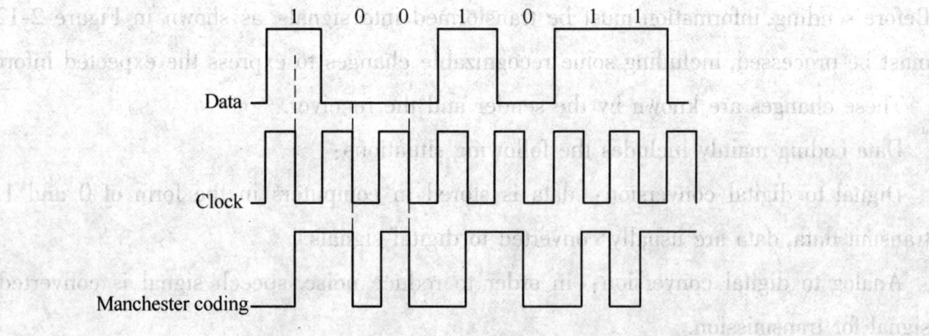

Figure 2-11 Manchester Encoding

Different code encoding implementation can be seen in the following Table 2-3.

Table 2-3 Code encoding implementation

Types	Encoding mode
Two-level code	1: Negative level 0: Positive level
Differential code	1: Level transition occurs at the start of the bit 0: No level transition occurs at the start time of the bit
Bipolar alternating reverse code	1: Positive level or negative level alternately indicates two levels 0: 0 value level
Manchester encoding	1: Transition from negative level to positive level in the middle position of the bit 0: Transition from positive to negative level in the middle of the bit
Differential manchester code	A transition occurs in the middle of the bit, but this does not indicate data information 1: No bit transition occurs at the start time of the bit 0: Level transition occurs at the start of the bit
Miller code	1: Level transition occurs in the middle of the bit 0: No level transition occurs in the middle of the bit. If followed by another 0, a level transition occurs at the end of the bit
Duobinary code	0: 0 Value level 1: Positive level or negative level, the two levels are alternately indicated, and the level polarity is changed when there are an odd number of 1 between two zeros

2.4.2 Analog data coding

Analog data coding transforms the spectrum of the digital signal into a spectrum suitable for transmission in the analog channel by modulating the spectrum transform. The analog data encoding uses an analog signal to express the 0, 1 state of the data. The amplitude, frequency and phase of the signal are parameters describing the analog signal. The analog data can be encoded by changing the three teachings. Amplitude-shift keying (ASK), frequency-shift keying (FSK), and phase-shift keying (PSK) are three coding methods for analog data coding.

2.4.3 Coding techniques

Before sending, information must be transformed into signals, as shown in Figure 2-12. The signal must be processed, including some recognizable changes to express the expected information.

These changes are known by the sender and the receiver.

Data coding mainly includes the following situations:

Digital to digital conversion: data is stored in computers in the form of 0 and 1. In order to transmit data, data are usually converted to digital signals.

Analog to digital conversion: in order to reduce noise, speech signal is converted into digital signal for transmission.

Digital to analog conversion: for example, using public telephone line to transmit data, computer generated digital signals should be converted into analog signals.

Analog to analog conversion: when analog signal is transmitted, the signal frequency is not suitable for the selected medium, and needs to be tuned to the appropriate frequency.

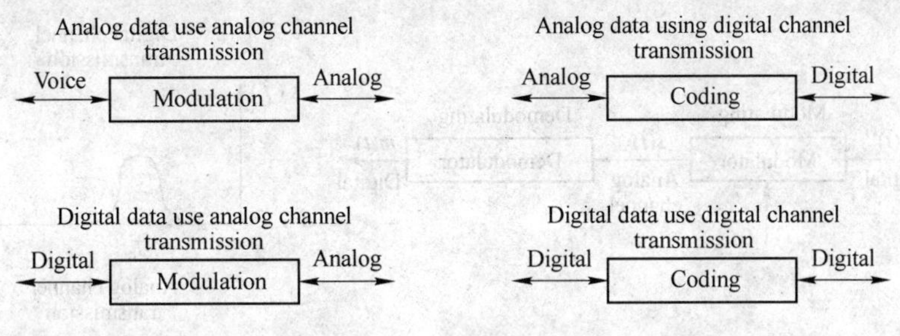

Figure 2-12 Analog transmission and digital transmission

2.4.3.1 Analog signal modulation of digital data

Baseband transmission is the way to send the encoded signal directly from the source to the channel. Baseband transmission requires wide bandwidth of the channel, and the waveform is greatly influenced by the distributed capacitance of the channel, so it is only suitable for short distance and internal data transmission. For long distance transmission, digital signals must be converted to analog signals that can be transmitted on a long distance channel. This is the frequency band transmission of digital signals, also known as carrier transmission. Modem is the main device for transforming digital signals and analog signals to facilitate the transmission of digital signals on analog channels. Therefore, the most important technology of frequency band transmission is modulation and demodulation.

A sine wave has three main characteristics: amplitude, frequency, phase. When we change one of them, we have another form of wave. By changing the characteristics of one aspect of electrical signals, it can be used to represent digital data, as shown in Figure 2-13. Any of the three characteristics of the wave can be changed in this way, so there are at least three mechanisms to transfer digital data to analog signals: amplitude shift keying, frequency shift keying, phase shift keying. In addition, the mechanism of combining amplitude and phase changes-quadrature amplitude modulation.

When the signal is transmitted by PSTN, the sine wave carrier of the 1000~2000Hz frequency range is used, and its amplitude, frequency and phase are modulated, and all of them can be used to transfer information. In analog transmission, the sending device generates a sine wave signal as the base frequency information loading signal, called the carrier signal.

There are mainly 3 methods of converting digital data to analog signal through modem, they are ASK, FSK and PSK, as shown in Figure 2-14.

ASK: Two binary values "0" and "1" are represented by the different amplitude of carrier sig-

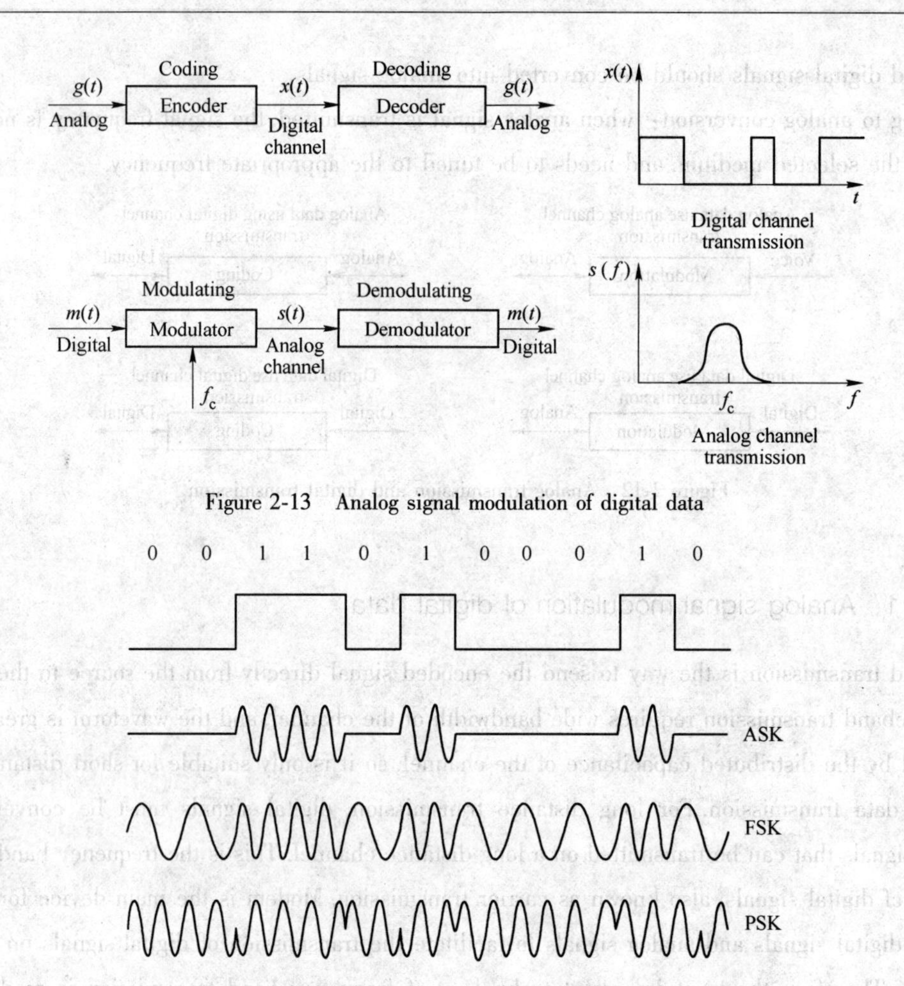

Figure 2-13 Analog signal modulation of digital data

Figure 2-14 ASK, FSK and PSK

nals (A_1 and A_2), and the carrier mode controls the amplitude A of the carrier by the digital signal sequence. ASK transmission technology is greatly affected by noise.

FSK: The two binary values "0" and "1" are represented by the different frequencies of the carrier signal (f_1 and f_2), and the oscillation frequency f of the carrier varies with the value of the digital sequence. FSK is mainly used for >1200bps transmission rate, and can also be used for 3 to 30MHz high frequency radio transmission and local broadcasting network. FSK is easy to implement and is mainly used for low speed transmission.

PSK: The two binary values "0" and "1" are represented by the different phases of the carrier signal (φ_1 and φ_2). PSK is mainly used for medium speed transmission. There are two ways to express it.

(1) absolute phase modulation: Two fixed phases represent digital information "0" and "1" respectively. Example: the sine wave of $\varphi_1 = 0$ represents "0", and the sine wave of $\varphi_2 = \pi$ represents "1".

(2) relative phase modulation: the phase of modulated signal is determined by the value of the

modulated phase and the standard digital information corresponding to the preceding digital information. That is: when the signal is "1", the phase of the modulated signal is unchangeable relative to the previous signal; when the signal is "0", the phase of the modulated signal relative to the previous signal is changed to π.

2.4.3.2 Analog signal conversion to digital signal

Because of the low distortion of digital signal, low error rate and high rate of data transmission, in the network, in addition to the digital signal produced directly by the computer, the digitalization of speech and image information has become the inevitable trend of development. Pulse code modulation (PCM) is the main way to simulate data digitalization, as shown in Figure 2-15 and Figure 2-16.

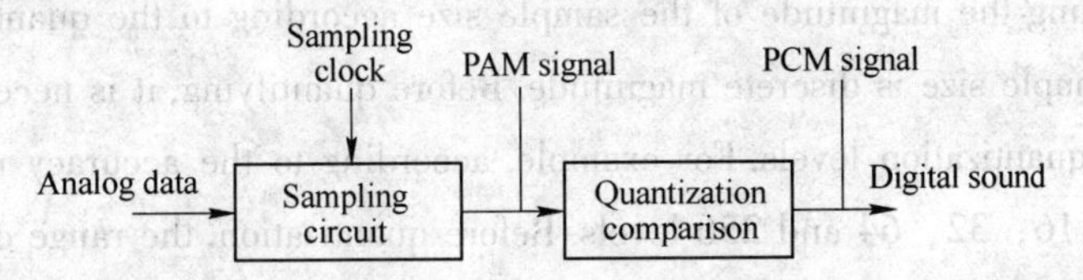

Figure 2-15 PAM and PCM

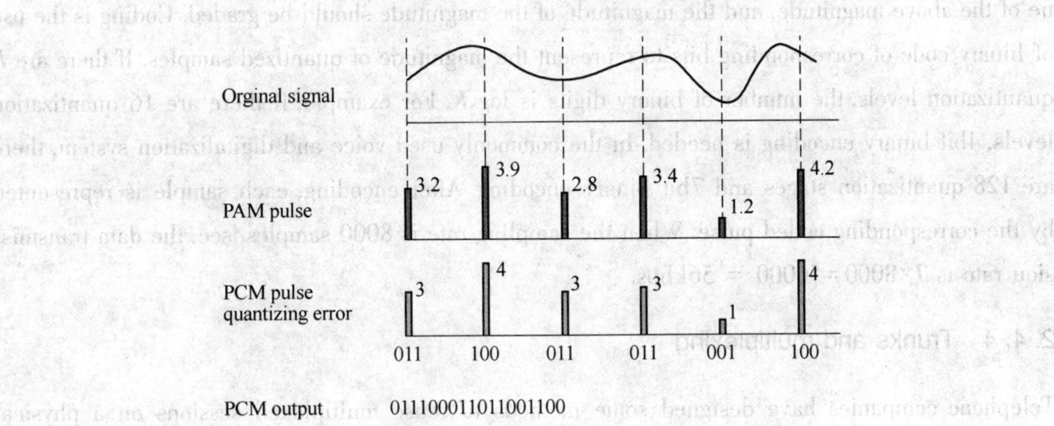

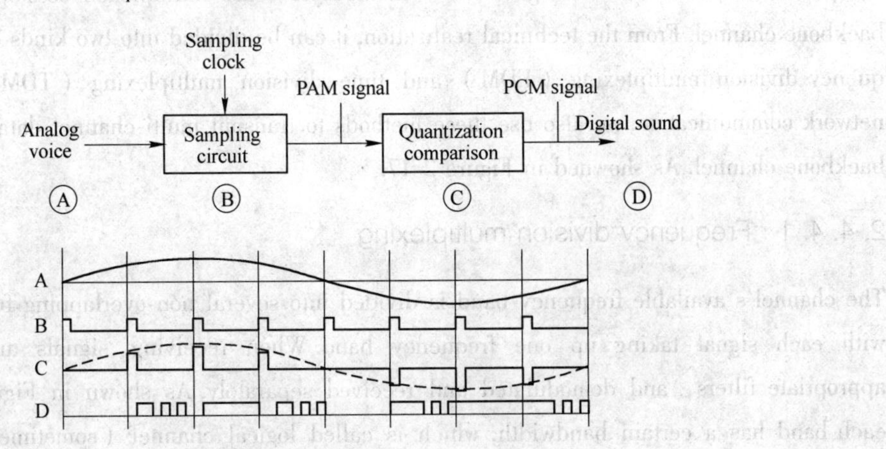

Figure 2-16 PCM conversion process

Pulse amplitude modulation (PAM): For the analog signal, we sample it, and then generate a series of pulses based on the results. However the pulse value is still amplitude, not digitalized, which is not suitable for data communication. PCM is needed to modulate the signal. PAM is the first step in PCM.

The typical application of PCM technology is speech digitization. Speech can be transmitted through telephone lines in the form of analog signals, but the voice signal must be digitized when the digital, text, graphics and images generated by the voice and the computer are transmitted simultaneously in the network. PCM operations include: sampling, quantization and coding.

Sampling theorem: a continuous variable analog data, with the highest frequency or bandwidth of F_{max}. If the sampling frequency is greater than or equal to 2 F_{max}, then the sampled discrete sequence can recover the original continuous analog signal without distortion. Quantization is the process of determining the magnitude of the sample size according to the quantization level. After quantization, the sample size is discrete magnitude. Before quantifying, it is necessary to divide the signal into several quantization levels. For example, according to the accuracy requirement, it can be divided into 8, 16, 32, 64 and 256 levels. Before quantization, the range of the magnitude of each level should be specified, and then the sample amplitude is compared with the magnitude value of the above magnitude, and the magnitude of the magnitude should be graded. Coding is the use of binary code of corresponding bits to represent the magnitude of quantized samples. If there are K quantization levels, the number of binary digits is $\log_2 K$. For example, if there are 16 quantization levels, 4bit binary encoding is needed. In the commonly used voice and digitalization system, there are 128 quantization stages and 7bit binary encoding. After encoding, each sample is represented by the corresponding coded pulse. When the sampling rate is 8000 samples/sec, the data transmission rate is $7 \times 8000 = 56000$ = 56kbps.

2.4.4 Trunks and multiplexing

Telephone companies have designed some methods to reuse multiplexed sessions on a physical backbone channel. From the technical realization, it can be divided into two kinds of methods: frequency division multiplexing (FDM) and time division multiplexing (TDM). In computer network communication, we also use these methods to transmit multi-channel data with a physical backbone channel. As showned in Figure 2-17.

2.4.4.1 Frequency division multiplexing

The channel's available frequency band is divided into several non-overlapping frequency bands, with each signal taking up one frequency band. When receiving, signals are separated by appropriate filters, and demodulated and received separately. As shown in Figure 2-18, FDM each band has a certain bandwidth, which is called logical channel (sometimes referred to as channel). In order to prevent the interference caused by the frequency coverage of adjacent

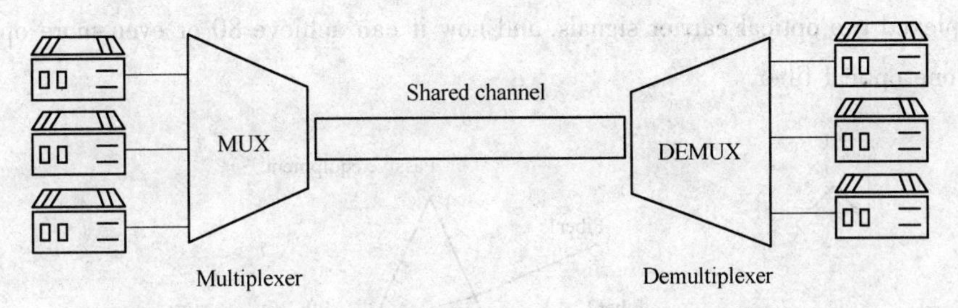

Figure 2-17 Transmitting multi-channel data with a physical backbone channel

channel signals, a certain "protection" band is set up between the two adjacent signal frequencies, but the corresponding spectrum of the protective band cannot be used to ensure that each frequency band is isolated from each other.

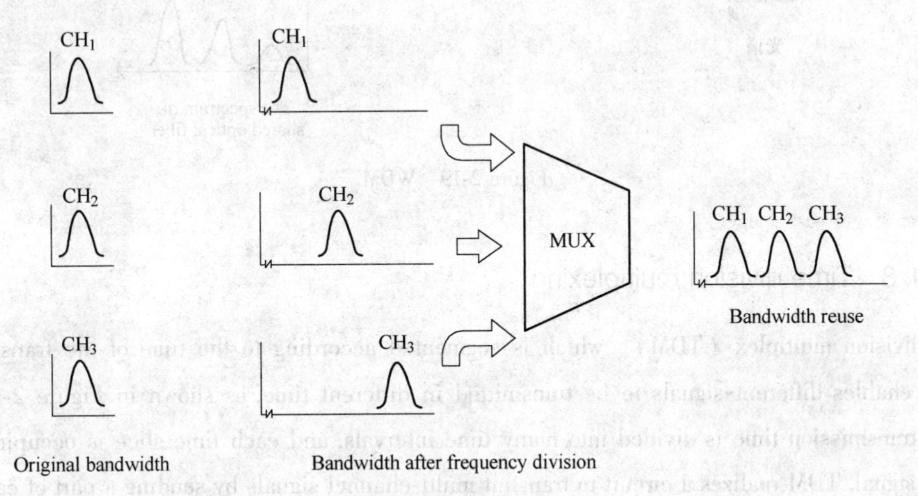

Figure 2-18 FDM

Frequency division multiplexing divides the frequency band resources of the circuit or space into multiple frequencies and assigns them to multiple users respectively. The data of each user terminal is transmitted through the sub-channel (frequency band) allocated to it, which is mainly used in the telephone and cable television (CATV) system. Typical examples are radio and TV, cable TV.

2. 4. 4. 2 Wavelength division multiplexing

Wavelength division multiplexing (WDM), also known as optical frequency division multiplexing, is an example of frequency division multiplexing used in an optical fiber channel. WDM uses the concept of frequency division multiplexing of traditional carrier phones, and uses a fiber to simultaneously transmit optical carrier signals which are very close to multiple frequencies, which multiplies the transmission capability of the fiber , as shown in Figure 2-19. WDM is mainly used in the communication system composed of all optical networks. Initially, one optical fiber multiplexed two optical carrier signals, and now it can achieve 80 or even more optical fiber signals on one optical fiber.

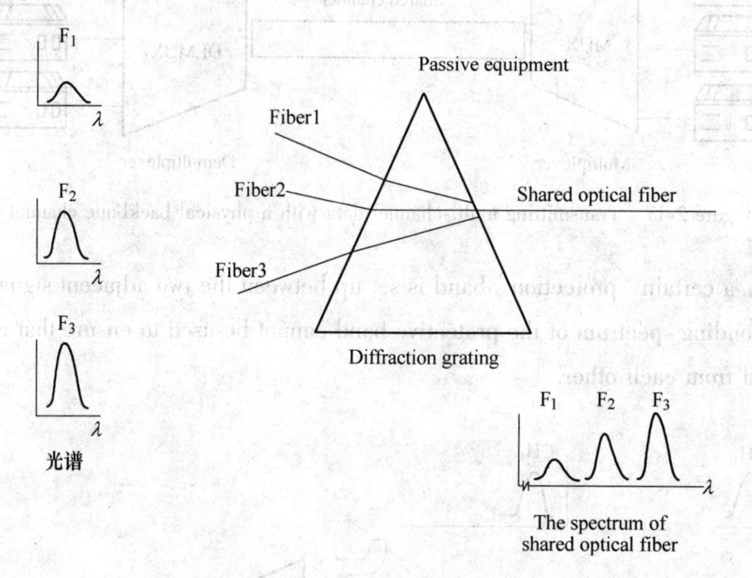

Figure 2-19 WDM

2.4.4.3 Time division multiplexing

Time division multiplex (TDM), which is segmented according to the time of the transmission signal, enables different signals to be transmitted in different time, as shown in Figure 2-20. The whole transmission time is divided into many time intervals, and each time slice is occupied by a single signal. TDM realizes a circuit to transmit multi-channel signals by sending a part of each signal across time. Table 2-4 shows the difference between time division multiplexing and frequency division multiplexing.

Table 2-4 Comparison with frequency division multiplexing

Time division multiplexing	There is only one signal at any time in the circuit. Digital signal is a finite number of discrete values, and is widely used in digital communication systems including computers. The shorter the slot length, the more sub-channels can be partitioned
Frequency division multiplexing	There are many different frequencies of signals at any time on the circuit. Frequency division multiplexing is suitable for analog communication systems. The wider the frequency band, the more sub-channels can be divided within the bandwidth

2.4.4.4 Synchronous time division multiplexing

Synchronous time division multiplexing (SDM) is a kind of fixed time slice allocation. Throughout

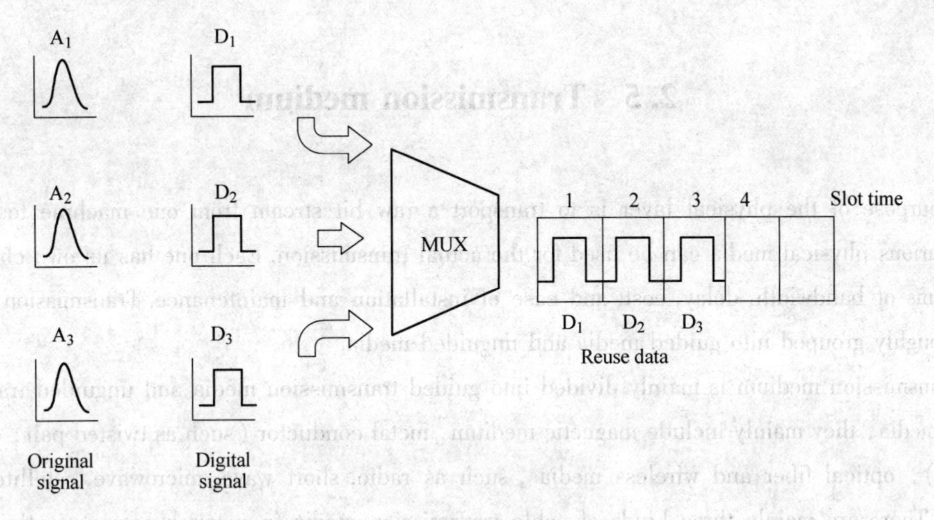

Figure 2-20　TDM

SDM, the time of transmission signal is continuously divided into specific time periods (one cycle) according to specific lengths. Each time segment is divided into equal time slots, and each time slot is allocated to various digital signals in a fixed way. Each digital signal is allocated to one time slot in every time period.

2. 4. 4. 5　Statistics TDM

Asynchronous time division multiplexing, statistical TDM, as shown in Figure 2-21, dynamically allocates time slot on demand. When a user has data to send, time slots can be allocated to it; when the user suspends sending data, no time slots is allocated to it. The idle time slot of the circuit can be used for data transmission of other users. Statistics TDM avoids idle time slots in each time period.

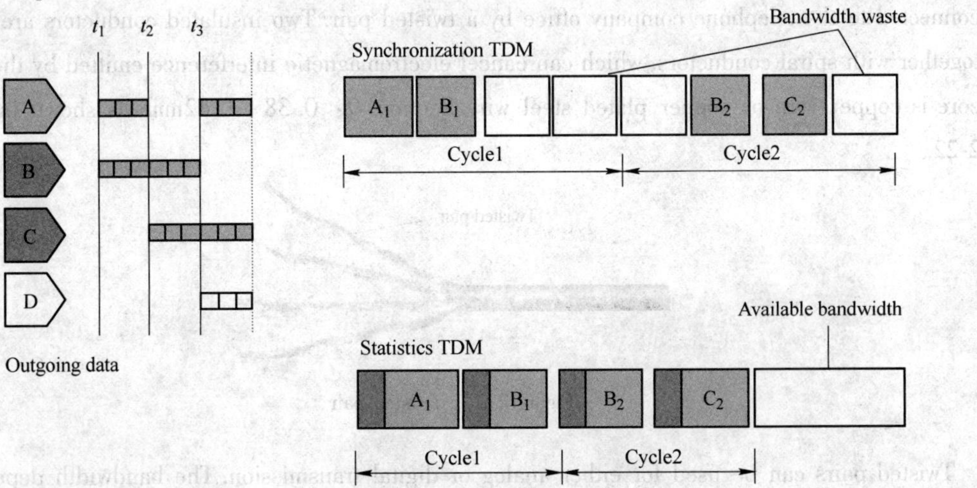

Figure 2-21　Statistical TDM

2.5 Transmission medium

The purpose of the physical layer is to transport a raw bit stream from one machine to another. Various physical media can be used for the actual transmission. Each one has its own character in terms of bandwidth, delay, cost, and ease of installation and maintenance. Transmission media are roughly grouped into guided media and unguided media.

Transmission medium is mainly divided into guided transmission media and unguided transmission media, they mainly include magnetic medium, metal conductor (such as twisted pair, coaxial cable), optical fiber and wireless media , such as radio, short wave, microwave, satellite, light wave. There are mainly three kinds of cable transmission media in network communication. They are twisted pair cable (UTP and UN-Shielded Twisted Pair), coaxial cable and fiber optic cable.

2.5.1 Magnetic media

One of the most common ways to transport data from one computer to another is to write them onto magnetic tape or floppy disks, physically transport the tape or the disks to the destination machine, and read them back in again.

(1) Advantages: simple, low cost, high bandwidth, high performance.

(2) Disadvantages: the delay characteristic is poor; transmission time is measured in minutes or hours not milliseconds.

2.5.2 Twisted pair

The most common application of the twisted pair is the telephone system. Nearly all telephones are connected to the telephone company office by a twisted pair. Two insulated conductors are twisted together with spiral conductors, which can cancel electromagnetic interference emitted by them. The core is copper wire or copper plated steel wire Section D: 0.38 ~ 1.42mm, as shown in Figure 2-22.

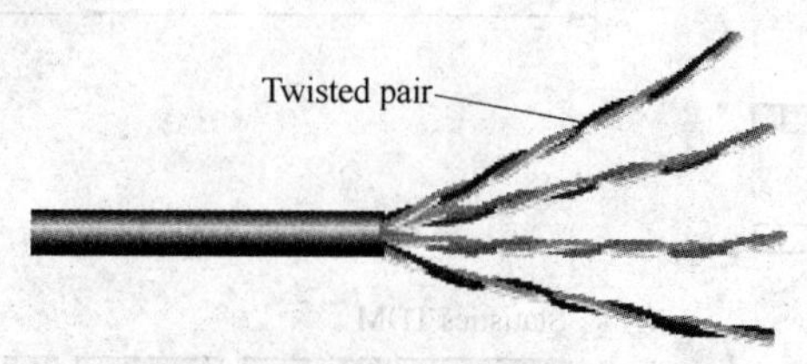

Figure 2-22 Twisted pair

Twisted pairs can be used for either analog or digital transmission. The bandwidth depends on the thickness of the wire and the distance traveled, but several megabits/sec can be achieved for a few kilometers in many cases. When the analog signal is transmitted, the maximum transmission distance is 15km; when transmitting the digital signal, it is 1~2km. It is simple and cheap, but it has poor security and high bit error rate.

2.5.2.1 UTP-unshielded twisted pair

The multiple twisted pair cables are gathered up and wrapped on a plastic reinforcement layer outside. All kinds of twisted pair cables have clear rules and regulations.

(1) Category 1 and Category 2 are Cables-voice and low speed data lines, which are suitable for voice and 4Mbps data transmission, and for telephone system wiring.

(2) Category 3 cables is suitable for data transmission with a maximum speed of 10Mbps. It is used for computer network.

(3) Category 4 cablesruns in, 20Mbit/s maximum data transmission rate.

(4) Category 5 cables is high speed data line which is used in the environment of maximum data transmission rate up to 100Mbps; used for computer network.

(5) Cables above Category 5 is high speed data lines, which bandwidth less than 1000Mbit/s, and fast Ethernet is the most common, cheap and high-quality network.

Among all of the kinds of twisted pair cables, Category 3 and 5 are important to computer networks. The Category 5 are similar to Category 3 pairs, but with more twists per centimeter and Teflon insulation, which results in less crosstalk and a better quality signal over longer distances, making them more suitable for high-speed computer communication.

2.5.2.2 STP-shielded twisted pair

The shielded sheath made of a pair of wire mesh shielding sheathed on a twisted pair line must be grounded when used. It can reduce interference and crosstalk. In addition, its cost is higher. Figure 2-23 shows the comparison between UTP and STP.

Types	Advantages	Disadvantages
UTP	Low price, easy to install and reconfigure; Mature technology and stability	The delay characteristic is poor; Less crosstalk; Poor confidentiality
STP	Mature technology and stability; Wider bandwidth than UTP	The shielded sheath made of a pair of wire mesh shielding sheathed on a twisted pair line must be grounded when used; It can reduce interference and crosstalk; High price

Figure 2-23 Comparison between UTP and STP

2.5.3 Coaxial Cable

Another common transmission medium is the coaxial cable (known to its many friends as just "coax" and pronounced "co-ax"). It has better shielding than twisted pairs, so it can span longer distances at higher speeds. Two kinds of coaxial cable are widely used. One kind, 50-ohm cable, is commonly used when it is intended for digital transmission from the start. The other kind, 75-ohm cable, is commonly used for analog transmission and cable television. This distinction is based on historical, rather than technical, factors (e.g., early dipole antennas had an impedance of 300-ohm, and it was easy to use existing 4 : 1 impedance-matching transformers). Starting in the mid-1990s, cable TV operators began to provide Internet access over cable, which has made 75-ohm cable more important for data communication.

The coaxial cable is made up of two concentric conductors inside and outside. The inner layer is a conductor and the outer layer is the metal cylinder concentric with the inner conductor, which is supported by the filler of the insulating material and ensures the internal and external two layers of coaxial. The outer layer has a reinforced layer of rubber or plastic, as shown in Figure 2-24. In order to maintain the correct electrical characteristics of the coaxial cable, the cable must be grounded, and there must be an end connector at both ends, which is used to weaken and absorb the reflection effect of the signal.

The construction and shielding of the coaxial cable give it a good combination of high bandwidth and excellent noise immunity. The bandwidth possible depends on the cable quality and length. Modern cables have a bandwidth of up to a few GHz. Coaxial cables used to be widely used within the telephone system for long-distance lines but have now largely been replaced by fiber optics on longhaul routes. Coax is still widely used for cable television and metropolitan area networks, however.

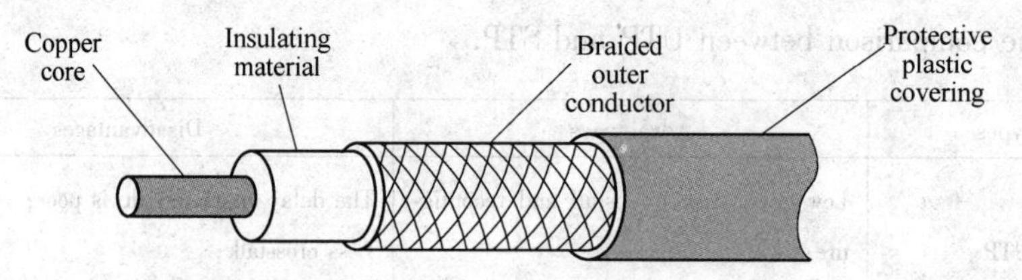

Figure 2-24 Coaxial cable

2.5.4 Fiber optics

Optical fiber is a kind of optical transmission medium, and it is a very thin fiber medium that can conduct light. Its radius is only a few microns to one or two hundred microns. The material for making optical fiber can be super pure silicon, synthetic glass or plastic. Because the visible frequency can reach 108MHz, the optical fiber transmission system has enough bandwidth. A fiber optic cable consists of a bundle of fibers.

Fiber optics is mainly composed by the fiber core, cladding and protection outer layer. The fiber core is the conductor part and the core part. It is the channel of light wave, made of glass fiber or plastic with excellent light conductivity. Cladding is mostly plastic or plastic coating, and the action is to reflect the light back to the fiber, while preventing interference from other light sources. The protection outer layer, mainly serves to protect, and there are gaps between it and the cladding, which can accommodate thin lines or bubble end, oil and so on.

The refractive index of the cladding and fiber layer is different. The optical fiber uses the difference of the refractive index to transmit the beam encoded by the signal by total internal reflection. When the incident angle of the light is greater than or equal to the critical value, the light will be completely confined in the optical fiber, and lossless propagation will take several kilometers. This is called the Refraction Theorem, as shown in Figure 2-25.

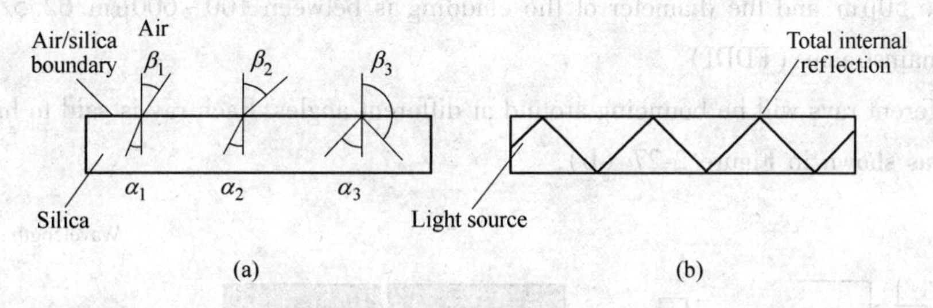

Figure 2-25 Refraction theorem

(a) Three examples of a light ray from inside a silica fiber impinging on the air/silica boundary at different angles; (b) Light trapped by total internal reflection

We choose the optical fiber transmission mode according to the way of light transmission between the optical fiber and cladding. It can be divided into single mode propagation and multimode propagation, as shown in Figure 2-26. Single mode propagation uses specific material and highly concentrated light source to limit the emitted light to a very low level, while multimode propagation means multiple beams of light propagage from light sources through different optical paths. Each kind of fiber needs different physical properties.

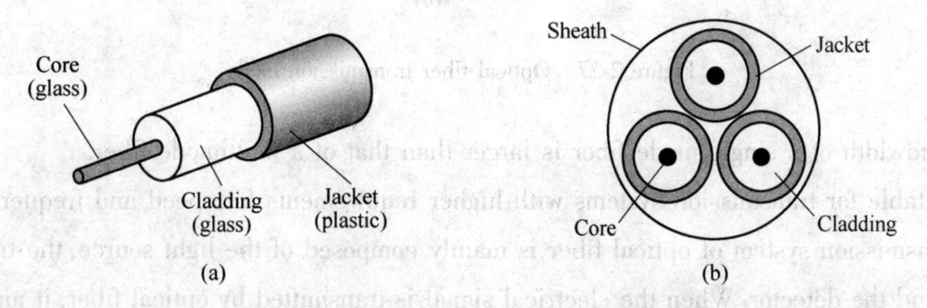

Figure 2-26 Typical optical fiber cable

(a) Side view of a single fiber; (b) End view of a sheath with three fibers

2. 5. 4. 1 Single mode

The beam travels in a straight line, with a single frequency and no refraction, and the core diameter is less than 10μm.

If the fiber′s diameter is reduced to a few wavelengths of light, the fiber acts like a wave guide, and the light can only propagate in a straight line, without bouncing.

In the manufacturing of a single mode fiber, the diameter and the very low density of the multi-mode fiber are much smaller than the multimode fiber, and the total reflection angle is close to 90 degrees so that the light propagating is basically horizontal, as shown in Figure 2-27 (a).

2. 5. 4. 2 Multiple mode

The beams propagate forward in waves, and coexist at various frequencies. The core diameter is mostly above 50μm, and the diameter of the cladding is between 100 ~ 600μm. 62. 5/125μm becomes the mainstream (FDDI).

Many different rays will be bouncing around at different angles. Each ray is said to have a different mode, as shown in Figure 2-27 (b).

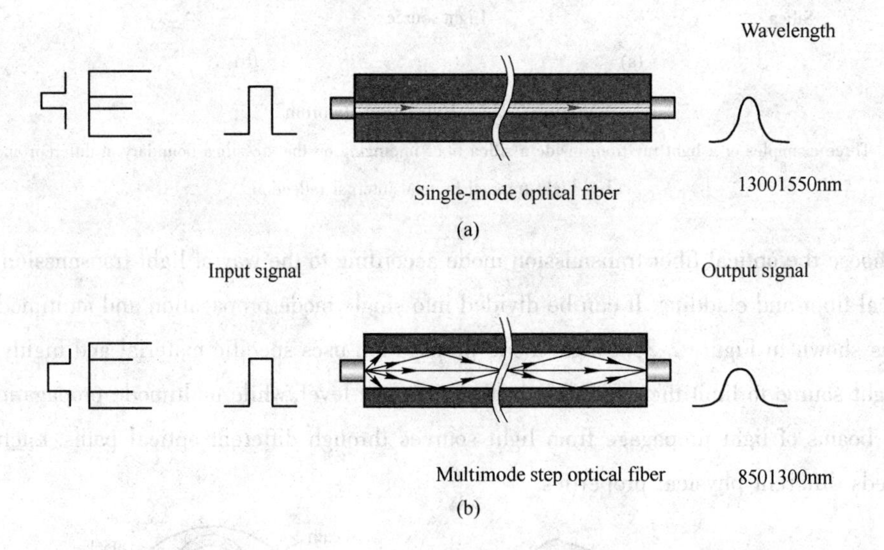

Figure 2-27 Optical fiber transmission mode

The bandwidth of a single mode fiber is larger than that of a multimode fiber.

It is suitable for transmission systems with higher requirements for speed and frequency.

The transmission system of optical fiber is mainly composed of the light source, the transmission medium, and the detector. When the electrical signal is transmitted by optical fiber, it must be converted to an optical signal at the transmitter side, and at the receiving end, it will also be restored by an optical detector to an electrical signal.

2. 5. 4. 3 Advantage

The transmission speed is very high, the bandwidth is very wide, and the capacity of transmitting information is huge;

The anti-interference ability is strong. The optical fiber is not affected by power interference and static interference. Even in the same optical cable, there is almost no interference between all optical fibers, so it is keeps data safe from outside signals that could otherwise interfere with the transmission.

2. 5. 4. 4 Shortcoming

High cost, difficulty in installation/maintenance and vulnerability.

2. 5. 4. 5 Conclusion

These are all " cable transmission" . According to the distance and the requirement of the communication rate, different cable media can be selected. Tabel 2-5 shows the performance comparasion between the above Transmission Medium. However, if the communication line is to pass through mountains, islands, and rivers, it is very difficult to lay the line and the cost is very high. At this time, we can consider the use of radio wave in the air. The spread of space to achieve a variety of communications.

Table 2-5 Performance comparison

class	Twisted pair	Coaxial line	Optical fiber
Cost	low	medium	high
Installation	easy	easy	complex
Bandwidth	155Mbit/s	500Mbit/s	>400Gbit/s
Decay	100m	1km	60km
Interference	bad	good	Very good

Because of the development of information technology and the rapid development of radio communication in the last decade, people can not only carry on mobile phone communication in motion, but also carry on computer data communication, which can not be separated from data transmission of the wireless channel.

2. 6 Wireless transmission

The biggest difference between wireless and wired transmission is that. Radio transmission does not use electricity or light but electromagnetic waves as the transmission medium. When wireless transmission is used, we need to configure the corresponding wireless transmitting and receiving devices. The most commonly used wireless transmission media are microwave, infrared and laser. Radio communication has been widely used in radio and television broadcasting. The relationship between electromagnetic spectrum and communication types is shown in Figure 2-28.

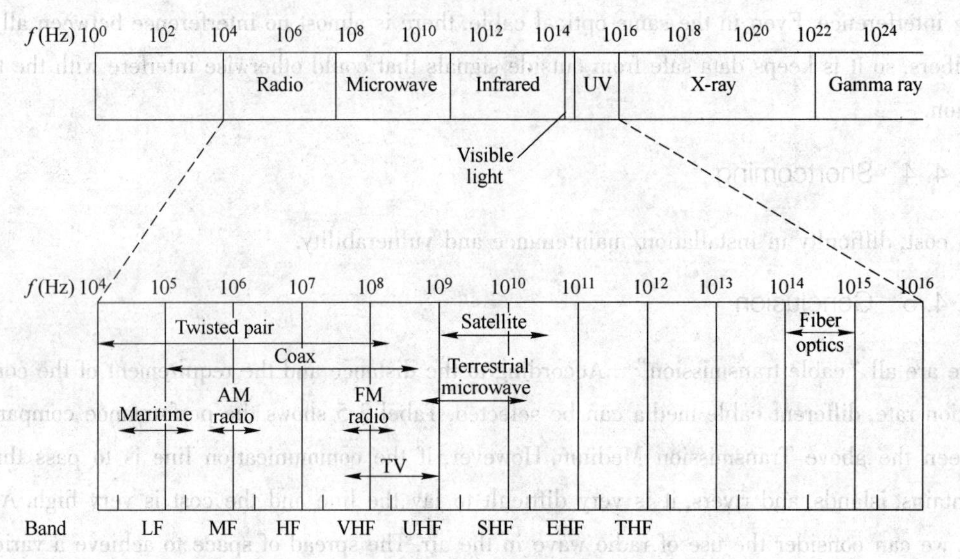

Figure 2-28 The electromagnetic spectrum and its use for communication

Wireless transmission mode mainly includes radio transmission, microwave transmission, infrared transmission and laser transmission.

2.6.1 Radio transmission

Radio transmission, also known as wireless power transmission, refers to the conversion of electrical energy into other forms of relay energy (such as electromagnetic field energy, laser, microwave, mechanical waves, etc.) through the transmitter, after a certain distance of space transmission, and then radio energy is transmitted by converting the relay energy into electrical energy through the receiver.

Radio waves are easy to generate, can travel long distances, and penetrate buildings easily, so they are widely used for communication both indoors and outdoors; radio waves also are omnidirectional, meaning that they travel in all directions from the source, so that the transmitter and receiver do not have to be carefully aligned physically.

The properties of radio waves are related to frequency. At low frequencies, radio waves pass through obstacles well, but the power falls off sharply with distance from the source. At high frequencies, radio waves tend to travel in straight lines and bounce off obstacles. They are also absorbed by rain. At all frequencies, radio waves are subject to interference from motors and other electrical equipment.

2.6.2 Microwave transmission

Different from modern communication network transmission methods such as coaxial cable communication, optical fiber communication and satellite communication, microwave communication uses microwave directly as a medium, and does not require a solid medium. When the distance between two points is unobstructed, microwave transmission can be used. The use of microwave for communication has a large capacity and good quality and can be transmitted over a long distance. Therefore, it is an important communication means of the national communication network, and is also generally applicable to various dedicated communication networks.

An electric wave used for spatial transmission is an electromagnetic wave that propagates at a speed equal to the speed of light. Radio waves can be classified and named according to frequency or wavelength. We refer to electromagnetic waves with frequencies above 300MHz as microwaves. Since the propagation characteristics of each band are different, it can be used in different communication systems. For example, the medium wave mainly propagates along the ground and has a strong diffraction ability, which is suitable for broadcasting and maritime communication. The short wave has strong ionospheric reflection capability and is suitable for global communication. Ultrashort wave and microwave have poor diffractive ability and can be used as line-of-sight or over-the-horizon relay communication.

Due to its wide frequency bandwidth and large capacity, microwave communication can be used for transmission of various telecommunication services, such as telephone, telegraph, data, fax and color TV, which can be transmitted through microwave circuits. Microwave communication has good disaster resistance. It is generally unaffected by natural disasters such as floods, windstorms and earthquakes. However, microwaves are transmitted over the air and are susceptible to interference. The same frequency cannot be used in the same direction on the same microwave circuit. Therefore, the microwave circuit must be constructed under the strict management of the radio management department. In addition, due to the characteristics of the linear propagation of microwaves, there should be no high-rise blockage in the direction of the radio beam. Therefore, the urban planning department should consider the planning of urban space for microwave channel so that communication is not affected by the barriers of high buildings.

Problems

2-1 What problems does the physical layer solve? What are the main characteristics of the physical layer?

2-2 Try to explain the following nouns: Data, Signal, Analog signal, Digital signal, Baseband signal, Band-pass signal, Symbol.

2-3 Try to explain the following nouns: simplex communication, half-duplex communication, full-duplex communication, asynchronous transmission and synchronous transmission.

2-4 What are the characteristics of the baseband signal and the broadband signal.

2-5 Please give examples of single-pass communication, half-duplex communication, and full-duplex communication.

2-6 What is asynchronous transmission, why asynchronous transmission is used in computer networks?

2-7 Among the following cases, the half-duplex transmission method is ( ).

A. Analog TV B. Walkie-talkie C. Phone D. Radio station

2-8 Among the following encoding methods, the method that provides synchronization information is ( ).

A. Manchester code B. NRZ code

C. NRZ inversion code D. Four-level natural code

2-9 In the following statement about the transmission rate, the correct one is ( ).

A. The unit of baud rate is baud, and the unit of bit rate is bit

B. The baud rate and bit rate are equivalent in rate

C. In serial transmission, when two-state modulation is used, the baud rate and bit rate are numerically equal

D. The baud rate and the bit rate are two values that are completely unrelated

2-10 What are the commonly used wired transmission media, and which types of signal transmission do they adapt to.

2-11 What are the commonly used wireless transmission media, and which types of signal transmission do they adapt to.

2-12 In the following statement, the correct one is ( ).

A. In an optical fiber, the refractive index of the cladding is greater than the refractive index of the core

B. Multimode abrupt fiber has a larger bandwidth than single mode fiber

C. The core of the multimode graded fiber has a constant refractive index

D. Fiber has strong anti-electromagnetic interference performance

2-13 Which of the following technologies is used to transmit digital signals?

A. Frequency division multiplexing

B. Synchronous time division multiplexing

C. Statistical time division multiplexing

D. B and C

2-14 In remote terminal communication, the role of the transmitter modem is ( ).

A. Turn the TTL level to RS-232-C level

B. Turn the analog information into a digital information

C. Turn the RS-232-C level to TTL level

D. Turn digital information into analog information

2-15 Explain the difference between baud rate and bit rate.

2-16 Try to encode data 100100111010100 using Manchester code, differential Manchester code, non-return-to-zero inversion code, and non-return-to-zero level code.

2-17 For the data 11100010, draw its Manchester coded map.

2-18 Suppose a transmission channel has a bandwidth of 50Hz. If the channel reaches 300kb/s, what is the required signal-to-noise ratio for this channel?

2-19 What are the factors that limit the transmission rate of data in the channel? Can the signal to noise ratio be increased arbitrarily? What is the significance of the Shannon formula in data communication? What is the difference between "bits/second" and "symbols/second"?

2-20 Suppose that a channel is limited by the Nyquist criterion to a maximum symbol rate of 20000 symbols per second. If amplitude modulation is used and the amplitude of the symbol is divided into 16 different levels for transmission, how high a data rate (bit/s) can be obtained?

2-21 Suppose you want to transmit 64kbit/s of data over a 3kHz bandwidth telephone channel (no error transmission). How high is the signal-to-noise ratio (represented by the ratio and decibel)? What does this result mean?

2-22 Calculate with the Shannon formula, assuming a channel bandwidth of 3100Hz and a maximum information transmission rate of 35kbit/s, how many times should the signal-to-noise ratio S/N be increased if the maximum information transmission rate is increased by 60%? If the signal-to-noise ratio S/N is increased to 10 times on the basis of the calculation just now, can the maximum information rate increase by 20%?

2-23 Compute the Fourier coefficients for the function $f(t) = t\ (0 \leqslant t \leqslant 1)$.

2-24 A noiseless 4kHz channel is sampled every 1 msec. What is the maximum data rate?

2-25 If a binary signal is sent over a 3kHz channel whose signal-to-noise ratio is 20dB, what is the maximum achievable data rate?

2-26 What are the advantages of fiber optics over copper as a transmission medium? Is there any downside of using fiber optics over copper?

2-27 It is desired to send a sequence of computer screen images over an optical fiber. The screen is 2560 × 1600 pixels, each pixel being 24bits. There are 60 screen images per second. How much bandwidth is needed?

2-28 Is the Nyquist theorem true for high-quality single-mode optical fiber or only for copper wire?

2-29 Ten signals, each requiring 4000Hz, are multiplexed onto a single channel using FDM. What is the minimum bandwidth required for the multiplexed channel? Assume that the guard bands are 400Hz wide.

2-30 Three packet-switching networks each contain n nodes. The first network has a star topology with a central switch, the second is a (bi-directional) ring, and the third is fully interconnected, with a wire from every node to every other node. What are the best-, average-, and worst-case transmission paths in hops?

2-31 Why use channel multiplexing technology? What are the commonly used channel multiplexing techniques?

2-32 Try to write the full text of the following English abbreviations and explain them briefly.

FDM, TDM, STDM, WDM, DWDM, CDMA, SONET, SDH, STM-1, OC-48.

2-33 The propagation rate of the signal on the transmission medium is 2×10^8 m/s and the propagation distance is 1000km. Try to calculate the transmission delay and propagation delay for the following two cases, and analyze the characteristics of the delay in both cases.

(1) The data length is 10^7 bit and the data transmission rate is 100kbit/s;

(2) The data length is 10^3 bit and the data transmission rate is 1Gbit/s.

3 Computer Network Foundation

Goals:

(1) Mater the OSI model and five-layer protocol.

(2) Understand the composition of the internet.

(3) Understand the network classification.

3.1 Overview

The function of computer networks mainly reflects four aspects: data transmission、sharing of resources、the reliability and availability of computers and distributed processing.

Data transmission is the most basic function of computer networks. It is used to quickly transfer various types of information between computers and terminals, computers and computers, including text letters, news messages, consulting information, photo materials, newspaper layouts, etc. By using this feature, units or departments scattered in various regions can be connected by computer networks for unified deployment, control and management.

Sharing of resources is the most attractive function of computer networks. "Resources" refers to all software, hardware, and data resources in the network. "Shared" means that users on the network are able to enjoy these resources in part or in whole. For example, the database of some regions or affiliates (such as airline tickets, hotel rooms, etc.) can be used throughout the network; some external devices such as printers can be user-oriented and can be used in places where these devices are not available. If resource sharing cannot be achieved, each region needs to have a complete set of software, hardware and data resources, which will greatly increase the investment cost of the whole system.

Computer networks improve the reliability and availability of the computer. Improved reliability makes it possible for computers to rely on each other. And improved availability balances the burden of computers and avoids overly busy and idle states.

Computer networks can easily to carry out distributed processing. When a computer is overburdened, or when the computer is processing a certain job, the network can transfer the new task to the idle computer to balance the load of each computer and improve the real-time processing problem; large comprehensive problems can be handed over to different computers for different parts of the problem, making full use of network resources, expanding the processing power of computers and enhancing practicality; a computer system combined by multiple computers working collaboratively and processing in parallel. Which is much cheaper than purchasing a high-performance mainframe separately.

3.2 Network topology

Network topology refers to the physical form of nodes connection in the network. Common types are star, tree, ring, bus shape, irregular shape, integrity, etc., as shown in Figure 3-1. There are two types of nodes in the network: one is the transfer node that acts by controlling and forwarding information in network communication, including SPC exchanges, communication processors, hubs and terminal controllers; the other is the end node which is both a source node and a sink node of communication, including computer hosts and terminals. The link is the connection between the two nodes. The topology design of a computer network has great influence on network performance, system reliability and communication cost.

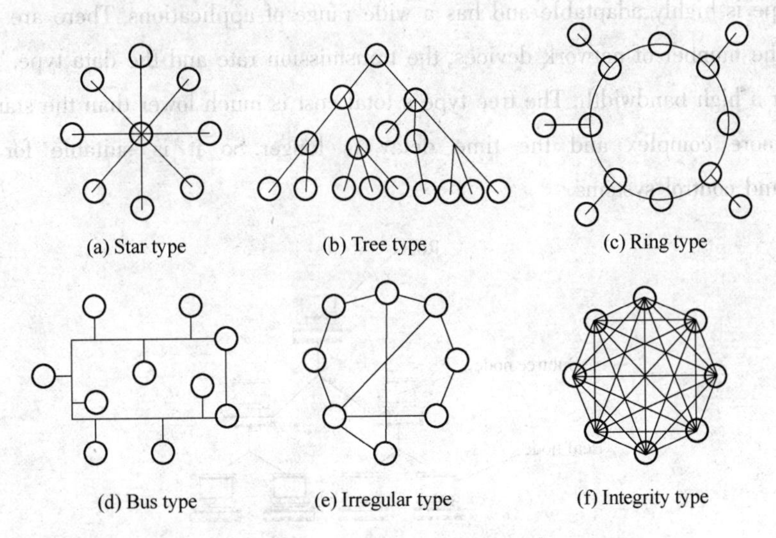

Figure 3-1 Common network topology

3.2.1 Star type

The star type includes the central node and the slave node. For the powerful computer, the central node has dual functions, which are data processing and storage and forwarding. For SPC exchanges or concentrator, the central node only connects from each node. The star usually has a simple structure and the network is easy to build and the early ethernet was a bus type, but now because of the Hub, the switch is physically star-shaped. However, it is logically a bus type.

As show in Figure 3-2 Star, when a channel or a slave node is faulty, it will not affect other parts' work, but when the central node fails, the whole network will stop working. The central processing unit needs high-speed processing performance and high reliability, and its cost is high.

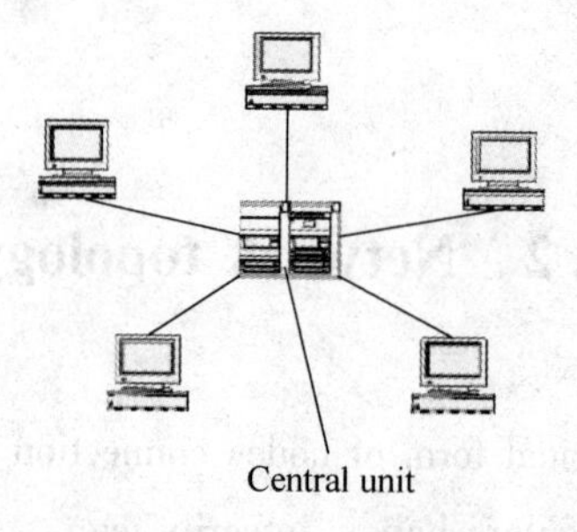

Figure 3-2 Star type

3. 2. 2 Tree type

The tree type is a stratified structure, an expansion of the star type, as shown in Figure 3-3. There is only one forwarding node in the star network, but both the root node and the sub-tree node are the forwarding nodes.

The tree type is highly adaptable and has a wide range of applications. There are not many restrictions on the number of network devices, the transmission rate and the data type. The tree type also allows for a high bandwidth. The tree type's total cost is much lower than the star type, but its structure is more complex and the time delay is longer. So it is suitable for hierarchical management and control systems.

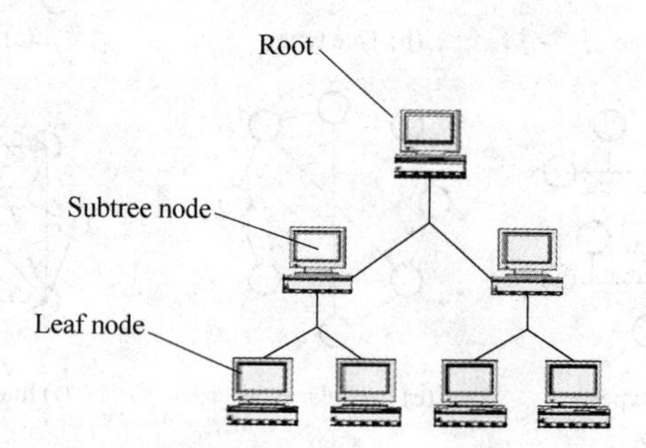

Figure 3-3 Tree type

3. 2. 3 Bus type

A common bus type as shown in Figure 3-4 is used to connect all workstations (hosts) and other shared devices (file servers, printers, etc.) through corresponding hardware interfaces. Early 10M coaxial cable is a typical bus type. Now Ethernet can connect to the hub and switch. Physically, it is a star type, but logically it is a bus type. This is determined by the mechanism.

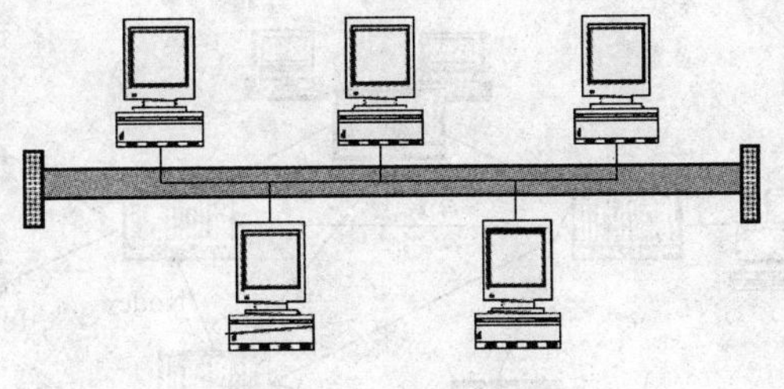

Figure 3-4 Bus type

The total cost of the bus type is much lower the star type, but its structure is more complex and the time delay is longer. Mainly it is used in LAN, such as Ethernet.

3. 2. 4 Ring type

Each host or terminal passes through the ring interface to form a closed ring.

The initial installation of the ring type is relatively easy, and the location of fault diagnosis is relatively accurate. The ring type, as shown in Figure 3-5, utilizes point-to-point connection and one-way transmission; it is suitable for optical fiber transmission. But the reliability is poor, and it is difficult to reconfigure. It is mainly used in local area networks, such as FDDI.

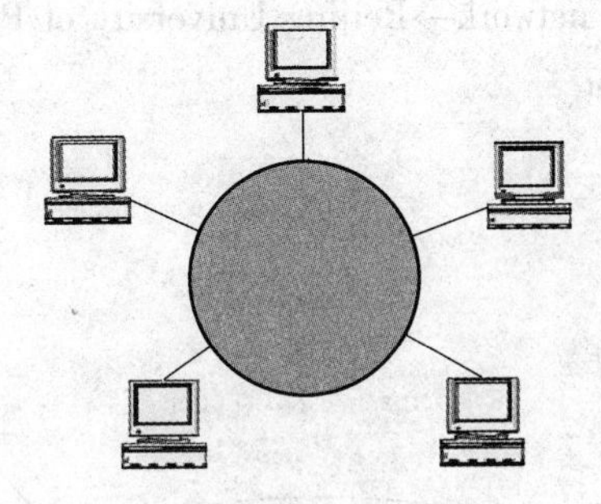

Figure 3-5 Ring type

3. 2. 5 Reticulation type

The reticulation (irregular and complete) type, as shown in Figure 3-6, is connected with computers located at different places by channels, and its shape is arbitrary. In a reticulation network, all nodes can be wired to all other nodes in the topology to form a "local area network." The difference between a reticulation network and a general network architecture is that all nodes can communicate data through multiple hops, although they are usually not mobile devices.

Although the nodes connection in a reticulation network take an irregular form, its reliability is relatively high relatively. It is mainly used in large wide area network.

With the wide application of the microcomputer, a large number of microcomputers are connected to the WAN through the LAN, and the interconnection between LAN and WAN, WAN and WAN is realized through the router.

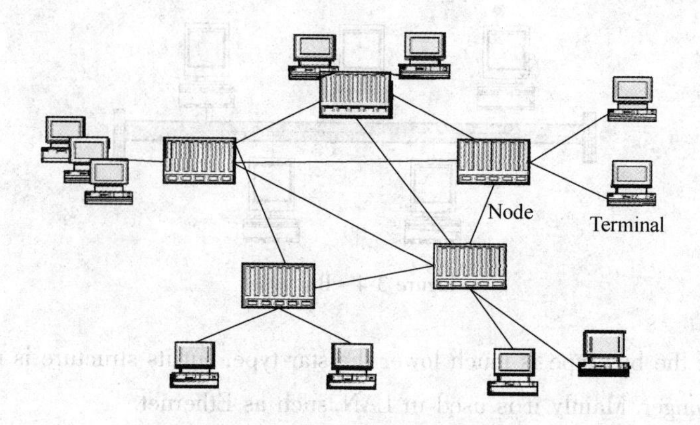

Figure 3-6 Reticulation type

To connect to the Internet, the user's computer needs to join the backbone network through the campus network, the enterprise network, or the ISP, and the regional backbone network is connectal to high-speed backbone networks among conntries through the national backbone network. In this way, a large and hierarchical interconnection network will be formed by routers, as shown in Figure 3-7.

Example: computer→campus network→Beijing University of Posts and Telecommunications→Tsinghua University→the Internet.

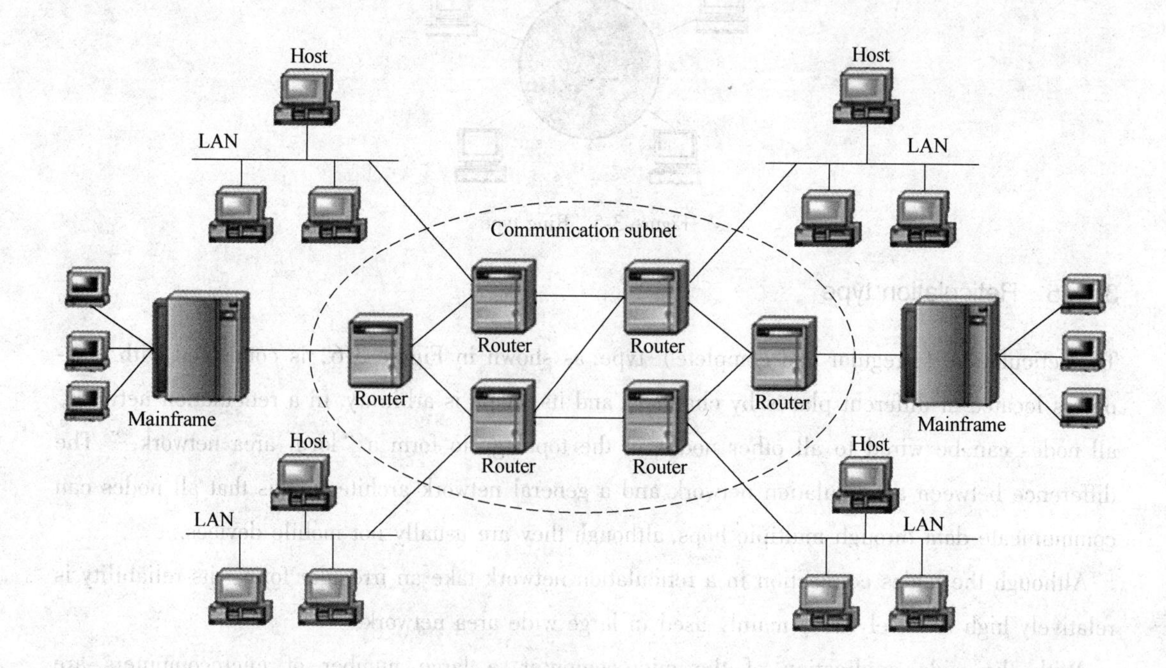

Figure 3-7 Schematic diagram of large and hierarchical interconnection

3. 3 Computer network hardware

3. 3. 1 The classification of computer network

Computer network classification standards are diverse. According to the topology structure, there are star networks, tree networks, bus networks, ring networks, mesh networks and so on. According to the scope of use, there are public networks and private networks. According to the way of information exchange, there are circuit switching networks, packet switching networks and integrated switching networks. According to the mode of communication, there are twisted pair networks, coaxial cable networks, optical fiber networks, wireless networks and satellite networks. According to the transmission technology, there are broadcast network and point-to-point networks. According to the network size, there are local area networks, metropolitan area networks and wide area networks. In the following section, we give a brief introduction to network hardware.

3. 3. 2 Classification by transmission technology

3. 3. 2. 1 Broadcast network

Packets can be sent by any machine and received by all other machines. The address field of the packet indicates which machine should be the receiver.

The broadcasting system shown in Figure 3-8 uses a special code in the address field to send packets to all targets. Every machine on the network receives and handles packets. This mode of operation is called broadcasting. Some broadcast systems also support the sending function to a subset of the machine, which is known as multicasting. If packets are sent to only one computer in the network, it is called unicasting.

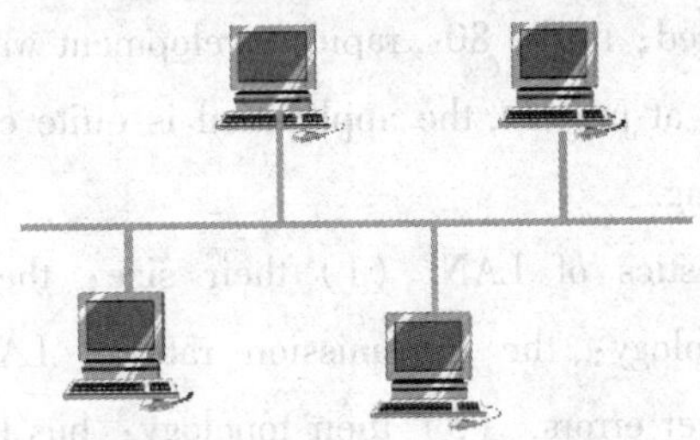

Figure 3-8 Broadcast network

3. 3. 2. 2 Point-to-point network

The point to point network consists of multiple connections between machines, and the routing algorithm is important.

In short, when the network is small or in a local range, we usually use broadcast mode, and for the network with a large scale or wide coverage area, we usually use point-to-point mode.

3. 3. 3 Classification by network scale

According to the network size, there are personal area network, local area network, metropolitan area network, wide area network, wireless networks and internetworks. As the Figure 3-9 shows, different types of networks are used for different interconnected processor distances.

Interprocessor distance	Processors located in same	Example
1m	Square meter	Personal area network
10m	Room	Local area network
100m	Building	Local area network
1km	Campus	Local area network
10km	City	Metropolitan area network
100km	Country	Wide area nework
1000km	Continent	Wide area nework
10000km	Planet	The Internet

Figure 3-9 Classification of interconnected processors by scale

3. 3. 3. 1 Local area networks

For local area networks (LAN), privately-owned networks, geographic range differs usually from tens of meters to a few kilometers. It belongs to a small network built by a department or unit. Broadcasts are widely used in LAN.

In the late 70s, LAN appeared; in the 80s, rapid development was achieved; in the 90s, it entered the period of prosperity; at present, the application is quite common, reflecting the demand for information resources sharing.

There are three characteristics of LAN. (1) their size: the LAN is restricted in size. (2) their transmission technology: the transmission rate of LAN reaches 10 to 100Mbps, bringing out low delay and fewer errors. (3) their topology: bus type, star and ring type all can be used in LAN. For example, there are various types of topology for broadcast LAN. For bus, only one machine at any time can be sent to the main station, and centralized or distributed arbitration mechanism is used to resolve conflicts. Commonly, IEEE802. 3 Ethernet, a bus based broadcast network, uses distributed control, and the rate is 10Mb/s or 100Mb/s. For ring, the arbitration mechanism solves the simultaneous access to the ring network. IEEE802. 5, an IBM token ring network based LAN, the rate is 4Mb/s or 16Mb/s.

It is also possible to divide one large physical LAN into two smaller logical LANs. An occasion where a LAN needs to be divided is when the layout of the network equipment does not match the organization's structure. For example, the engineering and finance departments of a company might have computers on the same physical LAN because they are in the same wing of the building, but it might be easier to manage the system if engineering and finance logically each had its own network virtual LAN or VLAN. In this design each port is tagged with a "color", say green for engineering and red for finance. Two switches then forward packets so that the computers attached to the green ports are separated from the computers attached to the red ports. Broadcast packets sent on a red port, for example, will not be received on a green port, just as though there were two different LANs. We will cover VLANs at Chapter 5.

3.3.3.2 Metropolitan area networks

Metropolitan area networks (MAN) —are extended LANs which use a similar technology, and cover no more than a city or region. They are networks that can cover a city range.

At first, the main application of the MAN is to connect multiple LANs in the interconnected city. Now, the city network can be used to carry out different types of business, including the transmission of all kinds of real-time data, voice and videos.

MAN, based on a large LAN, usually uses similar technology to LAN, as shown in Figure 3-10, One of the main reasons why MAN is listed alone is that there is already a standard: the Distributed queue dual bus (DQDB), that is, IEEE802.6 which is made up of two buses, and all computers are connected to it. Because MAN and Ethernet, Token Ring, and Token Bus are all 802.x standard, so it is suitable for general LAN technology.

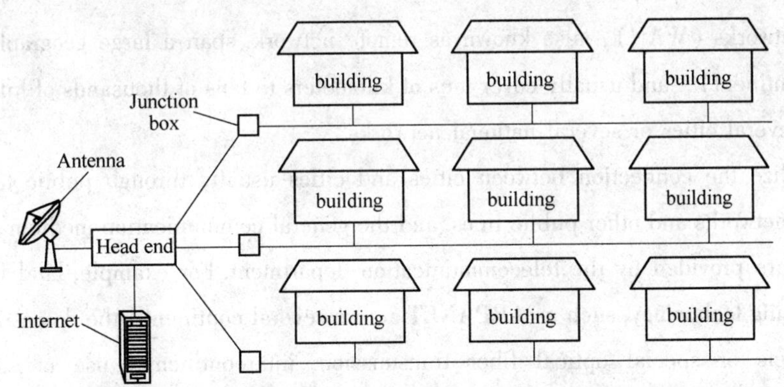

Figure 3-10 A metropolitan area network based on cable TV

MAN can use different physical media, such as optical fiber and twisted pair (the most widely used is optical fiber). Because MAN can provide multiple services and support various communication devices, its application will be more and more common.

The DQDB Metropolitan area networks structure is shown in Figure 3-11 DQDB Metropolitan area networks structure. The upper bus is used when the destination computer is on the right side of the sender. Conversely, the lower bus is used when the destination computer is on the left side of the sender.

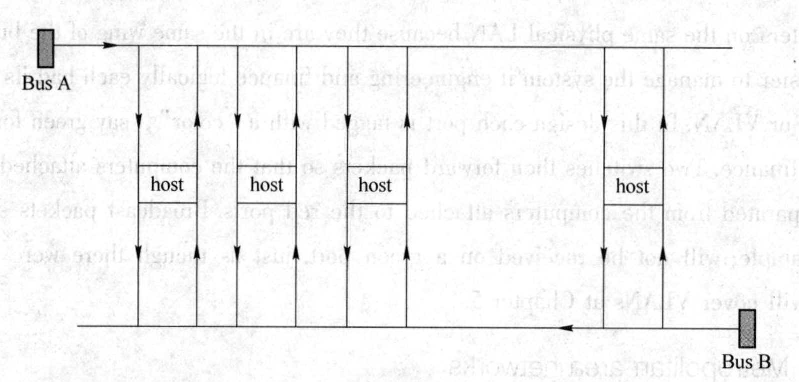

Figure 3-11 DQDB metropolitan area networks structure

Compared with LAN, MAN has many characteristics, such as a wide coverage area (up to 100km), high transmission speed (>50Mbps), a low bit error rate ($<10^{-9}$) and a large number of accommodating stations (>500). Additionally, the transmission media is fiber. MAN can be used not only for private networks but also for public networks, but LAN can only be used as a private network. LAN can only transmit data between PC machines, but MAN can not only transmit burst data services, but also transmit constant speed services and real-time services. MAN can work effectively in a variety of application environments.

3.3.3.3 Wide area networks

Wide area networks (WAN), also known as remote network, span a large geographical area (a country or continent), and usually cover tens of kilometers to tens of thousands of kilometers, such as crossing several cities or several national networks.

WANs realize the connection between cities and cities usually through public telephone networks, cable networks and other public lines, and the general communication medium and transmission devices are provided by the telecommunication department. For example, land networks uses packet switching technology, such as ARPANET across several continents, the domestic use of telephone channels or special optical fiber transmission, intercontinental use of satellite channels. Satellite networks uses satellite channels to realize packet switching, multiple access and broadcast distributed communication channels. Packet radio networks, in the relatively small range, uses common radio channel communication.

The development of WAN started in the 60s. The typical representative is ARPANET of the US Department of Defense. Chinese WAN includes mainly:

(1) CHINANET;

(2) CERNET;
(3) CSTNET;
(4) UNINET;
(5) CNCNET;
(6) CIETNET;
(7) CMNET;
(8) CGWNET;
(9) CSNET.

The Internet can be regarded as a special case of WAN.

Obviously, there are different characteristics of LAN, MAN and WAN, as shown in Table 3-1. Additionally, the topology of LAN and WAN is different. When computers, peripherals and communication equipment are connected within a geographical scope of more than ten kilometers, we usually use LAN, which provides a high rate and a low bit error rate while ensuring conveniece and simplicity. When a wide range is required, usually reaching dozens of kilometers, hundreds of kilometers, or even farther, we use WAN, which covers for the distances but with a low rate, a relatively high bit error rate and a more complex mechanism.

Table 3-1 Different characteristics of LAN, MAN and WAN

Class	Coverage	Transmission technology	Management	Whether to support broadcasting
LAN	Within a few kilometers	Usually broadcasting	Private network	Yes
MAN	Within 100 kilometers	The technology of LAN	Private network/ Public network	Yes
WAN	A country or continent	Need to forward or exchange devices	Private network/ Public network	No
Ethernet	Global	Need to forward or exchange devices	Public network	No

3.3.4 Wireless networks

Wireless networks can be classified to system interconnection named Bluetooth (as shown in Figure 3-12(a)), wireless LANs(as shown in Figure 3-12(b))and wireless WANs. Wireless networks have a wide range of applications, such as in portable office, trucks, taxis, buses and repairing persons for keeping in contact with home, even in the military. But there are some shortcomings of wireless networks. (1) its capacity is lower than wired LANs; (2) error rates are often much higher; (3) the transmissions from different computers can interfere with one another.

Figure 3-12 Bluetooth configuration(a) and wireless LAN(b)

3.4 Computer network software

Two computer systems that communicate with each other have to work in a highly coordinated way, and this "coordination" is rather complicated. In order to design such a complex computer network, a "Stratification" approach was proposed as early as the initial ARPANET design. "Stratification" can transform large and complex problems into smaller local problems, and these smaller local problems are relatively easy to research and deal with.

The OSI reference model is an important network architecture. As long as we follow the OSI standard, a system can communicate with any other system that follows the same standard anywhere in the world. But OSI failed in the market. OSI experts have no commercial driving force when they finish the OSI standard. The implementation of OSI protocol is too complex and inefficient. The establishment period of OSI standards is too long, so the equipment produced by OSI standards can not enter the market in time. The level division of OSI is not very reasonable, and some functions are repeated at multiple levels.

Therefore the legal international standard OSI has not been recognized by the market. The international standard TCP/IP is used now most widely. TCP/IP is often referred to as the international standard.

3.4.1 The necessity of protocol and hierarchies

Data exchange in computer networks must obey the rules agreed before. These rules clearly specify the data format and the synchronization problem. A network protocol, as simple as a protocol in the real world, is a rule, standard or convention for data exchange in a network. Furthermore, the network protocol is mainly composed of the following three components:

(1) Syntax: the structure or format of data and control information.

(2) Semantics: what kind of control information do we need to send, what actions to do and how to respond.

(3) Synchronization: order description of the order of event implementation.

Assuming that Host 1 sends a file through the network to the Host 2, it is possible to divide the work to be done as follows.

The first type of work is directly related to the transfer of documents. For example, the file transfer application on the sender should make sure that the file management program at the receiving end is ready to receive and store documents. If the file format is different, the two parties must coordinate a consistent file format, so the two hosts takes the file transfer module as the highest level. The rest of the work is responsible for the following modules.

However, we do not want the file transfer module to do all the details of the work, which makes the file transfer module too complicated. As shown in Figure 3-13, a communication service module can be set up to ensure that file and file transfer commands are reliably exchanged between the two systems. That is, We should let the file transfer module located above utilize the services provided by the communication service module below. We can also see that if the file transfer module located above is replaced with an e-mail module, the e-mail module can also utilize the reliable communication service provided by the communication service module below it.

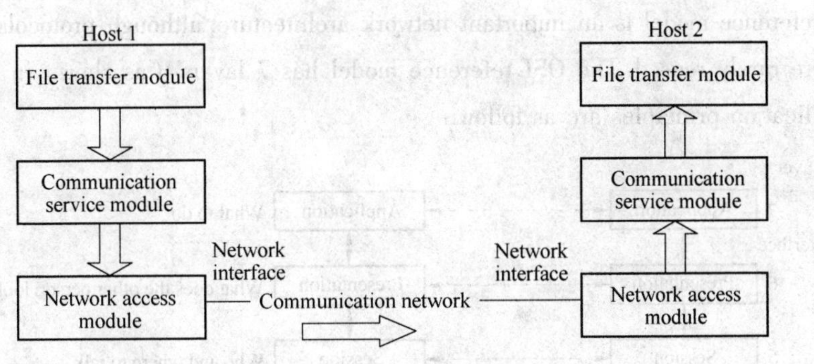

Figure 3-13 Example of hierarchy division

Similarly, the network access module is responsible for doing things related to network interface details. For example, the frame format and the maximum length of the frame are specified.

Stratification as shown in the examples above can bring about a number of benefits:

(1) Each layer is independent. A layer does not need to know how its next layer is implemented, but only needs to know the services provided by the layer through the interface between the layers. Since each layer implements only a relatively independent function, it is possible to decompose a problem that is difficult to handle into a number of smaller problems that are easier to handle. In this way, the complexity of the whole problem is reduced.

(2) The flexibility is good. When any layer changes, such as due to technical changes for example, as long as the interface relationship between the layers remains unchanged, the layers above or

below this layer are not affected. In addition, the services provided for a layer can be modified. When a service provided by a layer is no longer needed, it can even be cancelled.

(3) Its structure can be split open. Each layer can be implemented using the most suitable technology.

(4) Easy to implement and maintain. This structure makes it easy to implement and debug a large and complex system because the entire system has been broken down into several relatively independent subsystems.

(5) It can promote standardization work. Because the function of each layer and the services it provides are precisely described.

If the number of layers is too small, the protocol of each layer is quite complex. Too many layers will encounter more difficulties in describing and synthesizing system engineering tasks of all functions. So the number of layer should be appropriate.

The architecture of a computer network is the collection of all layers and its protocols. In other words, the architecture is the precise definition of the functions of the computer network and its components. Implementation is a problem that depends on what kind of hardware or software to perform these functions under the premise of this architecture. The architecture is abstract, but the implementation is concrete, which is the actual operation of computer hardware and software.

The OSI reference model is an important network architecture, although protocols related to the OSI model are rarely reused. The OSI reference model has 7 layers (as shown in Figure 3-14), and its stratification principles are as following:

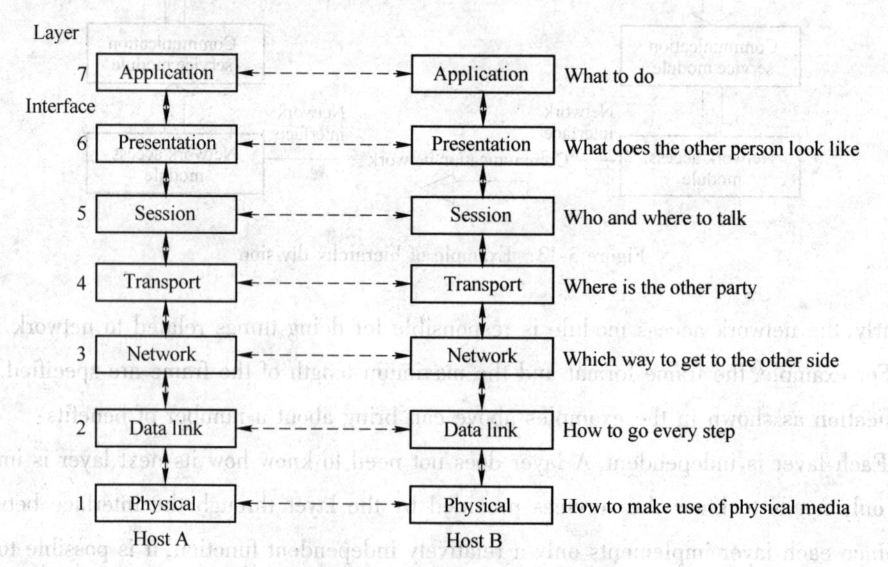

Figure 3-14 OSI reference model

(1) According to the abstract of different levels;

(2) Each layer should achieve a defined function;

(3) The selection of functions for each layer should help to formulate international standards for network protocols;

(4) The choice of boundary should minimize the amount of traffic across the interface;

(5) The number of layers should be enough to avoid mixing different functions in the same layer, but not too much, or else the architecture will be too large.

3.4.1.1 Physical layer

The physical layer does not refer to a specific transmission medium, but only refer to how to make use of the physical media. Its main function is to provide a physical connection for the data link layer, to ensure that information is entered into the channel and taken down by the receiver, so that the bit stream can be transparently transmitted. its data unit is bit.

A repeater belongs to a network interconnecting device at the physical layer. The main function of the repeater is to extend the transmission distance of the cable and to increase the number of nodes connected on the network segment. There are two types: electrical signal repeater and optical signal repeater. The repeater does not change the functionality of the network in any way, and on the segment it has the same data rate, protocol, and address.

A hub is a device that connects multiple Ethernet twisted pairs or fiber collections to the same physical medium. A hub is a physical layer that operates in the OSI model. It can be thought of as a multi-port repeater, and if it detects a collision, it will submit a blocking signal. There are two types of hubs: Passive Hub and Active Hub.

Notice that the physical layer is the lowest level of OSI, and is the interface between network physical devices. It is not the transmission medium, but only the physical devices: the data terminal device DTE (such as displays, PC, workstations, etc.) and the data circuit end connection device DCE (such as a modem).

Typical problems solved by the physical layer include: (1) The voltage level of "0" and "1"; (2) How long each one lasts; (3) Full duplex synchronization; (4) How to establish initial connections and terminate connections; (5) Mechanical interface, Pins; (6) Transmission medium.

3.4.1.2 Data link layer

The data link layer refers to how to go every step. Its main function is to establish, maintain and dismantle links between two adjacent nodes, and to transform unreliable physical links into error free data links through error control and flow control, which can improve link reliability and provide an error free data link logically for network communication. And its data unit is frame. Each frame includes a certain amount of data and some necessary control information.

A bridge is the primary interconnecting device at the data link layer. Through the bridge, you can divide a large LAN into multiple network segments, or interconnect two or more LANs into one large extended network. That is, the bridge is used to connect different network segments of the same type in the local area network, and the interconnection uses networks with different transmission rates and different transmission media. A bridge belongs to a store-and-forward device that acts on the physical layer and the data link layer. Frame interruption occurs between two network segments of the same type, and data frames are stored or forwarded between the local area networks.

A switch is a machine used to exchange messages. Most of them are link layer devices (Layer 2 switches), which can perform address learning and exchange messages in the form of store-and-forward.

3.4.1.3 Network layer

The network layer refers to is about choosing which way to get to the other side. From the source to the destination, there are multiple intermediate nodes and many ways to go. The network layer's main function is responsible for selecting an optimal path from it. That is to say, it determines how the packet is routed from source to destination. And its data unit is packet.

Problems to be dealt with in the network layer include:

(1) Routing. This is the main problem to be solved, considering factors such as optimality, fairness, simplicity, adaptability and stability.

(2) Congestion control. It aims at the problem of resource congestion in the network.

(3) Network addressing. Routers are used to connect two or more different logical networks.

(4) Network interconnection. When packets reach their destination cross a network, the addressing methods, the limit of packet length, and the protocols of the two networks are different, and now the network interconnection problem needs to be solved.

The router works in the third layer of the OSI protocol. Its main task is to receive data packets from a network interface and decide to forward to the next destination address according to the destination address contained therein.

3.4.1.4 Transport layer

The transport layer is about to where is the other. It is the core layer, which is located between the top level of OSI (5~7 layers, user oriented) and the low level (1~3 layers, network oriented), as shown below in Figure 3-15.

The transport layer is the level of a host to host. Its main functions include: (1) make best use of the network resources according to the characteristics of the communication subnet, and (2) establish a transmission channel between the session layer of the source host and the destination host in a reliable and economical way for transmitting the message transparently. The data unit is message. The basic function is to accept data from the session layer, split it up into smaller units if needed, pass these to the network layer, and ensure that the pieces all arrive correctly at the other

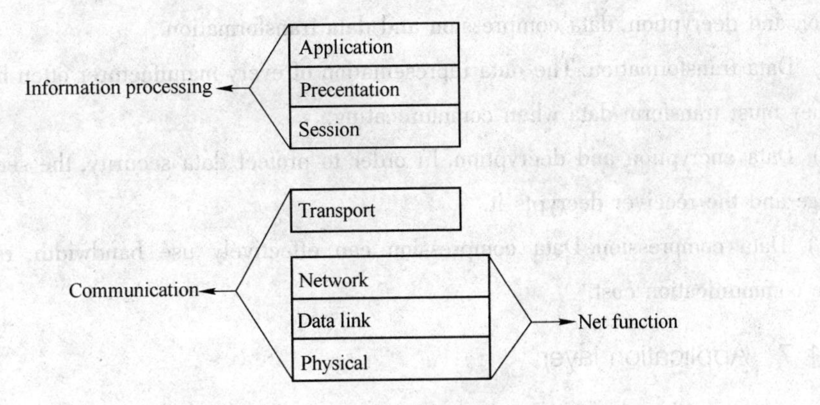

Figure 3-15 The transport layer location

end. The transport layer makes the session layer free from the influence of hardware technology change.

3. 4. 1. 5 Session layer

The session layer refers to who should speak and where to speak. The main function is managing dialogue control and user interface. Although the session layer does not participate in the specific data transmission, it manages the data transmission. The conversation layer establishes organizes and coordinates interaction activities between two communication applications. Its specific functions include the following:

(1) The establishment and release of session connection. Both sides of the conversation must be approved and have the right to participate in the conversation.

(2) Session synchronization service. It ensures that the whole session activity does not need to be repeated when the transport connection fails, and allows the session users to set the synchronization points freely in the transmitted data and gives synchronization numbers.

(3) The conversational interaction management. It coordinates and controls the interaction of multiple users, ensuring that actions are performed sequentially without confusion.

The session layer′s data unit is message. The layers above the session layer are application oriented. The session layer acts as a connection, adding some coordination management dialogues on the basis of transportation services, providing better services for the upper level entities. The layers below the session layer are communication oriented, providing transparent and reliable data transmission for the upper level of the two session entities, so that the session entity does not consider the communication problem during the session.

3. 4. 1. 6 Presentation layer

The presentation layer refers to what does the other person looks like. The main function is to handle the data representation problem between the two sides of communication, including data encryption and decryption, data compression and data transformation.

(1) Data transformation. The data representation of every manufacturer often has its own rules, and they must transform data when communicating.

(2) Data encryption and decryption. In order to protect data security, the sender encrypts the message and the receiver decrypts it.

(3) Data compression. Data compression can effectively use bandwidth, reduce delay and reduce communication cost.

3.4.1.7 Application layer

Obviously, the application layer refers to what to do. As the highest level of the OSI model, the application layer is a direct user oriented layer. The main function of the layer is to provides access to the process and the OSI environment by providing distributed information services to the application and system management processes. It contains a variety of protocols that are commonly needed. Its main work is to provide network services and solve the problem of how these services enable users to know and how to make users use.

The common network services include file service, e-mail service, virtual terminal service, network management service, printing service, integrated communication service, directory service, security service and so on.

3.4.2 Five-layer protocol architecture

TCP/IP is the four layers architecture: application layer, transport layer, Internet layer and network interface layer. But the bottom layer of the network interface is not specific. Therefore, a compromise is often adopted, that is to combine the advantages of OSI and TCP/IP, and to adopt an architecture with five layers of protocols as Figure 3-16 shows.

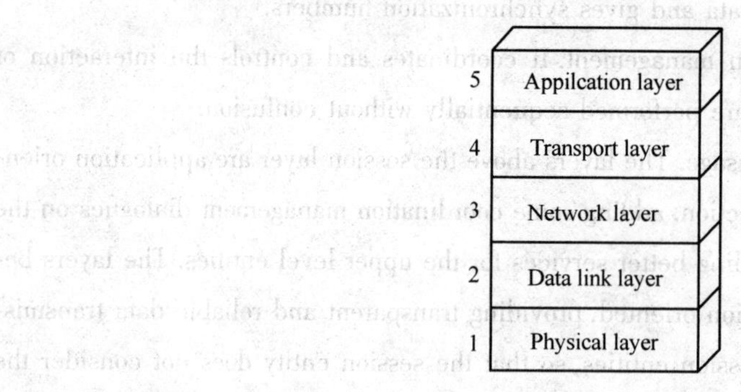

Figure 3-16 Five layers reference model

Figure 3-17 illustrates the changes experienced by the application process's data as it travels between layers assuming that Host 1 is connected to Host 2 by a router. It is assumed that the application process AP1 of the Host 1 transmits data to AP2 of Host 2. AP1 now hands over its data to

Layer 5 (application layer) of the Host 1. The fifth layer adding the necessary control information H_5 becomes the data unit of the next layer. After receiving the data unit, the fourth layer (transport layer) adds the control information H_4 of this layer, and then transfers to the third layer (network layer), thereby becoming the data unit of the third layer; so on and so forth. However, after the second layer (data link layer), the control information is divided into two parts, added to the header (H_2) and the tail (T_2) of the data unit in this layer respectively. And layer 1 (physical layer) is no longer added with control information because it is a bit stream. Please note that the bit stream should be transmitted from the header.

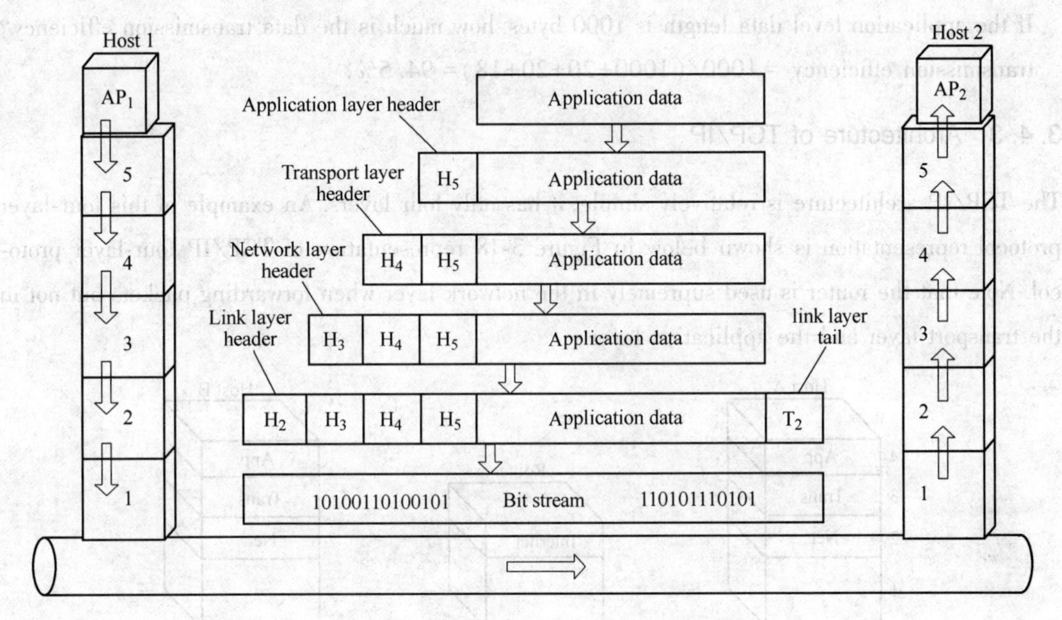

Figure 3-17 Data transfer between layers

When this series of bit streams is transmitted from Host 1 via the router to the destination station Host 2, it rises from the first layer of Host 2 to the fifth layer. Each layer performs the necessary operations based on the control information, and then strips the control information and hands over the remaining data units of the layer to a higher layer. Finally, the data sent by the application process AP_1 is handed over to the application process AP_2 of the destination station.

A simple example can be used to describe the above process. There is a letter that passes down from the top. Every time you go through a layer, pack a new envelope and add the necessary address information. After the letter containing the multiple envelopes is sent to the destination station, from the first layer, open the envelope in each layer and hand over to the upper layer. After reaching the top level, take the letter from the sender.

Although the application process data can be sent to the destination through the complicated process shown in Figure 3-17, these complex processes are blocked for the user, so that the application process AP_1 feels as if the data is directly delivered to AP_2. Similarly, any two identical levels (e. g. at layer 4 of the two systems) seem to transfer the data (i. e. the data unit plus the control letter) horizontally. This is the communication between the so-called "peer layers" . The various protocols mentioned before are actually the rules when passing data between peers.

Example: Data at the application level with a length of 100 bytes is sent to the transport layer, and 20 bytes of TCP header must be added. When you send it to the network layer, you need 20 bytes of IP header. Finally, the Ethernet transport to the data link layer is added, plus header and tail 18 bytes. Try to find the transmission efficiency of the data.

$$\text{Transmission efficiency} = 100/(100+20+20+18) = 63.3\%$$

If the application level data length is 1000 bytes, how much is the data transmission efficiency?

$$\text{transmission efficiency} = 1000/(1000+20+20+18) = 94.5\%.$$

3.4.3 Architecture of TCP/IP

The TCP/IP architecture is relatively simple, it has only four layers. An example of this four-layer protocol representation is shown below in Figure 3-18 representation of TCP/IP four-layer protocol. Note that the router is used supremely in the network layer when forwarding packets but not in the transport layer and the application layer.

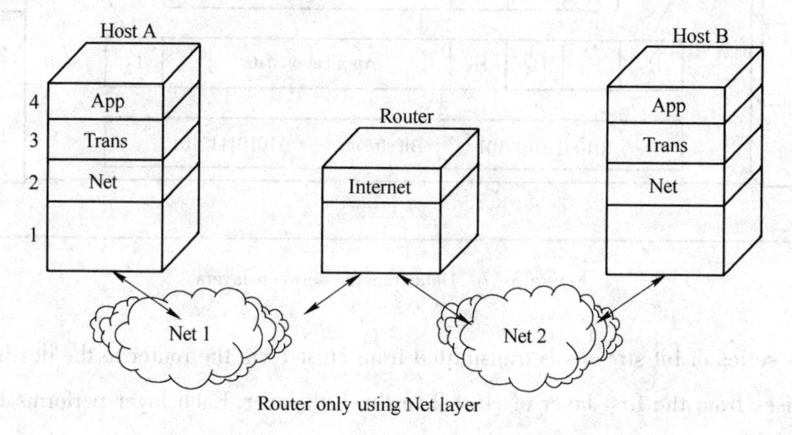

Figure 3-18 Representation of TCP/IP four-layer protocol

Another method is to draw a specific protocol to represent the TCP/IP protocol family (as shown in Figure 3-19). Its characteristics are that the upper and lower ends are large and the middle is small: the application layer and the network interface layer have many protocols, the middle IP layer is small, and the various protocols of the upper layer are aggregated downward into an IP protocol. This TCP/IP protocol family, much like the shape of an hourglass timer, shows that the TCP/IP protocol can serve a wide variety of applications (so-called everything over IP), while the TCP/IP protocol also allows IP protocols to be used in a variety of applications. Various networks are constructed to run on the Internet (so-called IP over everything). Because of this, the Internet was able to grow to this global scale today.

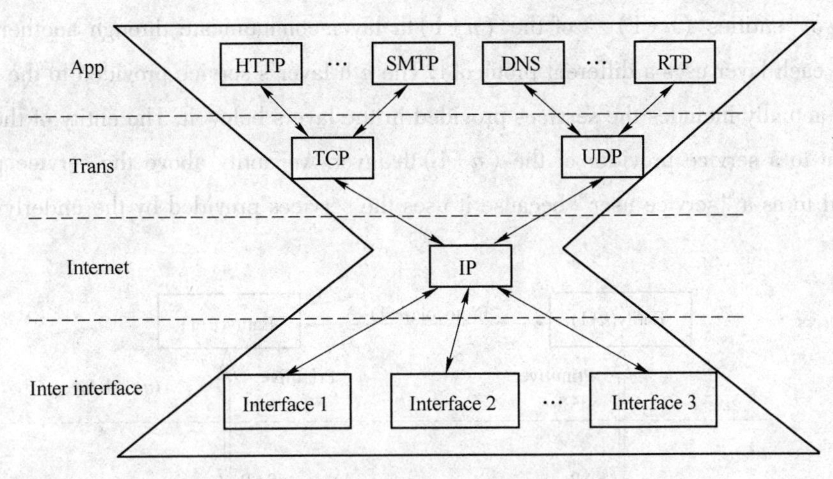

Figure 3-19 Hourglass timer shape TCP/IP protocol family

3.4.4 Entities, protocols, services, and service access points

Entity represents any hardware or software process that can send or receive information. In many cases, an entity is a specific software module.

A protocol is a set of rules that control the communication between two peers. The grammatical rules of the protocol define the format of the information exchanged, while the semantics rules of the protocol define the operations to be performed by the sender or receiver. For example, under what conditions data must be retransmitted or discarded.

Under the control of protocol, communication between two peer entities enables the layer to provide services to the next level. To achieve this protocol, we need to use the services provided by the lower level.

First, the implementation of the protocol guarantees the ability to provide services to the next level. Entities using this layer of service can only see services but cannot see the following protocols. The following protocols are transparent to the users above.

Second, the protocol is "horizontal", that is, the protocol is the rule to control the communication between peers. But the service is "vertical", that is, the service is provided by the lower layer to the upper layer through the inter-layer interface. In addition, functions don't perform in one layer are called services. Only those functions that can be seen "visible" by a higher level of entity can be called "services". The services provided by the low layer must be exchanged with the lower layer for some commands, which are called service primitives in the OSI.

The interaction between two adjacent entities in the same system is called SAP (Service Access Point), which is an abstraction. The relationship between any two adjacent layers can be summarized as shown in Figure 3-20 the relationship between any two adjacent layers below. It should be noted here that the two "entities (n)" of the n th layer communicate via "protocol (n)", while the two "entities $(n+1)$" of the $(n+1)$th layer communicate through another "protocol $(n+1)$" [each layer uses a different protocol]. The nth layer's service provided to the $(n+1)$ th layer above actually includes the services provided in the layers below it. The entity of the nth layer is equivalent to a service provider of the $(n+1)$th layer. An entity above the service provider is also referred to as a "service user" because it uses the services provided by the underlying service provider.

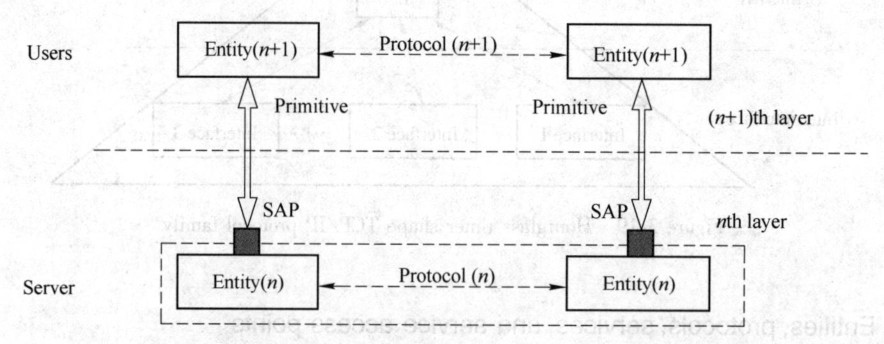

Figure 3-20 The relationship between any two adjacent layers

There are two services provided to the upper layer by the lower layer. One is connection-oriented service, taking the telephone system as a model to establish connections—to use connections and release connections. The other is connectionless service, taking the postal system as a model. Each message has a complete destination address, and each message is independent of other packets, passing through the selected route.

Service primitives: when the service user (n+1 entity) interacts with the service provider (n entity), the exchanged necessary information is used to notify the service user to take some action or to report the action of the service provider's peer entity to the service user. There are four types of service primitives, as shown in Figure 3-21.

(1) Request: an entity wants the service to do some work.

(2) Indication: an entity is to be informed of an event.

(3) Response: an entity wants to respond to an event.

(4) Confirm: the response to an earlier request has come back.

Examples of connection services:

(1) CONNECT. request: Calling the party service user to request to establish a connection.

(2) CONNECT. indication: The called party service provider reports to its service users that there is a connection establishment request.

(3) CONNECT. response: The called party service user indicates the acceptance of the connection request.

(4) CONNECT. confirm: The caller service provider notifies its service that the user connection has been established.

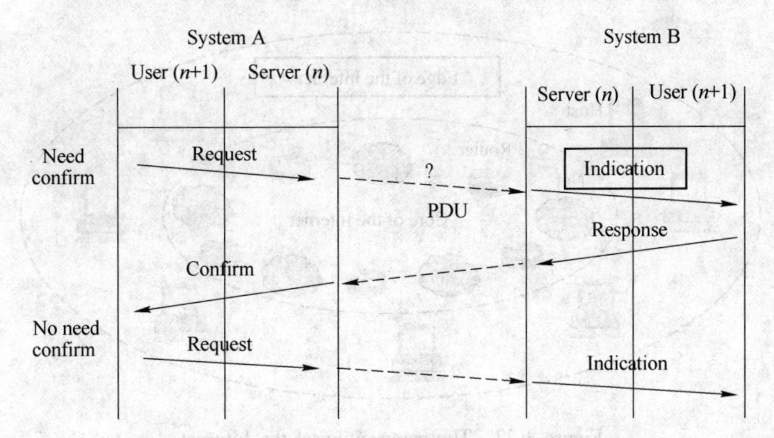

Figure 3-21 An example of the service primitives

(5) DATA. request: The service user requests its service provider to transmit the data to each other.

(6) DATA. indication: Service providers report services to users, and data arrive.

(7) DISCONNECT. request: Service users request to release connections.

(8) DISCONNECT. indication: The service provider notifies the service user that the other party has released the connection.

A connection oriented service example-comparison with a telephone system:

(1) CONNECT. request: Dial Aunt Millie's phone number;

(2) CONNECT. indication: Her phone rings;

(3) CONNECT. response: She picks up the phone;

(4) CONNECT. confirm: You hear the ringing stop;

(5) DATA. request: You invite her to tea;

(6) DATA. indication: She hears your invitation;

(7) DATA. request: She says she would be delighted to come;

(8) DATA. indication: You hear her acceptance;

(9) DISCONNECT. request: You hang up the phone;

(10) DISCONNECT. indication: She hears it and hangs up too.

3.5 The composition of the Internet

Although the topology of the internet is very complicated and covers the whole world geographically, it can be divided into the following two parts according to the way the Internet works, as shown in Figure 3-22.

(1) The edge part. It is composed of all hosts connected to the Internet. This part is directly

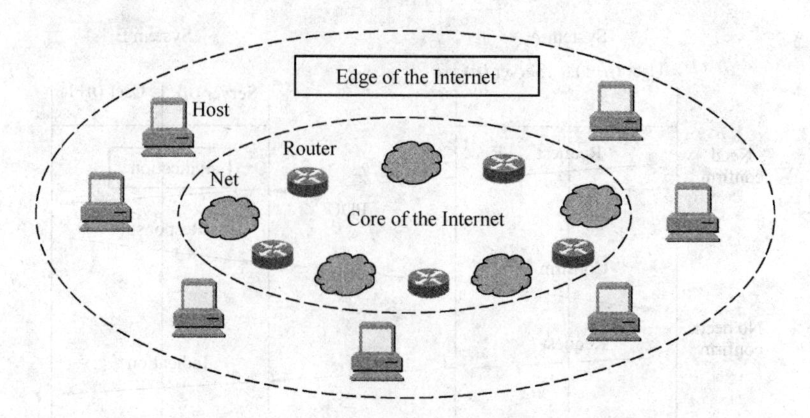

Figure 3-22 The composition of the Internet

used by users to communicate (transmit data, audio or video) and share resources.

(2) The core part. It is composed of a large number of routers connected to these networks. This part is to provide services for the edge part (providing connectivity and exchange).

3.5.1 The edge part of the Internet

The edge part of the Internet is all the hosts connected to the Internet. These hosts are also called end system. "Host A and Host B communicate", in fact, refers to: "a program running on Host A communicate with another program running on Host B". That is to say, one process of Host A communicates with another process on Host B, or it is referred to as "communication between computers" simply.

The means of communication between programs running in the end system of the network edge can usually be divided into two categories: Client/Server mode (C/S mode) and peer-to-peer mode (P2P).

3.5.1.1 Client server mode

Both client and server refer to application processes involved in communication. The client server mode (C/S mode) describes the service and serviced relationship between processes. The client is the requester of the service, and the server is the provider of the service. As the following Figure 3-23 C/S mode shown below, Client A sends a request service to Server B, and the Server B provides service to Client A.

The characteristics of client software include:

(1) After being invoked by users, it initiates communication (request service) to remote servers when planning to communicate. Therefore, the client program must know the address of the server program.

(2) No special hardware and complex operation system.

The characteristics of server software include:

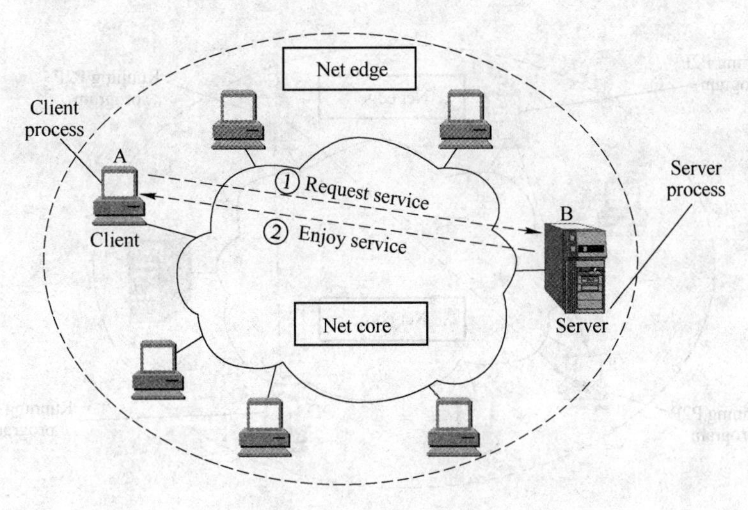

Figure 3-23 C/S mode

(1) A program specially designed to provide certain services, which can process multiple remote or local clients' requests at the same time.

(2) After the system starts, it calls automatically and runs continuously, passively waitting for and receiving communication requests from clients everywhere. Therefore, the server program does not need to know the address of the client program.

(3) Generally, strong hardware and advanced operation system support are needed.

After the communication relationship between the client and the server is established, the communication can be two-way, and both the client and the server can send and receive data.

3.5.1.2 Peer-to-peer mode

Peer-to-peer mode (P2P) refers to the fact that the two hosts do not distinguish the service requester or the service provider when communicating. As long as two hosts are running P2P software, they can perform equal and peer-to-peer communication. Both sides can download shared documents that have been stored in the hard disk. As Figure 3-24 P2P mode shows, Host C, D, E and F all run a P2P program, so they can perform peer-to-peer communication. Essentially, P2P still use the C/S mode, but every host in P2P is both a client and a server. For example, when Host C requests D's service, C is the client and D is the server. But if C provides services to F at the same time, C will play the role of server at the same time.

3.5.2 The core part of the Internet

The core part is the most complex part of Internet, because the core part of the network provides connectivity to a large number of hosts on the edge of the network, so that any hosts in the edge part can communicate to other hosts (that is, to transmit or receive various forms of data).

The router plays a special role in the core part of network. The router is a key component to real-

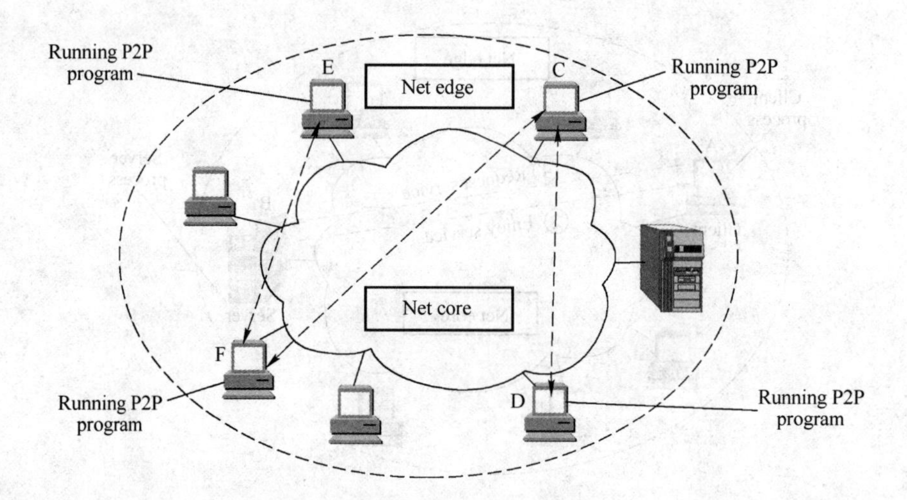

Figure 3-24 P2P mode

ize packet switching. Its task is to forward the received packets, which is the most important function of the core part of the network.

3.5.2.1 Main features of circuit switching

Shortly after the phone was introduced, people found it unrealistic to have all the phones connected in pairs. As Figure 3-24 shows, the two phones can connect to each other with only one pair of wires. But if five telephones are connected, ten pairs of wires are required, as shown in Figure 3-26 five telephones are directly connected to each other. Obviously, the N phones connected require $n(n-1)/2$ pairs of wires. When the number of telephones is large, the number of wire pairs required by this connection method is directly proportional to the square of the number of telephone. Then people realized when the number of telephone sets increases, switches will be used to complete the whole network exchange task, as shown in the following Figure 3-27.

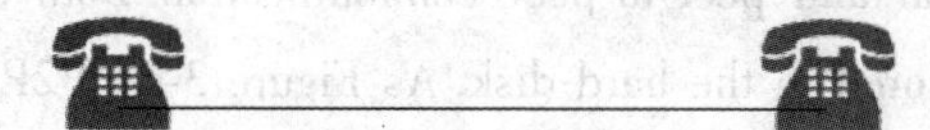

Figure 3-25 Two telephones are directly connected

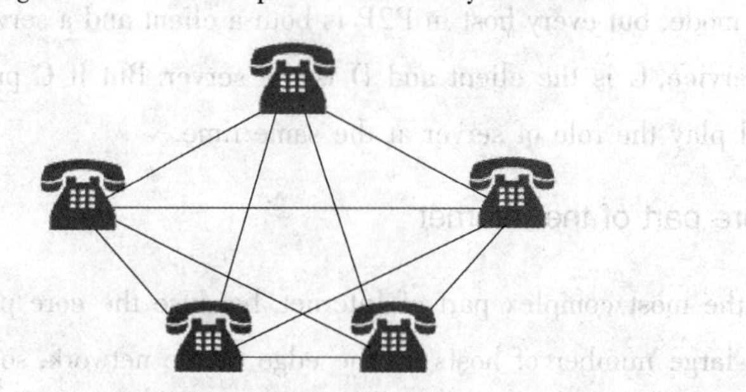

Figure 3-26 Five telephones are directly connected to each other

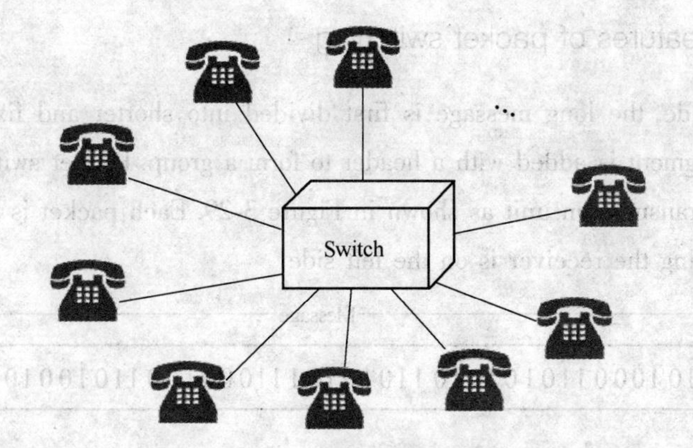

Figure 3-27 Connecting many telephones with a switch

Here, the meaning of switching is to transfer, that is to connect telephone line to another line. From the perspective of the allocation of communication resources, "switching" means dynamically distributing the transmission lines resources in a certain way. Before using a circuit connection, you must first dial up to request to establish a connection. When the called user hears the ringing tone sent by the switch and picks up the phone, a connection is established from the calling end to the called end, which is a dedicated physical channel. This connection guarantees the communication resources required for the two parties to talk, and these resources are not occupied by other users when communicating with each other. After that, both the calling and called parties can call. After the call is finished, the switch releases the dedicated physical channel that was just used. Circuit switching must be connection-oriented and go through three stages: establishing a connection, signaling communication and releasing the connection.

For example, as the following Figure 3-28 shows, A and B talk on a connection between them through four switches. C and D talk on a connection between them only through a local switch.

When using circuit switching to transfer computer data, the transmission rate of the line is often inefficient. Because the computer data is unexpected, this leads to low utilization of communication lines.

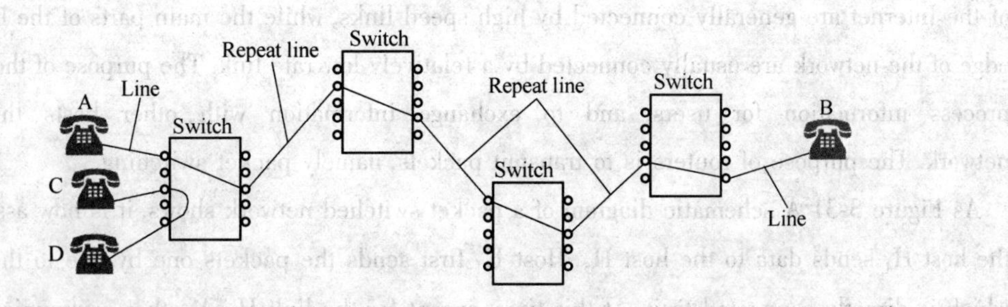

Figure 3-28 Example of circuit switching

3. 5. 2. 2 Main features of packet switching

At the transmitter side, the long message is first divided into shorter and fixed length data segments. Each data segment is added with a header to form a group. Packet switching network takes packet as the data transmission unit as shown in Figure 3-29. Each packet is sent to the receiving end in a row, assuming the receiver is on the left side.

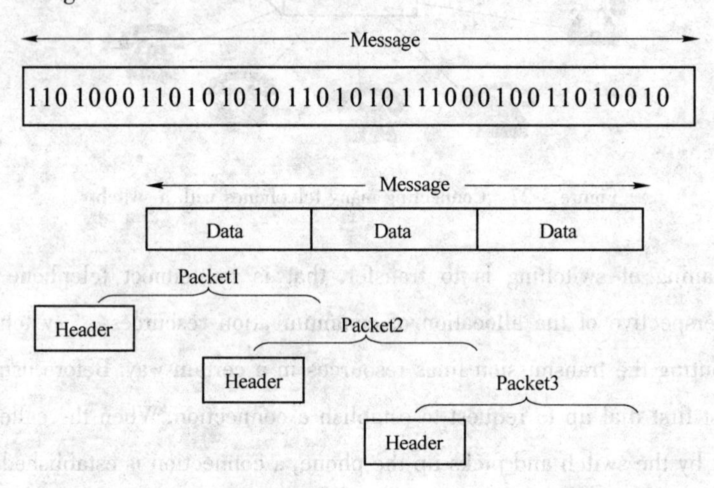

Figure 3-29 Transfer in the network by packet as the basic unit

The head of each packet contains control information such as the address. In the packet switching network, the node switches forward the packet to the next node switch according to the address information in the header of the packet. With this way of storage and forwarding, the final packet will arrive at the final destination.

After receiving the packet, the receiver is stripped of the header and reduced to a message. Finally, the received data is restored to the original message at the receiving end. Here we assume that there is no error in the transmission process of packets, and nor is it abandoned during forwarding.

As the following Figure 3-30 shows, the core of the Internet is composed of many networks and routers that interconnect them, and the host is at the edge of the Internet. Routers in the core part of the Internet are generally connected by high speed links, while the main parts of the host on the edge of the network are usually connected by a relatively low rate link. The purpose of the host is to process information for users, and to exchange information with other hosts through the network. The purpose of routers is to transmit packets, namely packet switching.

As Figure 3-31 A schematic diagram of a packet switched network shows, it is now assumed that the host H_1 sends data to the host H_5. Host H_1 first sends the packets one by one to the router A which is directly connected to it. At this time, except for the link H_1-A, other communication links are not occupied by both parties currently communicating. It should be noted that even the circuit

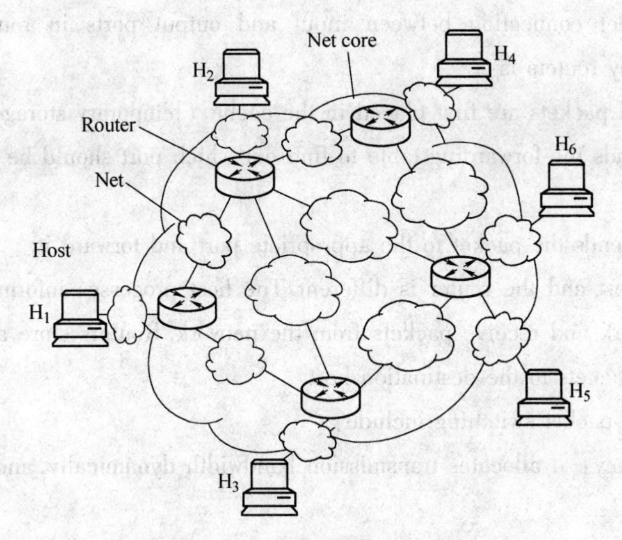

Figure 3-30 The core part of the router interconnects the network

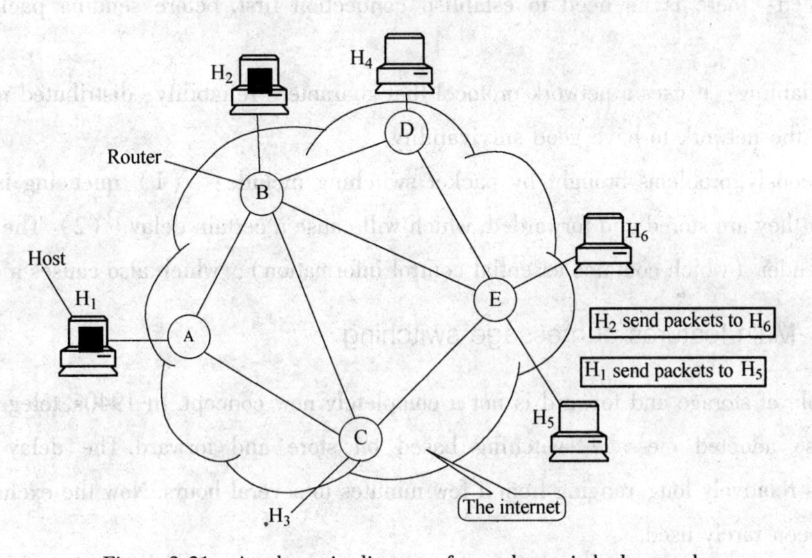

Figure 3-31 A schematic diagram of a packet switched network

H_1-A is only occupied when the packet is being transmitted on this link. The link H_1-A can still be used for packets sent by other hosts during the idle time between packet transmissions.

Router A puts the packet sent by host H_1 into the cache. It is assumed that the packet should be forwarded from the forwarding table of Router A to the link A-C. The packet is then transmitted to Router C. When a packet is being transmitted on the link A-C, it does not occupy resources of other parts of the network.

Router C continues to look up the forwarding table as described above, assuming that it should be forwarded to Router E. When the packet arrives at Router E, Router E finally hands the packet directly to Host H_5.

There is no direct connection between input and output ports in routers. The process of processing packets by routers is:

(1) The received packets are first placed in the cache (temporary storage).

(2) The route finds the forwarding table to find out which port should be forwarded to a destination address.

(3) The router Sends the packet to the appropriate port and forward it.

The role of the host and the router is different. The host processes information for users, send packets to the network and receive packets from the network. Routers store and forward packets, and finally deliver packets to the destination host.

The advantages of packet switching include:

(1) High efficiency: it allocates transmission bandwidth dynamically, and occupies communication link by stages.

(2) Flexibility: it groups transfer unit and looks up the route.

(3) Speed: there is no need to establish connection first, before sending packets to other hosts.

(4) Reliabling: it uses a network protocol that guarantees reliability; distributed routing protocol enables the network to have good survivability.

Simultaneously, problems brought by packet switching include: (1) queueing is needed for nodes when they are stored and forwarded, which will cause a certain delay. (2) The packet must carry the header (which contains essential control information), which also causes a certain cost.

3.5.2.3 Main features of message switching

The principle of storage and forward is not a completely new concept. In 1940s, telegraph communication also adopted message switching based on store and forward. The delay of message switching is relatively long, ranging from a few minutes to several hours. Now the exchange of messages has been rarely used.

Figure 3-32 shows the main difference between circuit switching, message exchange and packet switching. A and D are source and destination respectively, and B and C are intermediate nodes between A and D. At the bottom of the figure, the main features of the three exchange methods in the data transfer phase are summarized.

Circuit switching: The bit stream of the entire message continuously reaches the end point from the source point, as if it is transmitted in each pipe.

Message exchange: The entire message is first transmitted to the adjacent node, and all stored are searched for the forwarding table and forwarded to the next node.

Packet switching: A single packet (this is only part of the entire message) is sent to the adjacent node, stored and then looked up to the forwarding table and forwarded to the next node.

As can be seen from the figure, if a large amount of data is to be continuously transmitted and its

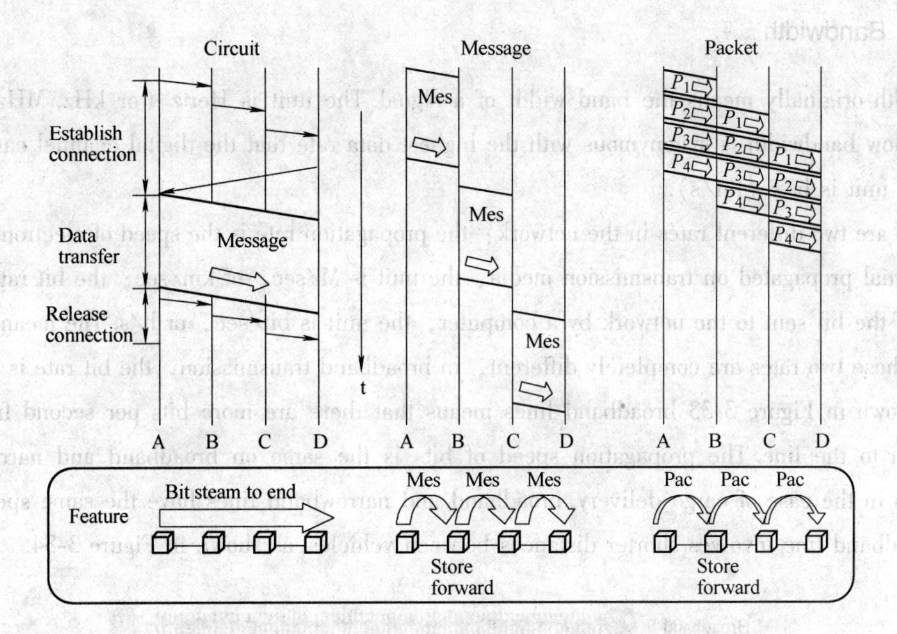

Figure 3-32 A comparison of three kinds of exchange

transmission time is much longer than the connection establishment time, the circuit switching transmission rate is faster. Message exchange and packet switching do not require pre-allocation of transmission bandwidth, and the transmission of burst data can improve the channel utilization of the entire network. Since the length of a packet is often much shorter than the length of the entire message, packet switching has a shorter delay than message exchange and it also has better flexibility.

3.6 The indexes of computer networks

Performance inderes measure computer performance in different ways. The six commonly used are described below.

3.6.1 Rate

Bit is the unit of data in computer, and it is also the unit of information quantity used in information theory. Bit comes from binary digit, which means a binary number. A bit is a 1 or 0 digit in binary number. Rate in network technology refers to the rate of data transfer, also known as data rate or bit rate. Rate is the most important performance index in computer networks. The unit of rate is b/s, or kb/s, Mb/s, Gb/s, etc. Rate often refers to the rate of fixed rate or nominal rate, not the rate at which the network is actually running.

3.6.2 Bandwidth

Bandwidth originally means the band width of a signal. The unit is Hertz (or kHz, MHz, GHz, etc.). Now bandwidth is synonymous with the highest data rate that the digital channel can transmit. The unit is b/s (bit/s).

There are two different rates in the network: the propagation rate is the speed of electromagnetic wave signal propagated on transmission media, the unit is M/sec, of km/sec; the bit rate is the speed of the bit sent to the network by a computer, the unit is bit/sec, or b/s. The meaning and unit of these two rates are completely different, in broadband transmission, the bit rate is higher.

As shown in Figure 3-33 broadband lines means that there are more bits per second from the computer to the line. The propagation speed of bits is the same on broadband and narrowband lines. As in the case of cargo delivery, broadband and narrowband lines have the same speed, but the broadband line involves shorter distances between vehicles, as shown in Figure 3-34.

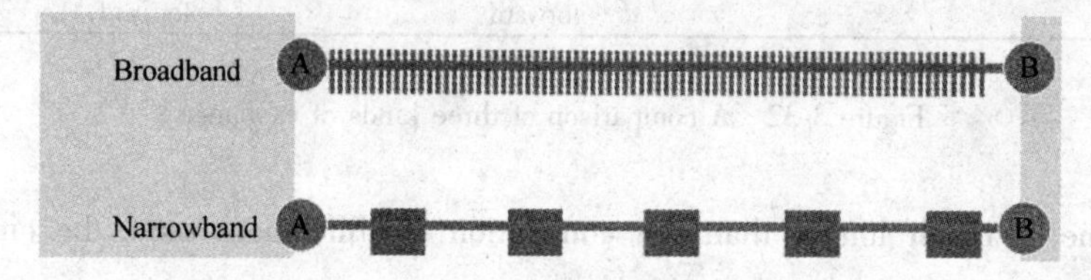

Figure 3-33 The bits in broadband and narrowband

Figure 3-34 An analogous illustration of broadband and narrowband same as metaphorical

A misconception is the belief that "broadband" is equivalent to "multi-lane". But in fact serial transmission is usually transmitted on a communication line, as shown in Figure 3-35.

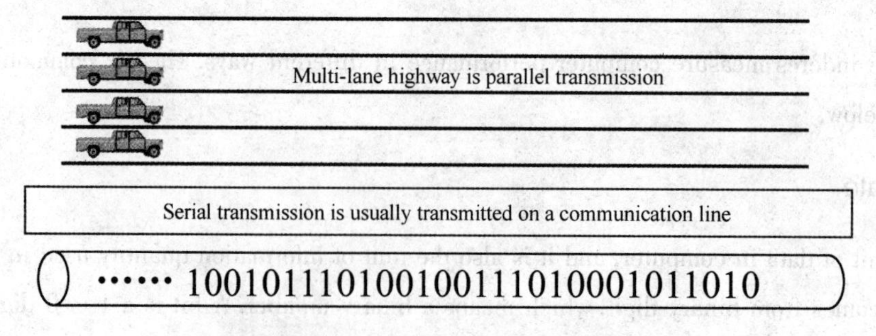

Figure 3-35 Serial transmission on a communication line

More commonly used bandwidth units are:

(1) A thousand per second, that is, kb/s (10^3 b/s).

(2) Megabytes per second, that is, Mb/s (10^6 b/s).

(3) Gigabit per second, Gb/s (10^9b/s).

(4) Terabit per second, Tb/s (10^{12} b/s).

Note: in the computer world, K = 2^{10} = 1024, M = 2^{20}, G = 2^{30}, T = 2^{40}.

On the time axis, the width of the signal is narrowed with the increase of the width as shown in Figure 3-36.

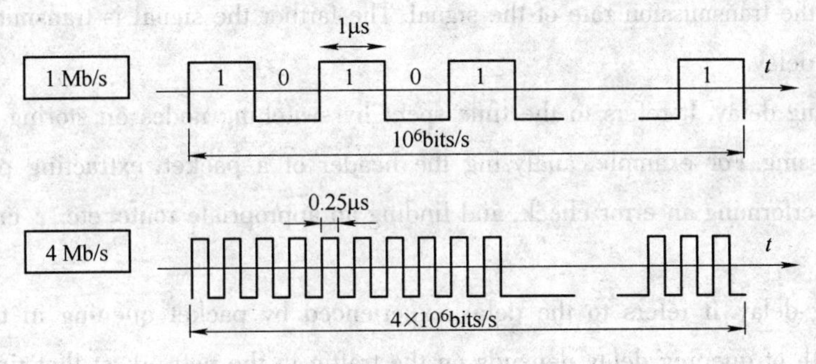

Figure 3-36 Digital signal flow over time

3. 6. 3 Throughput

Throughput represents the amount of data that is passed through a network (or channel and interface) in a unit time. Throughput is more often used for a measurement of the real world network in order to know how much data actually can be passed through the network. Throughput is limited by the bandwidth or the rated speed of the network. For example, for a 1 Gbit/s Ethernet, its rated speed is 1 Gbit/s, and then the integer value is also the absolute upper limit of the throughput of the Ethernet. Therefore, for a 1 Gbit/s Ethernet, the actual throughput may be only 100 Mbit/s, or even lower, and it does not reach its rated speed. Sometimes throughput can also be expressed in bytes or frames per second.

3. 6. 4 Delay

Delay or latency is the time that it takes for data (a message or packet, or even a bit) to travel from one end of the network (or link) to the other. The delay in the network consists of the following two parts.

(1) Transmission delay. It refers to the time that data blocks need to enter from the node to transmition media. That is, the time required from the first bit of the transmitted data frame to the last bit of the frame.

$$\text{Transmission delay} = \frac{\text{Data block length (bit)}}{\text{Transmission rate (bit/s)}}$$

(2) Propagation delay. It refers to the time spent when electromagnetic wave needs to propagate a certain distance in the channel.

$$\text{Propagation delay} = \frac{\text{Channel length (m)}}{\text{Propagation rate (m/s)}}$$

The signal transmission rate (i. e. transmission rate) and signal propagation speed on a channel are completely different concepts. The transmission delay occurs in the transmitter inside the machine and has nothing to do with the length of the transmission channel (or the distance traveled by the signal). However, the propagation delay occurs on the transmission channel media outside it, regardless of the transmission rate of the signal. The farther the signal is transmitted, the greater the propagation delay.

(3) Processing delay. It refers to the time spent by switching nodes on storing and forwarding necessary processing. For example, analyzing the header of a packet, extracting portions of data from a packet, performing an error check, and finding an appropriate route, etc., create a processing delay.

(4) Queuing delay. It refers to the delay experienced by packet queuing in the node cache queue. The length of queuing delay depends on the traffic in the network at that time.

The total delay experienced by data is the sum of transmission delay, propagation delay, processing delay and queuing delay.

Total delay = transmission delay + propagation delay + processing delay + queuing delay

In general, a network with a small delay is better than a network with a large delay. In some cases, a low-rate, small-latency network is likely to be better than a high-rate but large-latency network.

As Figure 3-37 the place of four kinds of delay shows, the place produced by four kinds of delay is different when data is sent from node A to node B.

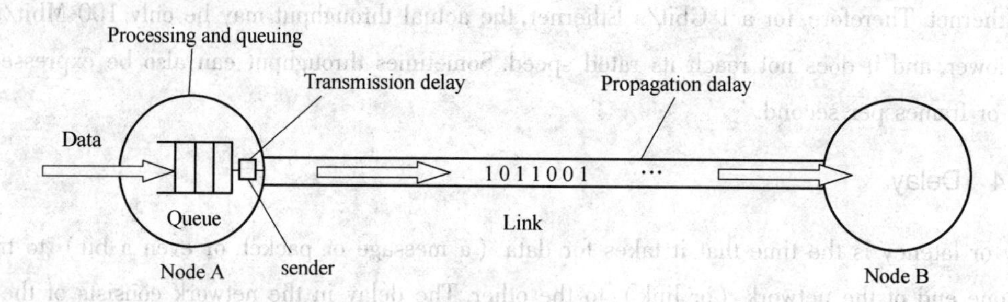

Figure 3-37 The place of four kinds of delay

For high-speed network links, we only increase the data transmission rate rather than the bit rate on the link. This improves link bandwidth and reduce data transmission delay.

Example: The transmission distance between the ends of the transceiver is 1000km. We try to calculate the transmission delay and propagation delay in the following two situations.

(1) The data length is 10^7bit, the data transmission rate is 100kbit/s, the propagation distance is 1000km, and the propagation rate of the signal on the media is 2×10^8m/s.

(2) Data length is 10^3bit, data transmission rate is 1Gbit/s, and the transmission distance and signal propagation rate on media are the same.

Answer:

(1) Transmission delay $=10^7/(100\times1000)=100$s

Propagation delay $=1000\times1000/(2\times10^8)=5\times10^{-3}s=5$ms

(2) Transmission delay $=10^3/(10^9)=10^{-6}$s$=1\mu$s

Propagation delay $=1000\times1000/(2\times10^8)=5\times10^{-3}s=5$ms

3. 6. 5 Bandwidth delay product

Bandwidth delay product (BDP) refers to the product of propagation delay and bandwidth. As Figure 3-38 bandwidth delay product dennonstrates, a cylindrical pipe represents a link, the length of the pipe is the propagation delay of the link, and the cross-sectional area of the pipe is the bandwidth of the link. Thus the delay bandwidth product represents the volume of this pipe, indicating how many bits the link can hold.

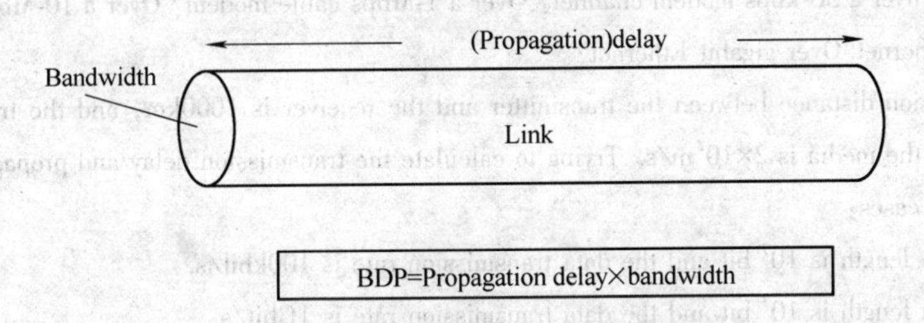

Figure 3-38 Bandwidth delay product

Example: It is assumed that the propagation rate of the signal on the media is 2.3×10^8m/s. The media length is:

(1) 10cm (network card).

(2) 100m (LAN).

(3) 100km (metropolitan area network).

(4) 5000km (wide area network).

Try to calculate the number of bits being propagated in the above media when the data rate is 1Mb/s and 10Gb/s.

Answer: propagation delay = channel length/electromagnetic wave in the channel transmission rate

Delay bandwidth = propagation delay×bandwidth

(1) $0.1/(2.3\times10^8)\times1\times10^6=0.000435$bit

(2) $100/(2.3\times10^8)\times1\times10^6=0.435$bit

(3) $100000/(2.3\times10^8)\times1\times10^6=435$bit

(4) $5\times10^6/(2.3\times10^8)\times1\times10^6=21739$bit

Problems

3-1 What are the reasons for using layered protocols? What is one possible disadvantage of using layered protocols?

3-2 What is the principal difference between connectionless communication and connection-oriented communication? Give one example of a protocol that uses (1) connectionless communication; (2) connection-oriented communication.

3-3 Which of the OSI layers and TCP/IP layers handles each of the following:

(1) Dividing the transmitted bit stream into frames.

(2) Determining which route through the subnet to use.

3-4 List two ways in which the OSI reference model and the TCP/IP reference model are the same. Now list two ways in which they differ.

3-5 The Internet is roughly doubling in size every 18 months. Although no one really knows for sure, one estimate put the number of hosts on it at 10 billion in 2018. Use these data to compute the expected number of Internet hosts in the year 2030. Do you believe this? Explain why or why not.

3-6 An image is 1024×768 pixels with 3 bytes/pixel. Assume the image is uncompressed. How long does it take to transmit it over a 56-kbps modem channel? Over a 1-Mbps cable modem? Over a 10-Mbps Ethernet? Over 100-Mbps Ethernet. Over gigabit Ethernet?

3-7 The transmission distance between the transmitter and the receiver is 1000km, and the transmission rate of the signal on the media is 2×10^8 m/s. Trying to calculate the transmission delay and propagation delay in the following two cases:

(1) The data length is 10^7 bit and the data transmission rate is 100kbit/s.

(2) The data length is 10^3 bit and the data transmission rate is 1Gbit/s.

What conclusions can be drawn from the above calculation results?

3-8 It is assumed that the propagation rate of the signal on the media is 2.3×10^8 m/s. Media length l is as follows:

(1) 10cm (Network Interface Card)

(2) 100m (LAN)

(3) 100km (MAN)

(4) 5000km (WAN)

Try to calculate the number of bits being propagated in the above media when the data rates are 1Mbit/s and 10Gbit/s.

3-9 Why does network architecture adopt hierarchical structure? Try to give examples of daily life that are similar to the idea of hierarchical architecture.

3-10 What is the difference between a protocol and a service? And what is the relationship between them?

3-11 What are the three elements of a network protocol? What do they mean?

3-12 Explain the following terms: protocol stack, entity, peer layer, protocol data unit, service access point, client, server, client-server mode.

3-13 Suppose you want to transfer a 1.5MB file over the network. Let the packet length be 1KB and the round trip time RTT = 80ms. There is also a time to establish a TCP connection before transferring data, which is 2×RTT = 160ms. Try to calculate the time required for the receiver to receive the last bit of the file in the following cases.

(1) The data transmission rate is 10Mbit/s, and data packets can be sent continuously.

(2) The data transmission rate is 10Mbit/s, but after each packet is sent, it has to wait for one RTT time to send the next packet.

(3) The data transmission rate is very fast, so the time required to send data can be ignored. However, it is stipulated that only 20 packets can be sent in each RTT time.

3-14 There is a point-to-point link with a length of 50km. If the transmission speed of data on this link is 2×10^8 m/s, how much bandwidth should the link have in order to make the transmission delay as large as the transmission delay of 100-Byte packets? What happens if you send a 512-Byte long packet?

3-15 There is a point-to-point link with a length of 20000km. The data transmission rate is 1kbit/s, and the data to be sent is 100bit. The transmission speed of data on this link is 2×10^8m/s. Assuming that we can see the bits transmitted on the line, try to draw the bits we see on the line (draw two pictures, one is when 100bits have just been sent and the other is after 0.05s).

3-16 Conditions are the same as above. But the data transmission rate is changed to 1Mbit/s. Compared with the results of the above questions, what conclusions can you draw?

3-17 Explain the following concepts:

(1) (N) Layer, (N) Subsystem, (N) Entity;

(2) Service and (N) Protocol;

(3) (N) Layer Service Access Points, (N) Service Primitives.

3-18 Which level of OSI deals with the following issues separately?

(1) Composition of bit streams into frame.

(2) Reliable transmission between two adjacent points in the network.

(3) Bit stream signal synchronization on transmission medium.

(4) Application process of reuse and decomposition.

3-19 An example of the concept of communication usage hierarchy in everyday life is given to illustrate peer-to-peer and peer-to-peer protocols.

3-20 The key points of network architecture with five layers of protocol are discussed, including the main functions of each layer.

3-21 Application layer data with length of 100 bytes is sent to transport layer, which needs 20 bytes of TCP header. To transfer to the network layer, add 20 bytes of IP header. Finally, the data is transmitted to the data link layer over Ethernet, assuming that the data portion of the Ethernet frame is 1500 bytes, plus 18 bytes of the first and last frame. Try to find the data transmission efficiency (assuming no consideration of the first part of the application layer).

3-22 Why does TCP/IP not define data link layer and physical layer?

3-23 What is the role of OSI and what levels are related to the communication subnet?

3-24 Draw the packing and unpacking process of protocol package transmission in OSI.

3-25 What are the layers of the TCP/IP reference model and which is the layer related to network interconnection?

3-26 What are the layers of LAN architecture and the characteristics of LAN?

4 The Data Link Layer

Goal:

(1) Understand methods of error detection and correction.

(2) Understand the elementary data link protocols.

(3) Master the sliding window protocols.

The data link layer is the second layer of the seven-layer OSI model. Located between the physical layer and the network layer, it can provide the functional and procedural means to transfer data between network entities and provide the means to detect and correct errors that may occur in the physical layer.

4.1 Data link layer design issues

The specific functions of the data link layer:

(1) Providing services to the Network Layer.

(2) Framing.

Putting the bits that are transmitted by the physical layer into a data frame.

(3) Error Control.

Handling errors in transmission.

(4) Flow Control.

Adjusting the speed of data frames, so that the slow receivers will not be overwhelmed by the fast sender.

4.1.1 Services provided to the network layer

We want to make it clear that "link" and "data link" are not the same thing. A link is a physical line (wired or wireless) from a node to an adjacent node without any other switching nodes. In the case of data communication, the communication path between the two computers often goes through many segments of links. The visible link is only part of a path.

A data link is another concept. When data needs to be transmitted on a line, there must be a physical line, in addition to some necessary communication protocols to control the transmission of the data (this will be discussed in later sections). If the hardware and software that implements these protocols are added to the link, it constitutes a data link. In order words.

Link: A physical line with point-to-point nodes but without any switching nodes in the middle;

Data link: Link plus control transmission rules; data transmission can be carried out on a data link.

Data link includes physical link and communication rules.

Data link is the digital channel provided by the link layer. A physical link is the passive point-to-point physical line provided by the physical layer. Communication rules are the data link layer protocol.

Data link management is responsible for establishing, maintaining and dismantling links between two adjacent nodes, and transforming unreliable physical links into error free data links by error control and traffic control, improving link reliability and providing a logically error-free data link for network communication.

Data link layer transmits data with a frame. Each frame includes a certain amount of data and some necessary control information.

The network layer data of the source node is reliably transmitted to the network layer of the adjacent destination node, by the data link layer, as shown in Figure 4-1.

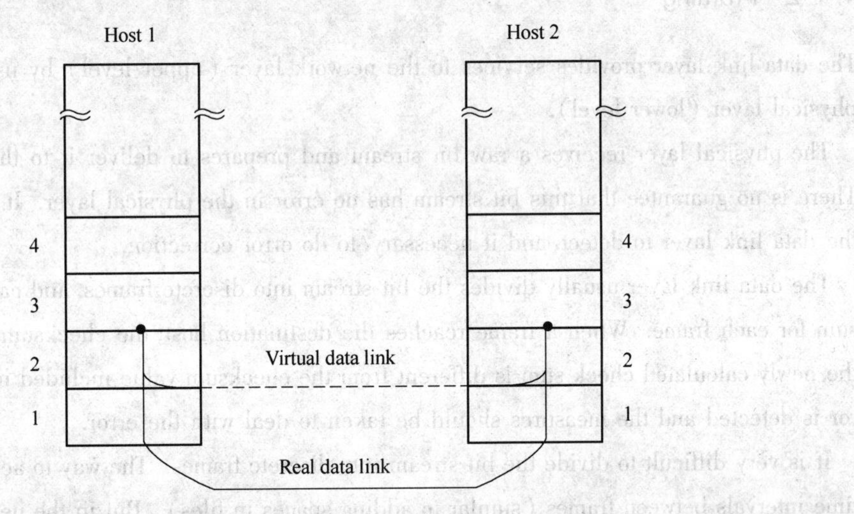

Figure 4-1 Data transmission

There are three basic services in the data link layer:

(1) Non-confirmed and connectionless service: The purpose of the node is not to be confirmed, and the error is responsible for the upper level.

The source machine sends independent frames to the destination machine, and the destination machine does not confirm the received frames. In advance, no connection is established, and there is no release afterwards. If the frame is lost due to the noise on the line, the data link layer does not recover it, and the recovery work is left to the upper level.

(2) Confirmed and connection-oriented service: The destination node has to confirm the received frame. The sending node can know whether the frame has arrived at the destination node safely and the mistaken frame retransmission.

The source machine sends independent frames to the destination machine, and the destination machine independently confirms each frame received. No connection is needed, and no release afterwards. But in this way, the sender can know whether the frame is arriving safely. If a frame is not reached within a specific time interval, it must be sent again.

(3) Confirmed service with connection: It is a reliable data services, providing the function of establishing, maintaining and releasing data links among network entities.

The source machine and the destination machine establish a connection before transmitting any data. Every frame sent on this connection is numbered, and the data link layer ensures that every frame sent is actually received. In addition, it guarantees that every frame is received once, and all the frames are received in the right order.

When a connectionless service is used, if the confirmation information is lost, it will cause a frame to be sent many times, so it will also receive many times. Connection-oriented services provide reliable transport of bit stream between network layers.

4.1.2 Framing

The data link layer provides services to the network layer (upper level) by using the services of physical layer (lower level).

The physical layer receives a raw bit stream and prepares to deliver it to the destination host. There is no guarantee that this bit stream has no error in the physical layer. It always transmits to the data link layer to detect, and if necessary, to do error correction.

The data link layer usually divides the bit stream into discrete frames, and calculates the checksum for each frame. When a frame reaches the destination host, the checksum is recalculated. If the newly calculated check sum is different from the checksum value included in the frame, the error is detected and the measures should be taken to deal with the error.

It is very difficult to divide the bit stream into discrete frames. The way to achieve it is inserting time intervals between frames (similar to adding spaces in files). But in the network, it is difficult to guarantee the accuracy of time and easy to lose or increase the time interval in transmission. Therefore, it is not reliable to determine frames which rely on timings. In order to solve the problem of reliable dividing bit stream, the following methods of frame division are proposed: byte count; flag bytes with byte stuffing; flags bits with bit stuffing and physical layer coding violation.

4. 1. 2. 1 Byte count

The number of bytes in a frame with a byte in the frame head.

Algorithm: we use a field in the frame header of each frame to mark the number of bytes in the frame.

Example: The number of bytes in frame 1, 2, 3, 4 are 5, 5, 8, 8 as shown in Figure 4-2.

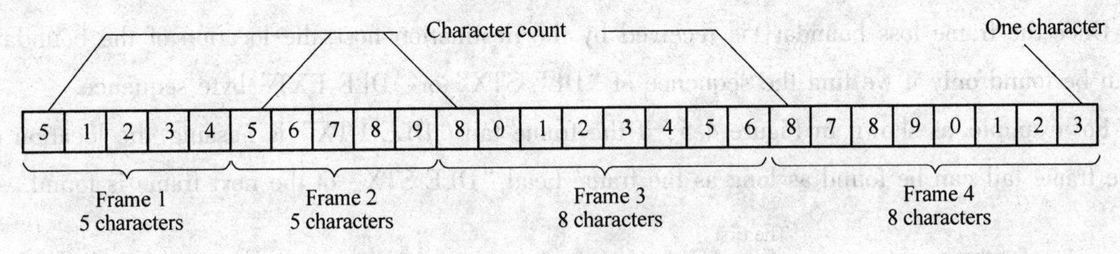

Figure 4-2 Byte count method

However, this method has its disadvantages, one of which is that the number of characters may be falsified due to transmission errors and can lead to some mistakes.

For example as shown in Figure 4-3, if the byte count value of the second frame should be 5, the transmission error is "tampered" to 7, causing the sender to not synchronize with the receiver and fail to determine the starting position of the next frame. At this point, if the destination host is known to be incorrect, it will still be unable to know where the next frame will start from the wrong frame. It is useless to send requests for retransmission to the source host, because the destination host does not know how many bytes to start from and to retransmit.

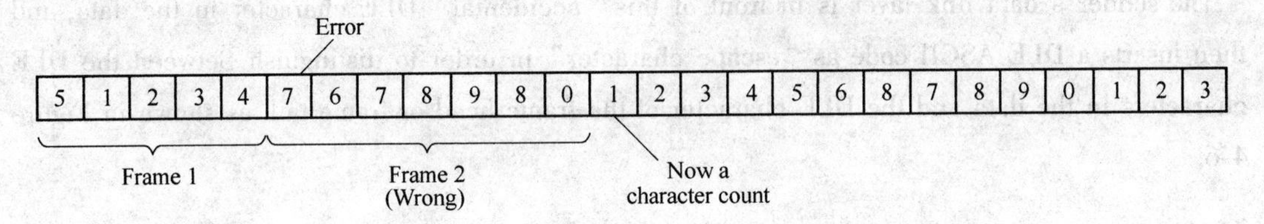

Figure 4-3 Example

4. 1. 2. 2 Flag bytes with byte stuffing

The control field of data frames is represented by a specific byte sequence DLE.

After the start of synchronous transmission marked by SYN, each frame is called the DLE STX with the ASCII character sequence as the frame head, and the DLE ETX as the frame tail, as shown in Figure 4-4.

DLE: Data Link Escape Character
STX: Start of Text
ETX: End of Text

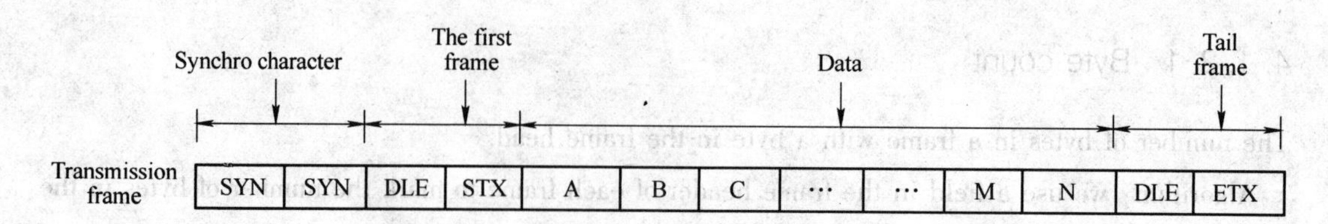

Figure 4-4 Flag bytes with byte stuffing

Once the frame loss boundary is received by the destination host, the location of the boundary can be found only if we find the sequence of "DLE STX" or "DLE EXT" byte sequence.

For example, as shown in Figure 4-5, if the frame tail "DLE ETX" is missing, the location of the frame tail can be found as long as the frame head "DLE STX" of the next frame is found.

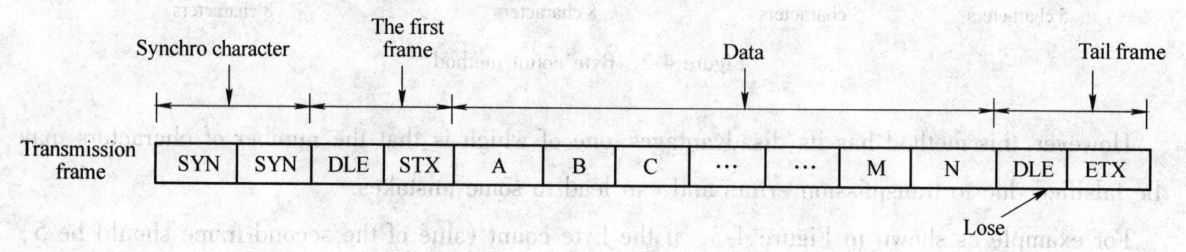

Figure 4-5 Lost DLE ETX

The problem of this method is that when data in the transmitted data appear the same as the "DLE STX" or "DLE EXT" character sequence, it will disturb the determination of frame boundary.

The sender's data link layer is in front of this "accidental" DLE character in the data, and then inserts a DLE ASCII code as "escape character" in order to distinguish between the DLE characters in the data and the DLE character of the frame head or frame tail, as shown in Figure 4-6.

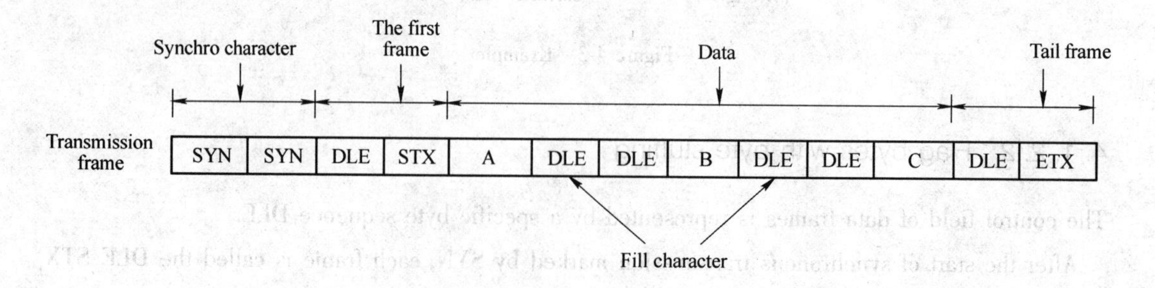

Figure 4-6 Transmission frame

The data link layer of the receiver loses the inserted DLE "escape character" before restoring the data to the network layer, so that it can be reduced to raw data.

Since the DLE characters in the data segment appear in pairs, the symbol character DLE for frame formation can be easily separated from the escape character DLE in the data, and the single DLE in the frame is the frame boundary character "DLE STX" or "DLE EXT", and the pairs of DLE are data.

Shortcoming: The frame boundary is completely dependent on the 8 bit character DLE. In fact, not all the character codes are 8bit (for example, UNICODE uses 16bit characters). Therefore, this method is not universal, nor can it be extended.

4.1.2.3 Flag bits with bit stuffing: HDLC

The first and tail marking method with filling position allows data frames to contain any length of bits, and also allow each character to encode any number of bits.

The header and tail labeling method with filling bits uses a unified data frame format to synchronize and determine the boundaries with specific bit sequences.

Algorithm: Each frame uses a special bit pattern, that is, 01111110 as the sign byte of start and end. When the sender's data link layer meets 5 consecutive "1" in the data, in order to distinguish between data and bit patterns, a "0" is automatically inserted into the output bit stream. This is the 0 bit stuffing.

When the receiver sees 5 consecutive "1" and the latter is "0", it automatically removes the "0" bit to restore it to the original data, as shown in Figure 4-7.

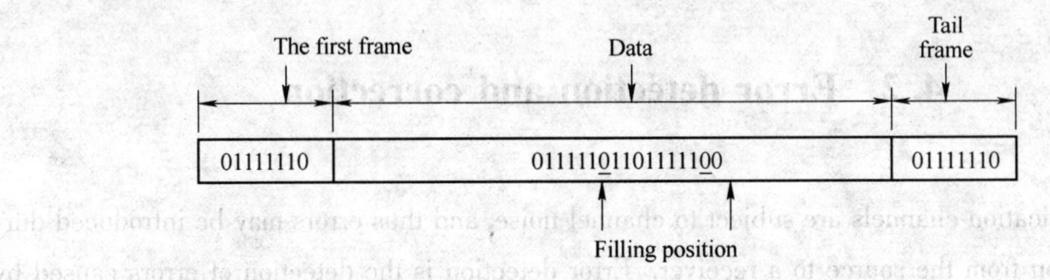

Figure 4-7 Filling position

The bit stuffing process is also completely transparent to the network layer. In the bit-stuffing-mechanism, the boundary between the two frames can be identified clearly by the logo mode. If the receiver loses frame synchronization, it only needs to scan the flag sequence in the input data stream.

4.1.3 Error control

Error control ensure that all frames can be delivered to the network layer of the target machine correctly.

The common method is to provide feedback to the sender of the receiver's reception;

The protocol requires the receiver to send a special control frame as an affirmation or negation of the input. If the sender receives an affirmative confirmation of a frame, it is known that the frame has arrived correctly; a denial confirmation means that a certain error has occurred and the corresponding frame must be retransmitted.

If the data frame is lost in the transmission process, the receiver will not send any response frame to the receiver because it does not receive the data frame. If the receiver is to receive the response frame and then send the next data frame, it is bound to wait forever and result in a deadlock. Similarly, if the receiver is to send the response frame loss will result in a deadlock.

A typical practice when a transmitter ends a data frame, a timeout timer is started, and if the response frame is still not received over the timing time set by the timer, the transmitter retransmits the data frame sent in front.

Generally, timing time is chosen to be slightly larger than "the average time required to send frames from receiving data to receiving frames".

4.1.4 Flow control

There also is a problem that the transmitting capacity of the sender may larger than that of the receiver flow control is applied to solve the problem by limiting the data traffic sent by the sender, so that its sending speed should not exceed the receiver's processing speed.

The protocol includes a set of rules defined completely, the rules describe the time the sender sends the next frame, and prohibits the frame from sending, before the recipient allows.

4.2 Error detection and correction

Many communication channels are subject to channel noise, and thus errors may be introduced during transmission from the source to a receiver. Error detection is the detection of errors caused by noise or other impairments during transmission from the transmitter to the receiver. Error detection techniques allow detecting such errors, while error correction enables reconstruction of the original data in many cases.

4.2.1 Methods to deal with errors

There are three ways to deal with errors:

(1) Error-correction code: also known as forward error correction; enough redundant information is added to every block that we want to send, so that the receiver can deduce what the characters that have been sent out.

(2) Error-detecting code: automatically request retransmission; only add enough redundant bits to inform the receiver that there is an error, but do not know what kind of error is, and then let the sender request retransmission.

(3) Mixed mode: the receiver adopts the error correction mix to correct a small amount of er-

ror automatically, while the error exceeding its correction ability is corrected by reissuing the original information.

4. 2. 2 Parity check

Parity check is the most simple error detection method. Its code rules are grouping the data element and attaching a check-bit after each group of data before the data is sent involving the check-bit, if the number of "1" in the group is even, it refers to even check. Conversely, it is about odd check. Using the same rule, if the data does not match the number of "1" from the sender, at the receiving end, there will be an error.

4. 2. 3 Cyclic redundancy check

Cyclic redundancy check (CRC) is cyclic redundancy check coding. First, the quantitative relationship between the binaries of the data to be sent is established, namely, the sender operates the binary numbers of the data frames according to certain rules to generate the binary check codes. Then the binary numbers are sent out together with the sent data. After receiving, the receiver checks the relationship between the binary numbers according to the same rules to judge whether there are errors in the transmission process.

The specific practice is that both the sender and receiver agree on a generating polynomial $G(x)$ (its highest and lowest order coefficients must be 1), the sender adds the check sum at the end of the frame, so that the polynomial of the frame with the check sum can be divided by the $G(x)$, after the receiver has received the data, the $G(x)$ is used to divide the polynomial, if there is a remainder, the transmission is wrong.

CRC is the most widely used error detection codes in computer networks and data communication, because of its low missing rate, and it is easy to realize.

The CRC code is also called a polynomial code. Any code consisting of binary digit strings can establish a one-to-one correspondence with a polynomial containing two coefficients: only 0 and 1.

For example, polynomial $x^5+x^4+x^2+x$, the corresponding code is 110110.

4. 2. 3. 1 The principle of cyclic redundancy check

Suppose the data to be transmitted is $M = 1010001101$ (total K bits). We add an bit redundant code for error detection after M.

The operation of 2^n multiplication M is performed by polynomial arithmetie module 2, which is equivalent to adding n "0" to M.

The k bits information to be sent corresponds to a $(k-1)$ power polynomial $K(x)$, and the r bits redundancy corresponds to a $(r-1)$ polynomial $R(x)$. The $(k+r)$ bits code involves information and redundancy corresponds to a $(n-1)$ polynomial

$$T(x) = x^r \cdot K(x) + R(x)$$

Coding process is solving $R(x)$ when knowing $K(x)$, in CRC code, we can find a specific r power polynomial $G(x)$ to achieve it. The remainder obtained by dividing $K(x)$ with $G(x)$ is $R(x)$.

4.2.3.2 Polynomial arithmetic modulo 2 (no borrow)

For polynomial division, as long as the corresponding coefficient can be divided.

For example, $K(x) = x^6 + x^4 + 1$ (1010001). If $r = 4$, $G(x) = x^4 + x^2 + x + 1$ (10111), then $X^4 \cdot K(x) = x^{10} + x^8 + x^4$ (10100010000). Then get the remainder $R(x)$ by division polynomial arithmetic modulo 2 as shown in follwing.

```
           1001111
10111 )10100010000
       10111
          11010
          10111
           11010
           10111
            11010
            10111
             11010
             10111
              1101
```

Here, 1101 of the last remainder is redundancy and $R(x) = x^3 + x^2 + 1$.

4.2.3.3 CRC reception checkout example

Suppose the bit stream from the sender to receiver is 11010110111, and the generated polynomial is $G(x) = x^4 + x + 1$. Please check if the data received by the receiver is correct.

Solution: by modulus 2 division. There is no carry addition and borrowing. Both the addition and the subtraction are equal to XOR.

The remainder is 0, indicating that the reception is correct as shown in Figure 4-8.

Example: The sending frame is 1101011011, the polynomial is $G(x) = x^4 + x + 1$, what is the data frame sent by the CRC?

Solution: $G(x)$ is 4 order, and the frame is 10+4 bits. The actual transmitted frame is: $T(x)$ = 11010110111110. When receiving check, if $T(x)$ can be divided by $G(x)$, it means that the transmission and reception are correct as shown in Figure 4-9.

Example: The sending frame is 1001001, the polynomial is $G(x) = x^4 + x + 1$, and the data

```
            1000101
11011 ) 11010110111
        11011
        -----
            11101
            11011
            -----
              11011
              11011
              -----
                  0
```

Figure 4-8 Example

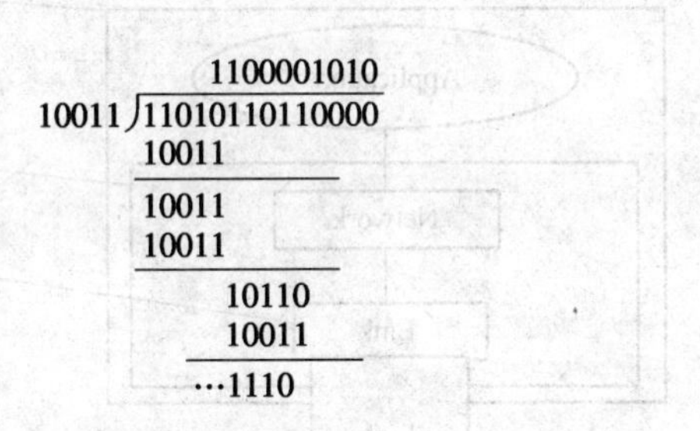

Figure 4-9 Example

frames sent by CRC are obtained. Its coefficient is 10011. The sender encodes the data before transmitting the data: dividing by modulo 2, dividing $xM(x)$ by $G(x)$, obtaining the quotient polynomial $Q(x)$ and the remainder polynomial $R(x)$;

Solution: $xM(x)$: 10010010000;

Quotient polynomial $Q(x)$: 10011;

Remainder polynomial $R(x)$: 1111;

Send polynomial $T(x)$: 10010011111.

CRC codes can be implemented by software, but usually CRC is encoded, decoded and erroneous by hardware. Mathematical analysis shows that only when $G(x)$ has some characteristics, various errors can be detected. In order to verify various error modes in different situations, several international standards of CRC generating polynomials have been studied.

The selection of the generator polynomial is the most important part of implementing the CRC algorithm. The most important attribute of the polynomial is its length (largest degree (exponent) +1 of any one term in the polynomial), because of its direct influence on the length of the computed check value.

Numerous varieties of cyclic redundancy checks have been incorporated into technical standards. There are three international standards:

CRC-12: $x^{12}+x^{11}+x^{3}+x^{2}+x^{1}+1$

CRC-16: $x^{16}+x^{15}+x^{2}+1$

CRC-CCITT: $x^{16}+x^{12}+x^{5}+1$

4.3 Elementary data link protocols

There are some assumptions of the communication model:

The physical layer, data link layer and network layer are independent processes. They communicate by sending information as shown in Figure 4-10.

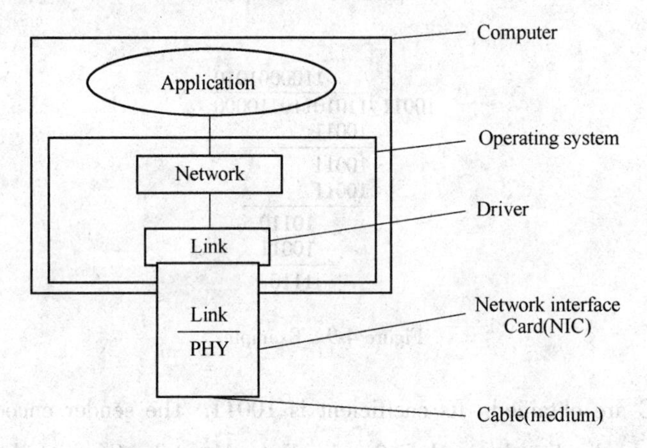

Figure 4-10 Implementation of physical, datalink, and network layers

Machine A hopes to send a long data stream to machine B in reliable and connection-oriented way.

In the sender, the data link layer receives a packet from the network layer, adds a data link layer head and tail on the packet, encapsulates it into a frame; and then uses the to-physical-layer to send the frame; the transmission hardware calculates and adds the checksum, and the data link layer software does not have to care about it.

In the receiver when a frame arrives, the hardware checks its checksum. If the checksum is incorrect, the data link layer is informed (event = checksum err); if the external frame is not damaged, the data link layer can be known (event = frame arrive), thus using from-physical-layer to obtain the frame, and then it would check the frame head control information. If everything is correct, the packet part is assigned to the network layer.

Note that the frame head is not given to the network layer. The purpose is to completely separate the protocols between the network layer and the data link layer.

The network layer obtains information from the transport layer, adding a network layer header, formly a packet to be sent to the data link layer. Then, we put the info field in the frame and sent it. When the frame reaches the destination, the data link layer extracts the packet from the frame and sends the packet to the network layer. In this way, the network layer can directly exchange packets between two machines.

In most protocols, it is assumed that the channel is not reliable, and sometimes the whole frame will be lost occasionally. In order to ensure recovery from this error, the sender's data link layer must start an internal timer or clock when sending the frame. If a reply is not received within a predetermined time interval, the data link layer will receive an interrupt signal, so that it can be processed accordingly.

4.3.1 An unrestricted simplex protocol

In the ideal work situation, the link is an ideal transmission channel. First, the transmission is completely reliable without error or loss, so there is not error control problem. Second, the receiver can accept and deliver the host regardless of the sending speed of sender, so there is not of flow control problem. As shown in Figure 4-11.

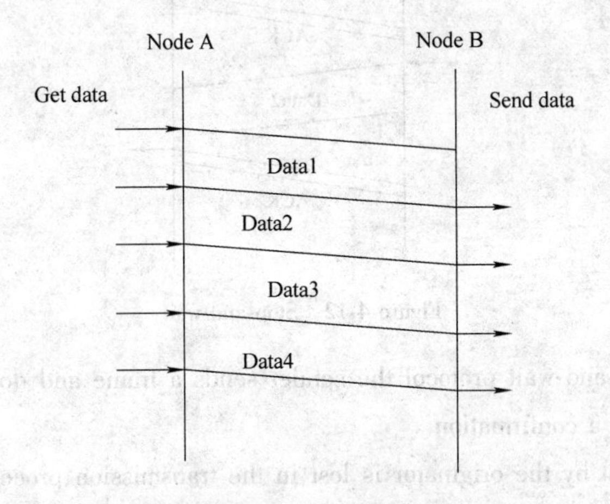

Figure 4-11 Ideal channel

An unrestricted simplex protocol includes the sending process and the receiving process.

The sending process is run on the data link layer of the source machine with infinite loop, and puts the data on line as fast as possible. The receiving process is to wait for frames to arrive, them get the new frames from the hardware buffer and assign them to variables. Finally, the data is transferred to the network layer, and the data link layer returns to the initial state and waits for the arrival of the new frame.

4.3.2 A simplex stop-and-wait protocol

In the non-ideal work situation, the link is an ideal transmission channel, and the transmission is completely reliable without error or loss, so there is not error control problem. However, the receiver cannot process the external data at an infinitely fast rate, so there is flow control problem.

The main question is how to prevent the sending process from sending data too fast, for the receiving process to process on time.

An available solution is for the sender to add a delay in protocol 1, which reduces the speed and ensures the recipient has the time to process. However, there are some disadvantages since bandwidth utilization is far below the best value.

The usual solution is to ask the receiver to provide feedback. The receiver is an boss; the sender sends a frame and then wait for the response from the receiver. The receiver receives the data frame and sends it to the host, and then sends a message (ACK) to the sender; the sender receives the message, and then sends the next frame of data, as shown in Figure 4-12.

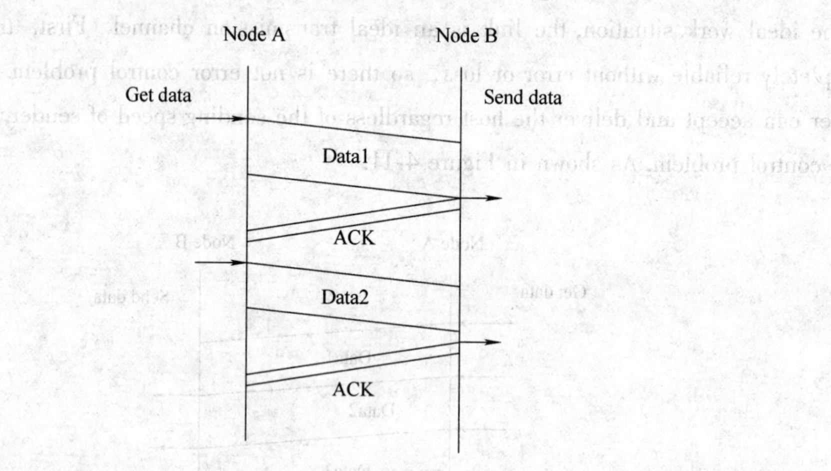

Figure 4-12 Stop-and-wait

It is named the stop-and-wait protocol the sender sends a frame and does not continue to send the frame, but waits for a confirmation.

If the data frame sent by the originator is lost in the transmission process, the receiver will not send any reply information to the sender because it has not received the data frame. In this case, if the sender wants to wait for the reply frame to send the next data frame, it will wait forever, and result in a deadlock. Similarly, if the response frame sent by the receiver is lost, a deadlock will also occur.

To solve the existence problem of "deadlock" as shown in Figure 4-13, the sender starts a timeout timer each time after sending a data frame. If it exceeds the timing time set by the timer, the receiver does not receive the recipient's response frame, the sender resends the previous data frame.

The timing time is slightly larger than the "average time from the end of sending frame to the end of receiving frame.

Most practical data link layer protocols do not use denial frames, and the confirmation frames have serial number *n*.

According to the expression of habit, ACK *n* means that "$n-1$ frame has been received, and now it expects to receive *n* frame".

For example, ACK1 means "0 frames have been received, and now the next frame is expected to

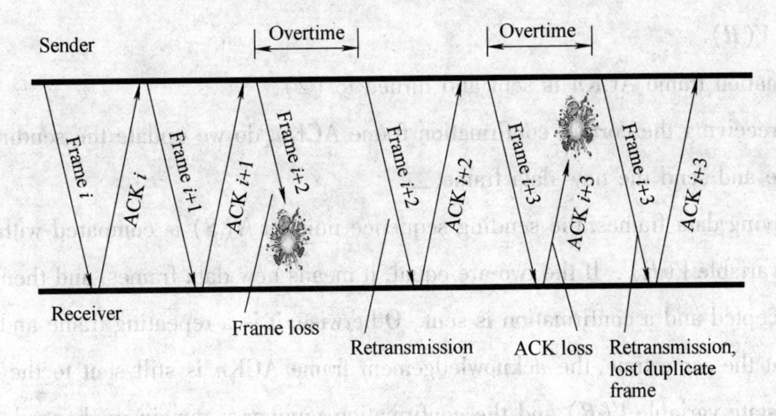

Figure 4-13 Solution of deadlock

be 1 frames". ACK0 means, "frame 1 has been received, and now the next frame we want to receive is frame 0".

In the stop-and-wait protocol, the sending node and the receiving node work together.

In the sending node:

(1) A data frame is fetched from the host and sent to the sending buffer.

(2) $V(S) \leftarrow 0$. $V(S)$: Sender state variable.

(3) $N(S) \leftarrow V(S)$. $N(S)$: A serial number of sending.

(4) The data frames in the send buffer are sent out.

(5) Set the timeout timer.

(6) Wait. {Wait for the first events in following (7) and (8).}

(7) Receive acknowledgement frame ACKn.

If $n = 1 - V(s)$, then:

Take a new data frame from the host and put it in the sending buffer;

$$V(S) \leftarrow [1 - V(S)], \text{ turn } (3).$$

Otherwise, the confirmation frame is discarded and turned to (6).

(8) If the timeout timer is reached, it turns to (4).

In the receiving node:

(1) $V(R) \leftarrow 0$.

(2) Wait.

(3) Receive a data frame;

If $N(S) = V(R)$, then execution (4);

Otherwise, the data frame is discarded and turned to (6).

(4) The data part in the received data frame is sent to the upper software, That is, the host in data link layer model.

(5) $V(R) \leftarrow [1 - V(R)]$.

(6) $n \leftarrow V(R)$.

The confirmation frame ACKn is sent and turned to (2).

Only after receiving the correct confirmation frame ACKn, do we update the sending state variable $V(S)$ once and send the new data frame.

When receiving data frames, the sending sequence number $N(S)$ is compared with the local receiving state variable $V(R)$. If the two are equal, it means new data frames, and then the new data frames are accepted and a confirmation is sent. Otherwise, it is a repeating frame and must be discarded. But at the same time, the acknowledgement frame ACKn is still sent to the sender, while the receiving state variable $V(R)$ and the confirmation number n remain unchanged.

Consecutive data frames with the same sending sequence number indicate that the sender has a timeout retransmission. Continuous acknowledgement frames with the same serial number indicate that the receiving terminals receive repeated frames.

After sending the data frame, the sender must temporarily retain the copy of the data frame in its sending buffer. In this way, the retransmission can be carried out in the case of error. We can only erase this copy when we confirm that the other party has received the data frame.

The practical CRC checker is done in hardware. It can automatically discard the detected error frames. Therefore, the so-called "throw away the wrong frame" is not felt for the upper software or users.

Sending end retransmission of the error data frame is automatic, so the error control system is often referred to as ARQ (Automatic Repeat request).

Stopping and waiting protocol ARQ is simple enough, but the utilization rate of the communication channel is lower; that is to say, the channel is far from being filled by data bits. To overcome this shortcoming, there are two other protocols, namely continuous ARQ and selective repeat ARQ.

4.3.3 A simplex protocol for a noisy channel

In the situation of non-ideal work, a link is not an ideal transmission channel, it results in unreliable transmission and frame may error and loss, so there is error control problem. The receiver can't process the external data at an infinitely speed, so there is flow control problem.

ARQ is an error-control protocol that automatically initiates a call to retransmit any data packet or frame after receiving flawed or incorrect data. When the transmitting device fails to receive an acknowledgement signal to confirm the data has been received, it usually retransmits the data after a predefined timeout and repeats the process a predetermined number of times until the transmitting device receives the acknowledgement.

ARQs are often used to assure reliable transmissions over an unreliable service. ARQ protocols reside in the data link or transport layers of the OSI model. TCP uses a variant of Go-back-N ARQ to ensure reliable data transmission over the Internet protocol. However, it does not guarantee delivery of data packets.

After sending a data frame, instead of stopping to wait for the confirmation frame, it can send several data frames continuously. If the acknowledgement frame is received at the time, the data frame can also be sent. Since the waiting time is reduced, the throughput of the entire communication is improved, as shown in Figure 4-14.

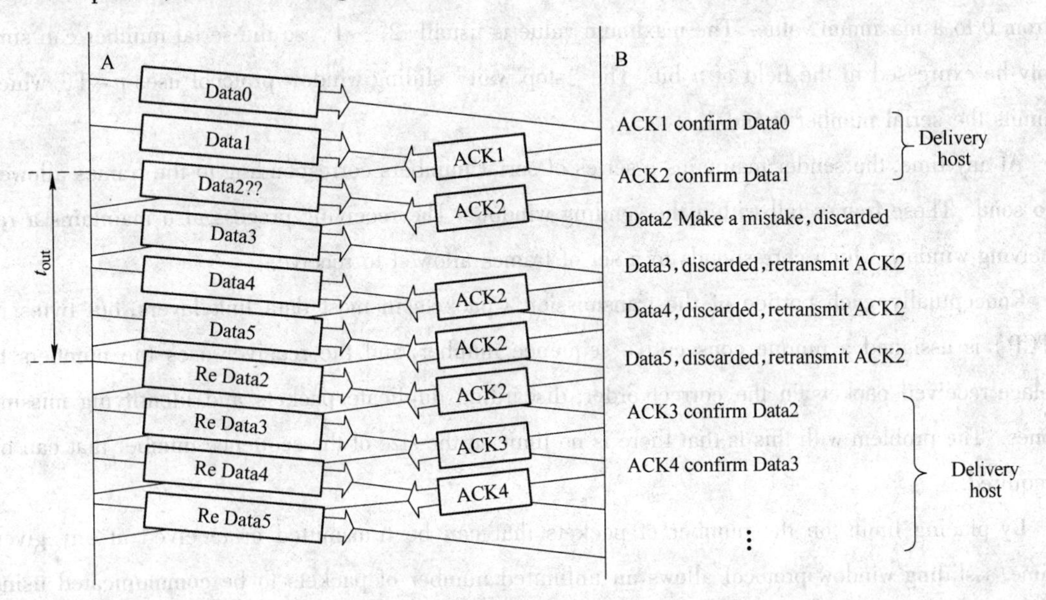

Figure 4-14 Continuous ARQ protocol

There are a few point to be noted.

The receiver only receives the data frame in sequence. Although the correct three frames are received after the error frame2, the receiver must discarded the frames, since the frame2 before the three frames has not been received. Although these non-error-free frames are discarded, the last acknowledgement frame that has been sent should be sent repeatedly to prevent the loss of the acknowledgement frame.

ACK1 means to confirm frame 0, and expects to receive the frame1 next. ACK2 means to confirm the frame1, and expects to receive the frame2 next time. And so on.

Node A sets up the timeout timer of the frame when sending out a data frame. If the acknowledgement frame is received within the set timeout time, the timeout timer is cleared immediately. However, if the acknowledgement frame is not received, beyond the set time the corresponding data frame must be retransmitted. and timeout time be reset.

When the frame2 is retransmitted before the acknowledgement frame2, node A has already completed frame5, but still must be returned to retransmit all the frame after frame2 (including frame2). Continuous ARQ, also known as Go-back-N ARQ, means that when errors occur, they must be retransmitted, and N frames should be returned, and then restart transmission.

4.4 Sliding window protocols

The three protocols mentioned above belong to the same protocol-sliding window protocol.

In all sliding window protocols, every frame to be sent contains a serial number, which ranges from 0 to a maximum value. The maximum value is usually $2^n - 1$, so the serial number can simply be expressed in the field of n bit. The "stop wait" sliding window protocol uses $n=1$, which limits the serial number 0 and 1.

At any time, the sender maintains a series of serial numbers corresponding to the frames allowed to send. These frames fall within the sending window. The receiving process also maintains a receiving window, which corresponds to a set of frames allowed to receive.

Conceptually, each portion of the transmission (packets in most data link layers, but bytes in TCP) is assigned a unique consecutive sequence number, and the receiver uses the numbers to place received packets in the correct order, discarding duplicate packets and identifying missing ones. The problem with this is that there is no limit on the size of the sequence number that can be required.

By placing limits on the number of packets that can be transmitted or received at any given time, a sliding window protocol allows an unlimited number of packets to be communicated using fixed-size sequence numbers. The term "window" on the transmitter side represents the logical boundary of the total number of packets yet to be acknowledged by the receiver. The receiver informs the transmitter in each acknowledgment packet the current maximum receiver buffer size (window boundary). The TCP header uses a 16 bits field to report the receive window size to the sender. Therefore, the largest window that can be used is 216 = 64 gigabit.

The sender and the receiver set the sending window and receiving window separately, the sending window is used to control the sender's flow; and the size of the sending window W_T represents that how many data frames at most can be sent without receiving the confirmation information, as shown in Figure 4-15.

At the receiving windows, only when the sending sequence number of the received data frame falls into the receiving window, the data frame can be allowed to be accepted. If the received data frames fall outside the receiving window, they will be discarded.

In the continuous ARQ protocol, the size of the receiving window is $W_R = 1$. The frame can only be received if the received frame number matches the receiving window. Otherwise, it should be discarded. Each receives a sequence number of the correct frame, and the receiver window moves forward (or to the right side) to slide the position of a frame. The confirmation of the frame is sent at the same time, as shown in Figure 4-16.

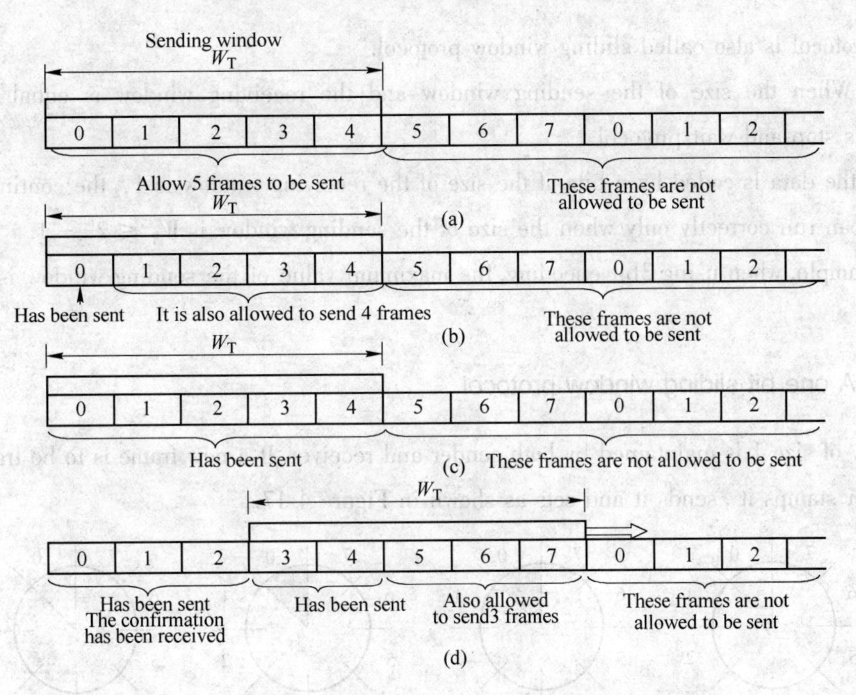

Figure 4-15 Sliding window in sender

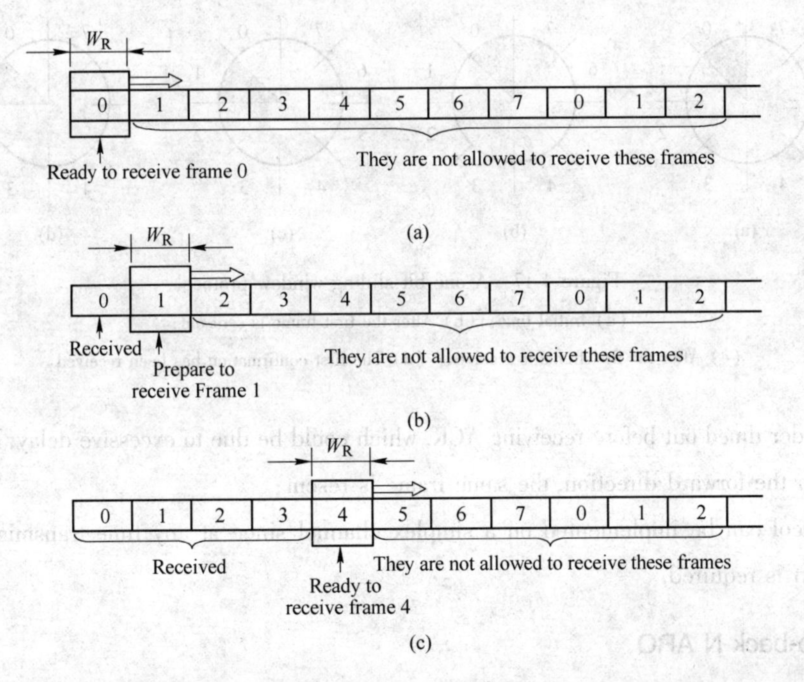

Figure 4-16 Example

The sliding window has some important characteristics:

(1) Only when the receiving window slides forward (at the same time, a confirmation is sent) can the sending window slide forward.

(2) The window of the sender and receiver keeps sliding forward according to the above rule,

so this protocol is also called sliding window protocol.

(3) When the size of the sending window and the receiving window is equal to 1, the protocol is stop-and-wait protocol.

When the data is coded by n bits, if the size of the receiving window is 1, the continuous ARQ protocol can run correctly only when the size of the sending window is $W_T \leqslant 2^n - 1$.

For example, when using 3bit encoding, the maximum value of the sending window is 7 instead of 8.

4.4.1 A one bit sliding window protocol

A window of size 1 is maintained by both sender and receiver. If a new frame is to be transmitted, the sender stamps it , sends it and sets as shown in Figure 4-17.

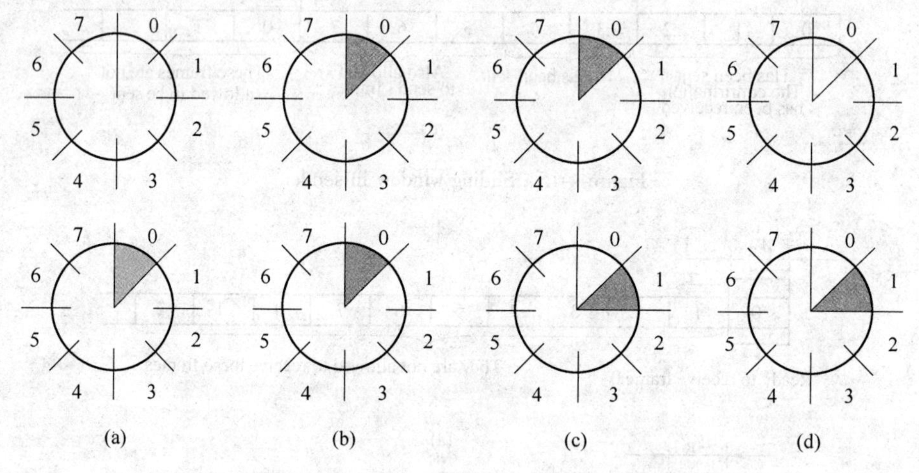

Figure 4-17 A one bit sliding window protocol

(a) Initial time; (b) After the first frame is sent out;

(c) After the first frame is received; (d) The first confirmation has been received

If the sender timed out before receiving ACK, which could be due to excessive delay, lost ACKs or lost frame in the forward direction, the same frame is resent;

The protocol can be implemented on a simplex channel since at any time transmission in only one direction is required.

4.4.2 Go-back-N ARQ

Go-back-N ARQ is a specific instance of the automatic repeat request (ARQ) protocol, in which the sending process continues to send a number of frames specified by a window size even without receiving an acknowledgement (ACK) packet from the receiver. It is a special case of the general sliding window protocol with the transmit window size of N and receive window size of 1. It can transmit N frames to the peer before requiring an ACK.

After sending a frame, the sender does not have to stop to wait for a response of from receiver, but can send several frames continuously. If one frame is lost or damaged in the middle of the frame stream, all subsequent frames are directly abandoned by the receiving process, and no confirmation is sent for the discarded frame. The transmission process starts with damaged frames or missing frames, and retransmit all the frames that are not confirmed. However, in the case of a high error rate, a lot of bandwidth will be wasted.

The premise of Go-back-N ARQ protocol is that in the continuous ARQ protocol, the maximum number of frames (issued without confirming) that can be sent continuously must be limited, then loop repeated use of a limited frame number.

For flow control, the sending window's size W_T indicates the number of data frames that can be sent continuously before receiving the information confirmed by the receiver, only the frame in the window can be sent.

The receiving window's size W_R indicates the maximum number of data frames that can be continuously received (only frames with serial numbers in the window can be received, others are discarded). The operation of the transmission window driven by the receiving window. The maximum value of the sending window is $W_T \leqslant 2^n - 1$ (n is the ordinal number).

4. 4. 3 Selective Repeat ARQ

The Go-back-N ARQ protocol improves efficiency because it involves continuously sending data frames. But when the error occurs, the efficiency is reduced due to the retransmission of the data frames that have been correctly transmitted after the error.

Selective repeat ARQ is part of the automatic repeat-request ARQ. With selective repeat, the sender sends a number of frames specified by a window size even without the need to wait for individual ACK from the receiver as in Go-back-N ARQ. The receiver may selectively reject a single frame, which may be retransmitted alone. The receiver accepts out-of-order frames and buffers them. We only retransmit the wrong data frames or timers' timeframes. This contrasts with other forms of ARQ, and ARQ must send every frame from that point again, as shown in Figure 4-18.

All the correct frames after the error frames are stored by the receiver's data link layer. When the sending process finally realizes a frame error, it only retransmit the error frame instead of all subsequent frames.

Namely, when data frames make mistakes or operate timeouts, if $W_R > 1$, data frames with serial number in the receiving window after the error frames are temporarily send.

It avoids retransmission of the correct frame and improve channel utilization. However, in the receiver, a certain amount of caching is needed.

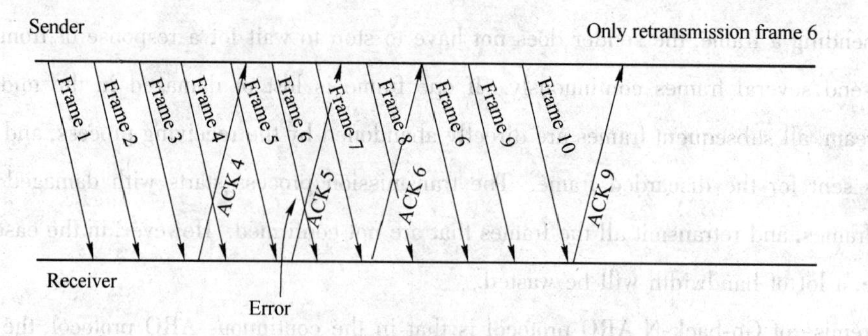

Figure 4-18 Selective repeat ARQ

The best value of the receiving window:

$$W_R < W_T$$
$$W_R < 2^{n-1}$$

4.5 Example data link protocols

4.5.1 HDLC—high-level data link control

High-level data link control (HDLC) is a data link layer protocol for data transmission and bit-oriented transmission on the synchronous network. There is an example as shown in Figure 4-19. It is formulated by the international organization for standardization (ISO). The original ISO standards for HDLC are as follows:

(1) ISO 3309-1979-Frame Structure.

(2) ISO 4335-1979-Elements of Procedure.

(3) ISO 6159-1980-Unbalanced Classes of Procedure.

(4) ISO 6256-1981-Balanced Classes of Procedure.

(5) The current standard for HDLC is ISO/IEC 13239: 2002, which has replaced all of the previous standards above.

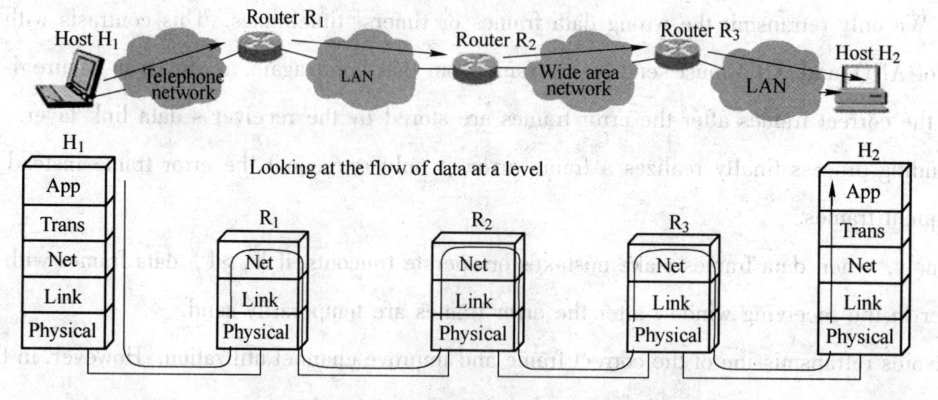

Figure 4-19 Example data link protocols

The protocols adopted at the high-level data link layer are basically divided into two categories.

4.5.1.1 Character oriented transmission control regulation

Character is used as the basic unit of information transmission, and a special 10-characters sequence is used to form necessary special control information for transmission control. The encoding is specified, such as ASCII, EBCDIC, etc.

It allows to use the synchronous and asynchronous transmission. Half duplex communication is often used. with check and stop-and-wait mode is used for sending control.

The binary synchronous communication procedure (BSC) in IBM is adopted for character-oriented transmission control procedure.

4.5.1.2 Bit-oriented transmission control regulation

Bit is used as the basic unit of transmission information, the transmission information can be an arbitrary combination of data bits, and the protocol data unit PDU has a fixed format called frame, which is transmitted in a frame. In the bit-oriented transmission control, data and control information of the message are completely independent, the sending made is continuous transmission with the CRC checking method. The HDLC for ISO using the bit-oriented transmission control regulation has been recognized by many countries as an international standard.

HDLC is a universal data link control protocol. When a data link is started, a specific operation mode is allowed. The so-called link operation mode, generally speaking, is a site in the main station mode of operation, or in the slave station mode of operation, or both.

The link used to control the destination is called the master station , and other stations controlled by the master station are called slave stations. The main station is responsible for organizing the data stream and recovering the errors on the link. The frame sent from the main station to the slave station is called the command frame, and the frame returned to the main station from the slave station is called the response frame.

Links with multiple stations usually use polling techniques. The stations polling other stations are called master stations, and each station in a point-to-point link can be a master station. The master station needs more logic functions than the slave station does, so when the terminal is connected with the host, the host is always the master station. In the case of multiple links connected by one station, the station may be a master station for some links and a slave station for others.

Some station can be combined with the functions of the master station and the slave station. This station is called a combined station. The protocol for the transmission of information between the combined stations is symmetrical, that is, the same transmission control function on the master and slave stations on the link. This is also called the balanced operation. This is a very important concept in the computer network. In contrast, if there is not same transmission control function on the master and slave stations, it is called unbalanced operations.

Link configurations can be categorized as following:

Unbalanced configuration: it is composed of one master station and one or more slave stations, supporting half duplex or full duplex communication. It is suitable for point-to-point link and multi-point links. In a multipoint link, there is a separate logical link between the master station and each slave station.

Balanced configuration: it is only suitable for point-to-point links, and consists of two complex stations. The composite station has both master and slave functions, supporting half duplex and full duplex communication.

There are three kinds of data transmission methods, among them, there are two kinds of data transmission methods in unbalanced configuration, as shown in Figure 4-20.

Normal Response Mode (NRM): only the master station can initiate data transmission, and the slave station can only send data after receiving the query order of the master station.

Asynchronous Response Mode (ARM): the slave station can send information and start data transmission without waiting for master station query.

There is only one kind of data transmission method in balanced configuration.

Asynchronous balanced mode (ABM): any combined station can start data transmission without the permission of another combined station.

4.5.1.3 HDLC frame structure

In HDLC, data and control messages are transmitted in standard format of frame. The frames in the HDLC are similar to the character blocks of the BSC, but the data messages and the control messages in the BSC protocol are transmitted independently, and the commands in the HDLC should be transmitted in a unified format. The complete frame of the HDLC consists of the flag field (F), the address field (A), the control field (C), the information field (I), the frame check sequence field (FCS), as shown in Figure 4-21.

Field F At the beginning and end of the frame, a special field called "01111110" which called field F is placed as the first and the last boundary of a frame.

It can solve the problem of frame synchronization, we correctly judge the beginning and end of a frame from the received bit stream.

Usually, the channel is still active for some time when the frame is not transmitted. In this state, the sender constantly sends the flag field, and it is considered that a new frame transfer has begun. The "0 bit stuffing" can be used to achieve transparent data transmission. At the sending end,

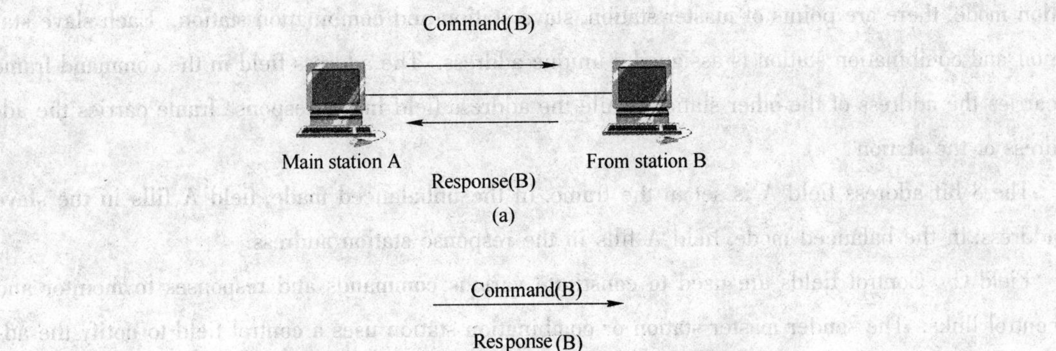

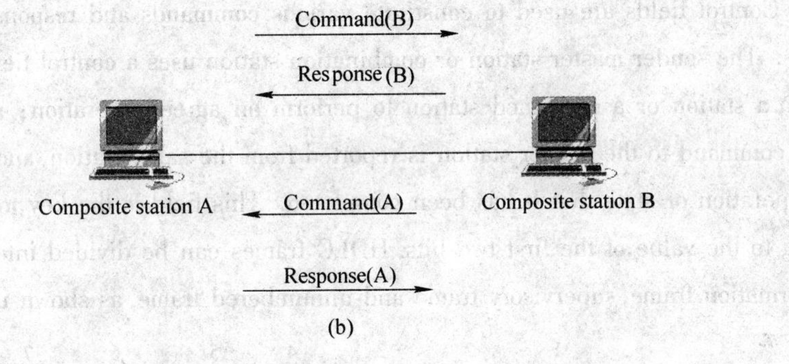

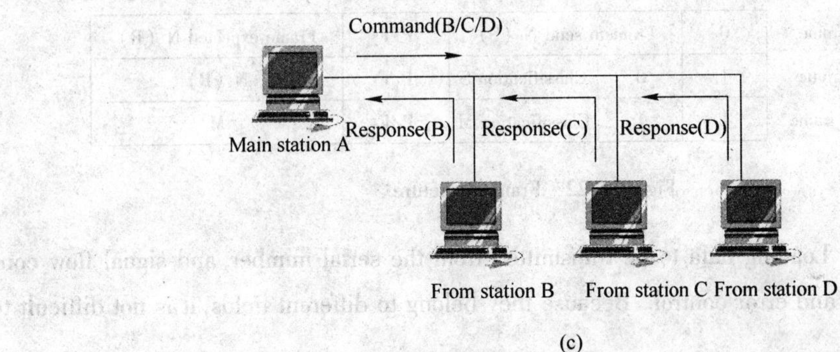

Figure 4-20 Two configurations of HDLC

(a) Unbalanced configuration; point to point; (b) Unbalanced configuration;

(c) Unbalanced configuration; Multipoint link

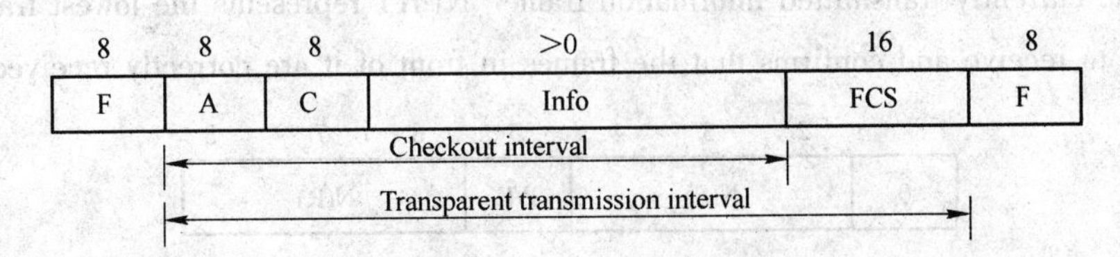

Figure 4-21 HDLC frame

when a string of bits has not been added to the flag field, the whole frame is scanned with hardware or software, and once the 5 "1" are found, then a "0" is filled after them, so that the F flag will not appear in the transmitted bit stream. No matter what kind of bit combination, it will not lead to a mistake in judging frame boundary.

Field A The content of the address field (Field A) depends on the way of operation. In operation mode, there are points of master station, slave station and combination station. Each slave station and combination station is assigned a unique address. The address field in the command frame carries the address of the other station, while the address field in the response frame carries the address of the station.

The 8 bit address field A is set in the frame. In the unbalanced made, field A fills in the slave address; in the balanced mode, field A fills in the response station address.

Field C Control fields are used to constitute various commands and responses to monitor and control links. The sender master station or combination station uses a control field to notify the addressed from a station or a combined station to perform an agreed operation; meanwhile, the response of a command to the master station is reported from the salve station, and reports the changes in the operation or state that it has been completed. This field is the key to HDLC.

According to the value of the first two bits, HDLC frames can be divided into three categories, namely, information frame, supervisory frame and unnumbered frame, as shown in Figure 4-22.

	1	2	3	4	5	6	7	8
Information frame	0	Frameto send N (S)			P/F	Frame expected N (R)		
Supervisory frame	1	0	Classification S		P/F	N (R)		
Unnum bered frame	1	1	Classification M		P/F	M		

Figure 4-22 Frame structure

Information frame Loading data to be transmitted from the serial number, and signal flow control, sequence control and error control. Because they belong to different fields, it is not difficult to identify them.

As shown in Figure 4-23, P/F (poll/final) the bit is used for asking and termimating. In the command frame, it is the query bit (P bit), and in the response frame, it is the stop bit (F bit) . The serial number N(S) and the N(R) are represented in 3 bits. N(S) represents the ordinal number of the currently transmitted information frame. N(R) represents the lowest frame number that is ready to receive and confirms that the frames in front of it are correctly received.

1	2	3	4	5	6	7	8
0	N(S)			P/F	N(R)		

Figure 4-23 Information frame

Supervisory frame The frame is used to provide control information to achieve ARQ while not using the piggy-back mechanism, it is used to supervise and control the frame transmission.

As shown in Figure 4-24, S=00 RR (receive ready) is ready to receive the next frame, confirm the frame N(R)−1 and the previous frames. It is equivalent to the acknowledgement frame ACK for flow control.

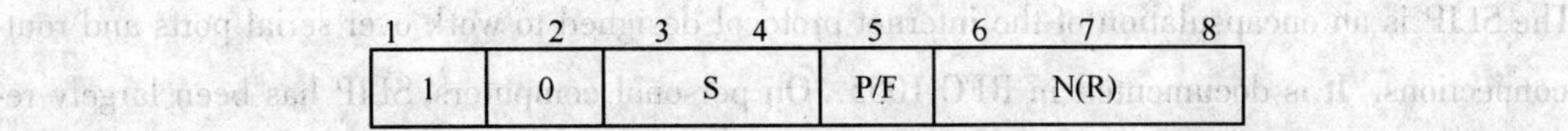

Figure 4-24 Supervisory frame

S=01 RNR (receive not ready), it means to suspend to receive the next frame, and confirm the frame N(R)−1 and all previous frames. It equal to ACK, it is used to but suspend reception for flow control.

S=10 REJ (reject), it denies frames from the beginning of N(R) and confirms the frame N(R)−1 and all previous frames. It is equivalent to denying frame NAK, which is used for error control in Go-back-N ARQ.

S=11 SREJ (selective seject), we only deny the frame N(R) and confirm N(R)−1 and previous frames. It is used to select error control in the selective repeat ARQ.

Unnumbered frame Provides link management functions.

As shown in Figure 4-25, the unnumbered frame provides additional data link control and management, such as the command and response for the establishment, release, recovery of the link, The unnumbered frame is not numbered, and can be sent at any time when needed without affecting the exchange sequence of a serial number of information frames. It uses 3, 4, 6, 7, 8 bits with the mark "M" to specify the type of the frame, it has 32 kinds of combinations, but currently only 15 kinds of UN numbered frames have been defined.

1	2	3 4	5	6 7 8
1	1	M	P/F	M

Figure 4-25 Unnumbered frame

Filed FCS The frame check sequence (FCS) is a 16-bit CRC-CCITT or a 32-bit CRC-32 computed over the address, control, and information fields. It provides a means by which the receiver can detect errors that may have been induced during the transmission of the frame, such as lost bits, flipped bits, and extraneous bits.

FCS is also used for error correction by return an affirmative response to an correct data frame. If a checkout error is detected, a negative response is sent and the transmitter is reissued, and if the frame lost is reissued over time.

4.5.2 The data link layer in the Internet

The data link layer in the internet has two famous protocols, Serial Line IP (SLIP) and point-to-point protocol (PPP).

The SLIP is an encapsulation of the internet protocol designed to work over serial ports and router connections. It is documented in RFC 1055. On personal computers, SLIP has been largely replaced by the PPP, which is better engineered, has more features and does not require its IP address configuration to be set before it is established. On microcontrollers, however, SLIP is still the preferred way of encapsulating IP packets due to its very small overhead.

The frame of SLIP is original IP package plus the tail flag 0xC0, character fill with 0xDB, 0xDC. It can optimize it this way that compress the header of TCP, IP, delete the same header and express different parts by the increment.

There are some problems: there is no error-free detection and verification; it only supports IP; it is unable to dynamically allocate IP addresses; it has no authentication, it is not suitable for dialing; it is not an Internet standard as there are multiple incompatible versions.

The PPP is commonly used as a data link layer protocol for connection over synchronous and asynchronous circuits, while it has largely superseded the older SLIP and telephone company mandated standards. The only requirement for PPP is that the circuit provided be duplex. Like SLIP, this is a full Internet connection over telephone lines via modem. It is more reliable than SLIP because it double checks to make sure that Internet packets arrive intact. It resends the damaged packets.

It is the formal standard of the Internet: which supports error detection, multiple protocols, IP addresses and allows authentication.

In PPP, the framing method of the tail of a frame and the head of next frame is clearly defined. This frame format also can handle error detection. When lines are no longer need the frames, pick them up, negotiate and choose, and release the link control protocol again——LCP (link control protocol). PPP uses the method of independent network layer protocol to negotiate the option of using the network layer for every supported network layer, the selected methods have different network control protocols NCP (network control protocol).

PPP is the most widely used data link layer protocol in the world now, when using a dial-up telephone line to access the Internet, users usually use the PPP protocol.

Figure 4-26 is a schematic diagram of a user's dialing into the net.

The PPP protocol was developed in 1992. After the revision in 1993 and 1994, the current PPP protocol has become the official standard of the Internet [RFC 1661]. The PPP protocol has three components: A method of encapsulating IP datagram to serial link; LCP and NCP.

As shown in Figure 4-27, the frame format of PPP is similar to HDLC. The symbol field F is still 0x7E (symbol “0x”), which means that the following character is expressed in sixteen binary system. The binary representation of sixteen binary 7E is 01111110. The address field A is only 0xFF. The address field does not actually work. The control field C is usually set to 0x03. PPP is byte oriented, and all PPP frames are integer bytes.

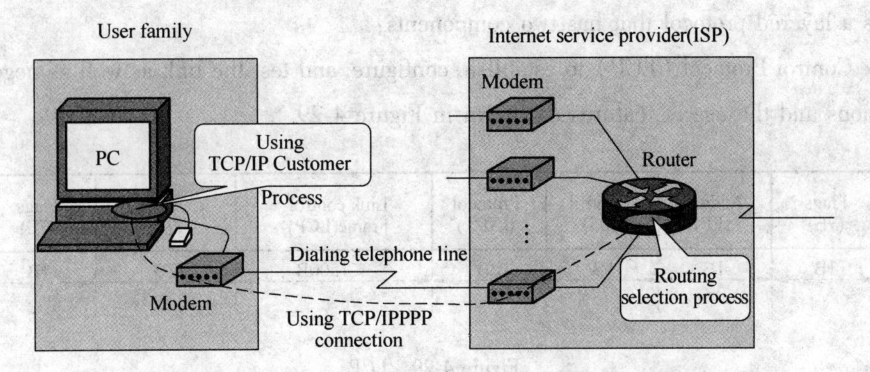

Figure 4-26 Schematic diagram of a user's dialing into the net

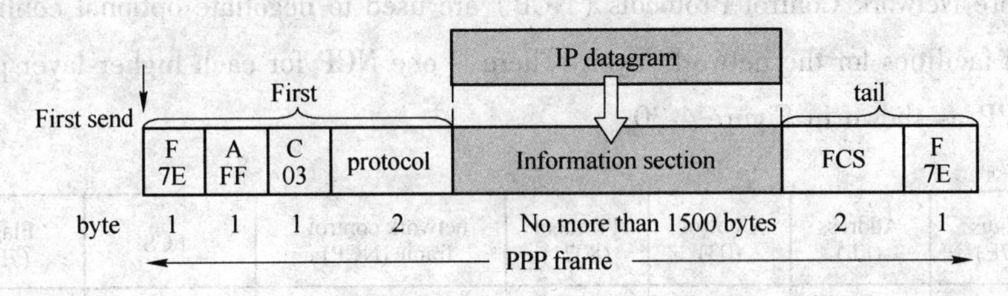

Figure 4-27 PPP frame

PPP has a 2 bytes protocol fields. When the protocol field is 0x0021, the information field of PPP frame is IP datagram. If the field is 0xC021, the information field is PPP link control data. If the field is 0x8021, it means network control data.

The PPP information frame format is shown in Figare 4-28.

Flags (7E)	Address (FF)	Control (03)	Protocol	Data field	FCS	Flags (7E)
1B	1B	1B	2B	≤1500B	2B	1B

Figure 4-28 Information frame format of PPP

Flag: 01111110

Address: The value is "FF" (11111111), indicating all the stations in the network. Receive the frame.

Control: The value is "03" (00000011).

Protocol: The length is 2 bytes. It identifies the type of network layer protocol data domain. The main types of network layer protocols commonly used are:

0021H—TCP/IP

0023H—OSI

0027H—DEC

Data field: Variable length.

PPP is a layered protocol that has two components:

A Link Control Protocol (LCP) to establish, configure, and test the link as well as negotiate settings, options and the use of features as shown in Figure 4-29.

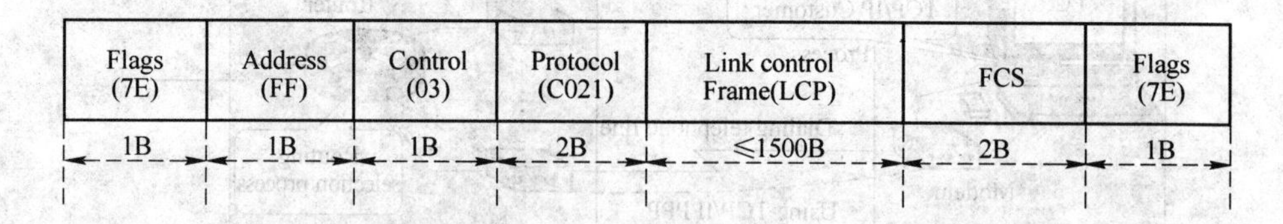

Figure 4-29 LCP

One or more Network Control Protocols (NCP) are used to negotiate optional configuration parameters and facilities for the network layer. There is one NCP for each higher-layer protocol supported by PPP as shown in Figure 4-30.

Flags (7E)	Address (FF)	Control (03)	Protocol (8021)	network control frame (NCP)	FCS	Flags (7E)
1B	1B	1B	2B	≤1500B	2B	1B

Figure 4-30 NCP

When PPP is used in synchronous transmission, protocol specifies hardware to complete bit stuffing (as HDLC does), and when it is used for asynchronous transmission, a special character filling method is used.

The reason why the PPP protocol does not use the serial number and the acknowledgement mechanism can be explained by the following considerations. The simpler PPP protocol is more reasonable when there is little probability of error in the data link layer. In the Internet environment, the information field placed in PPP is IP datagram. Reliable transmission of the data link layer cannot guarantee that the transmission of the network layer is also reliable. The FCS field of frame check sequence ensures no error acceptance.

As shown in Figure 4-31, when users dial in to access ISP, the router's modem confirms the dial-up and establishes a physical connection. The PC sends a series of LCP packets to the router (encapsulated into multiple PPP frames). These packets and their responses select some PPP parameters and complete network layer configuration. Then NCP assigns a temporary IP address to the newly connected PC machine, making the PC machine a host on the Internet. When communication is finished, NCP releases the network layer connection and restores the IP address that was originally assigned. Next, LCP releases data link layer connections. The final release is the connection of the physical layer.

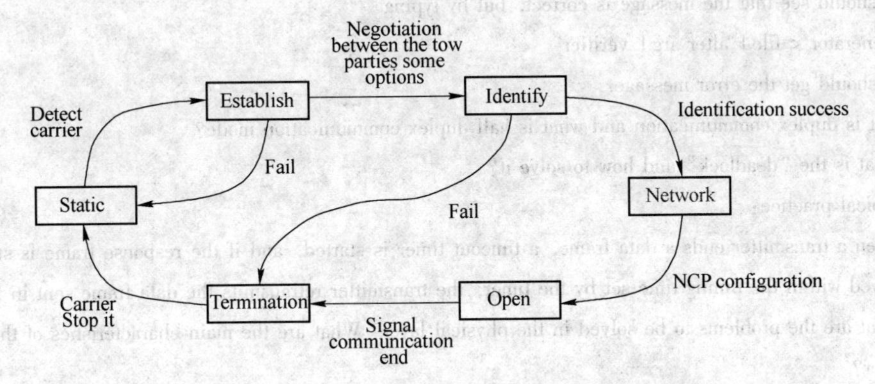

Figure 4-31 The state diagram of the PPP protocol

Problems

4-1 When bit stuffing is used, is it possible for the loss, insertion, or modification of a single bit to cause an error not detected by the checksum? If not, why not? If so, how? Does the checksum length play a role here?

4-2 Suppose that a message 1001 1100 1010 0011 is transmitted using Internet Checksum (4-bit word). What is the value of the checksum?

4-3 What is the remainder obtained by dividing x^7+x^5+1 by the generator polynomial x^3+1?

4-4 A bit stream 10011101 is transmitted using the standard CRC method described in the text. The generator polynomial is x^3+1. Show the actual bit string transmitted. Suppose that the third bit from the left is inverted during transmission. Show that this error is detected at the receiver's end. Give an example of bit errors in the bit string transmitted that will not be detected by the receiver.

4-5 In the discussion of ARQ protocol in Section 3.3.3, a scenario was outlined that resulted in the receiver accepting two copies of the same frame due to a loss of acknowledgement frame. Is it possible that a receiver may accept multiple copies of the same frame when none of the frames (message or acknowledgement) are lost?

4-6 Give at least one reason why PPP uses byte stuffing instead of bit stuffing to prevent accidental flag bytes within the payload from causing confusion.

4-7 What is the minimum overhead to send an IP packet using PPP? Count only the overhead introduced by PPP itself, not the IP header overhead. What is the maximum overhead?

4-8 The goal of this lab exercise is to implement an error-detection mechanism using the standard CRC algorithm described in the text. Write two programs, generator and verifier. The generator program reads from standard input a line of ASCII text containing an n-bit message consisting of a string of 0s and 1s. The second line is the k-bit polynomial, also in ASCII. It outputs to standard output a line of ASCII text with $n + k$ 0s and 1s representing the message to be transmitted. Then it outputs the generator program and outputs a message indicating whether it is correct or not. Finally, write a program, alter, that inverts 1 bit on the first line depending on its argument (the bit number counting the leftmost bit as 1) but copies the rest of the two lines correctly. By typing

generator < file | verifier

you should see that the message is correct, but by typing

generator < file | alter arg | verifier

you should get the error message.

4-9 What is duplex communication and what is half duplex communication mode?

4-10 What is the "deadlock" and how to solve it?

4-11 Typical practice:

When a transmitter ends a data frame, a timeout timer is started, and if the response frame is still not received within the timing time set by the timer, the transmitter retransmits the data frame sent in front.

4-12 What are the problems to be solved in the physical layer? What are the main characteristics of the physical layer?

4-13 What are the several basic concepts commonly used in data communication?

4-14 What are baseband transmission, frequency band transmission and broadband transmission?

4-15 What are modulation and coding technology, multiplexing technology and switching technology?

4-16 What is the difference between a data link and a link? What is the difference between "circuit connected" and "data link connected"?

4-17 The next layer of OSI to deal with the following problems, respectively? (1) divide the bitstream into frames; (2) identify which paths are used by the decision to reach the destination; (3) provide synchronization information.

4-18 The known CRC generating polynomial is $G(x) = X^4+X+1$, and the code word to be transmitted is 10110 (sent from left to right), and the check code is calculated.

4-19 A data communication system uses CRC verification, and the binary bit sequence of the generating polynomial $G(x)$ is 11001, and the binary bit sequence received by the destination node is 110111001 (including CRC cyclic check codes). Please judge whether there is an error in the transmission process. Why?

4-20 In the stop waiting protocol, why does the answer frame need no serial number (such as ACK0, ACK1, etc.)?

4-21 What is the frame format of HDLC? How to ensure the transparency of information? With a bit string of 10111110001111110010, after "0" bit insertion, what does it look like?

5 The Medium Access Control Sublayer

Goals:

(1) Understand the channel allocation problem.

(2) Understand the ALOHA and CSMA/CD.

(3) Master the standard 802 for LAN.

The computer network is divided into: point-to-point connection and broadcast channel. Medium access control is referred to as MAC. It can solve the problem of how to allocate the right to use the channel when the use of shared channels in the LAN generates competition.

When the use of the channel is competitive, the medium access control method can be used to control multiple nodes data transmitting and receiving in the public transmission medium and to allocate the right of the channel.

In IEEE 802 LAN/MAN standards, the medium access control (MAC) sublayer (also known as the media access control sublayer) and the logical link control (LLC) sublayer together make up the data link layer. Within the data link layer, the LLC provides flow control and multiplexing for the logical link, while the MAC provides flow control and multiplexing for the transmission medium.

5.1 The channel allocation problem

5.1.1 Static and dynamic channel allocation

In radio resource management for wireless and cellular networks, channel allocation schemes allocate bandwidth and communication channels to base stations, access points and terminal equipment. The objective is to achieve maximum system spectral efficiency in bit/s/Hz/site by means of frequency reuse, but still assure a certain grade of service by avoiding co-channel interference and adjacent channel interference among nearby cells or networks that share the bandwidth.

In a broadcast network, when a multiparty competition uses a channel, a single broadcast channel is allocated among multiple users through a protocol with the function of arbitration.

Channel-allocation schemes follow one of two types of strategy: static (fixed) channel allocation and dynamic channel allocation.

In fixed channel allocation (FCA) each cell is given a predetermined set of frequency channels. FCA requires manual frequency planning, which is an arduous task in time-division multiple access (TDM) and frequency-division multiple access (FDM) based systems since such systems are highly sensitive to co-channel interference from nearby cells that are reusing the same channel. Another drawback with TDM and FDM systems with FCA is that the number of channels in the cell remains constant irrespective of the number of customers in that cell. This results in traffic congestion and some calls being lost when traffic becomes heavy in some cells, and idle capacity in other cells.

A more efficient way of channel allocation would be dynamic channel allocation (DCA) in which signal channels are not allocated to cells permanently.

FDM is a traditional DCA method of multi-user allocation that is competitive. It is a simple and effective allocation scheme when the number of users is small, the number is fixed, and the traffic volume of each user is large. However, when users are more frequent and the number of traffic changes suddenly, they will waste resources. Even assuming that the number of users can be maintained at N, the method of dividing a single available channel into a fixed number of sub-channels is not efficient. The same problem exists in TDM.

None of the traditional static channel allocation methods are able to deal with communication bursts effectively.

5.1.2 LAN technology

A local area network (LAN), as shown in Figure 5-1 is a computer network that interconnects computers within a limited area such as a residence, school or office building. Ethernet and Wi-Fi are the two most common technologies in use for local area networks.

LAN has some basic features, the coverage is limited, and the distance is generally less than 10km. Its data transmission rate is relatively high; it may be 10Mbit/s, 100Mbit/s or 1000Mbit/s, 10Gbit/s Ethernet is being launched. The bit error rate is low, at about $10^{-8} \sim 10^{-10}$.

There are 3 aspects of the basic technology that can reflect the LAN features, network topology, transmission medium and media access control method.

5.1.2.1 Typical network topology

Network topology is the topological structure of a network and may be depicted physically or logically. Network topology can be used to define or describe the arrangement of various types of telecommunication networks, including command and control radio networks and computer networks. A wide variety of physical topologies have been used in LANs, including ring, bus, mesh and star as

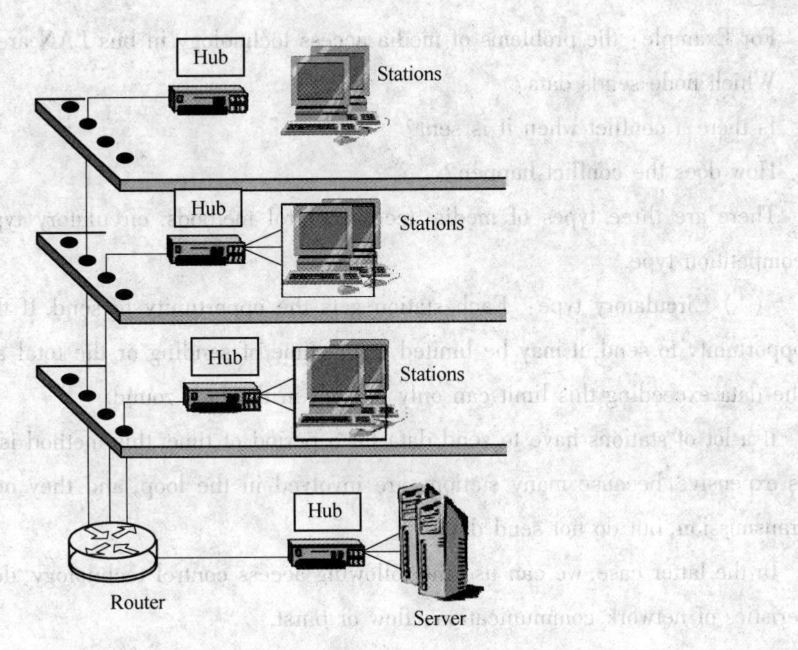

Figure 5-1 A local area network (LAN)

shown in Figure 5-2. Conversely, mapping the data flow between the components determines the logical topology of the network.

Bus/tree type: all nodes are directly connected to the bus channel.

Star type: all nodes are connected to the central node.

Ring type: nodes are connected to adjacent nodes through point-to-point links.

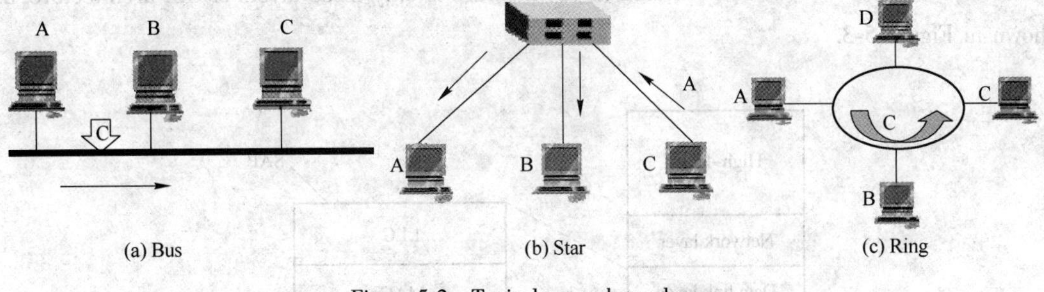

Figure 5-2 Typical network topology

5.1.2.2 Media access control method

Access refers to establishing connection between two entities and exchanging information.

Media access technology is one of the key technologies of LAN to study how to effectively utilize the problem of the channel. Its purpose is to adapt to the sudden characteristics of data transmission, and adopt appropriate reuse strategies to improve the utilization of the channel and reduce the delay in data transmission.

For Example: the problems of media access technology in bus LAN are:

Which node sends data?

Is there a conflict when it is sent?

How does the conflict happen?

There are three types of media access control methods: circulatory type、reservation type and competition type.

(1) Circulatory type: Each station gets the opportunity to send. If the workstation uses this opportunity to send, it may be limited to the time of sending or the total amount of data sent, and the data exceeding this limit can only be sent in the next round.

If a lot of stations have to send data for a period of time, this method is effective; conversely, it is expensive, because many stations are involved in the loop, and they only transmit the right of transmission, but do not send data.

In the latter case, we can use the following access control technology, depending on the characteristics of network communication: flow or burst.

(2) Reservations type: It is suitable for streaming communication, namely the long-time continuous transmission, such as voice communication, telemetry communication and long file transmission.

(3) Competition type: It is suitable for burst communication, namely the short-time sporadic transmission; it does not control the transmission rights of each workstation, but offers the opportunity freely by the various workstations. When the load is light or medium, the efficiency is very high and the load is very heavy.

LAN is a communication network, and its architecture is only three layers of OSI architecture, as shown in Figure 5-3.

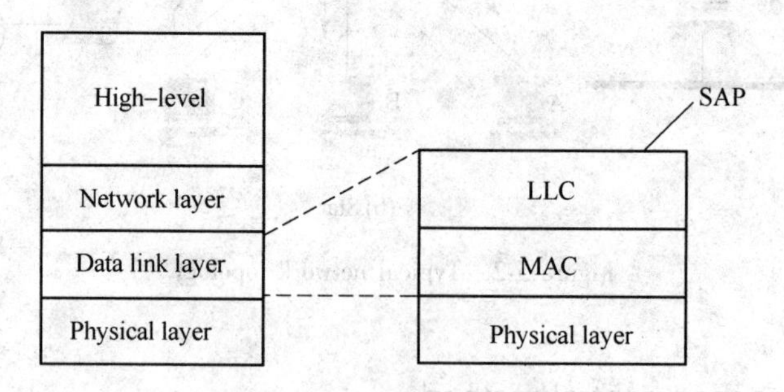

Figure 5-3 LAN reference model

Physical layer: it is an essential layer which is the same as the physical layer of the OSI reference model; it is responsible for the physical connection and the transmission of the bit stream in the media; its main task is to describe some of the characteristics of the transmission media structure.

Data link layer: it is closely related to network topology, channel (medium) access method and media type. In order to adapt to the development of technology and keep the interface between the data link layer and the upper layer as stable as possible, the IEEE 802 divides the LAN data link layer into two sublayers, as shown in Figure 5-4.

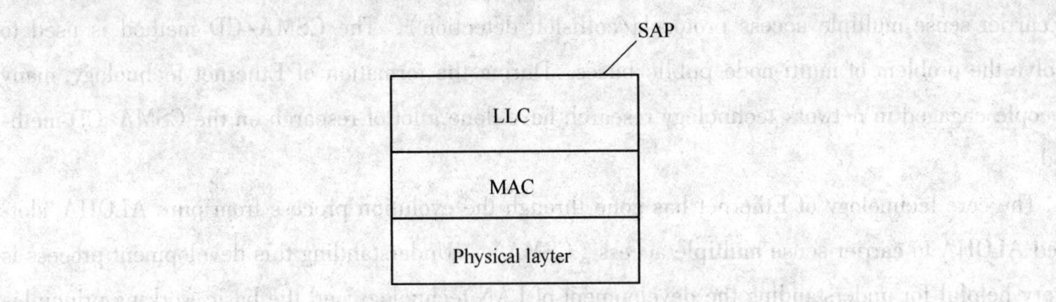

Figure 5-4 Part of OSI architecture

Network layer: it is used to broadcast communication does not need a routing function, so it is simplified into the service access point SAP of the upper protocol.

Data link layer: it is divided into two sublayer according to functions, LLC (logical link control) and MAC (medium access control), in order to distinguish hardware related parts and hardware independent parts according to their functions, so as to reduce the complexity of research and implementation.

MAC sublayer functions include framing/dismounting frames, realizing and maintaining MAC protocol, bit error detection and addressing. IEEE 802 stipulates that MAC has CSMA/CD, token bus and token ring.

Functions of the LLC sublayer include providing SAP to a high level, establishing/releasing logical connection, error control, and frame number processing.

5.2 Multiple access protocols

The data link layer is responsible for transmission of data between two nodes. Its main functions are data link control and multiple access control. The data link control is responsible for reliable transmission of messages over the transmission channel by using techniques like framing, error control and flow control. For data link control refer to-stop and wait ARQ.

If there is a dedicated link between the sender and the receiver then the data link control layer is sufficient; however if there is no dedicated link present, then multiple stations can access the channel simultaneously. Multiple access protocols are therefore required to decrease collision and avoid crosstalk. For example, in a classroom full of students, when a teacher asks a question and all the students (or stations) start answering simultaneously (send data at same time) then a lot of chaos is created (data overlap or data lost) then it is the job of the teacher (multiple access protocols) to manage the students and make them answer one at a time. Thus, protocols are required for sharing data on non-dedicated channels. This is multiple access protocols.

The core technology of Ethernet is the medium access control method. It is called CSMA/CD (carrier sense multiple access protocols/collision detection). The CSMA/CD method is used to solve the problem of multi-node public buses. During the formation of Ethernet technology, many people engaged in network technology research have done a lot of research on the CSMA/CD method.

The core technology of Ethernet has gone through the evolution process from pure ALOHA, slotted ALOHA to carrier sense multiple access (CSMA). Understanding this development process is very helpful for understanding the development of LAN technology and the basic working principles of Ethernet.

5.2.1 ALOHA

ALOHA was a pioneering computer networking system developed at the University of Hawaii. ALOHA used a new method of medium access (ALOHA random access) and experimental ultra-high frequency (UHF) for its operation, since frequency assignments for communications to and from a computer were not available for commercial applications in the 1970s. But even before such frequencies were assigned there were two other media available for the application of an ALOHA channel-cables & satellites.

In the 1970s, ALOHA random access was employed in the nascent Ethernet cable based network and then in the Marisat (now Inmarsat) satellite network.

5.2.1.1 Pure ALOHA

The ALOHA system was designed by the University of Hawaii in 70s for their ground wireless packet network.

As shown in Figure 5-5, the basic idea of pure ALOHA is that whenever users have data to send, they can send them at any time. When the information is sent, the transmission station waits

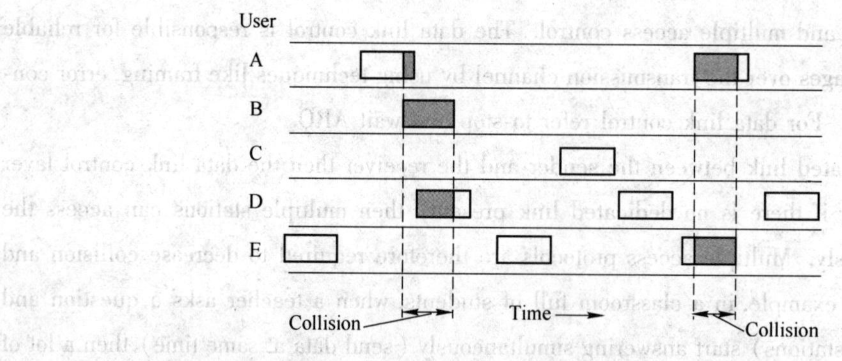

Figure 5-5 Pure ALOHA

for a period of time. If the response information of the receiving station is received within the waiting time, it is shown that the transmission is successful, otherwise, the data frame will be reissued.

In short, its main is that the transmission station waits for a random time to retransmit the frame to avoid conflict. Because multiple users share a common channel system in a way that may lead to conflict.

The throughput of the pure ALOHA channel is very low, and the utilization rate is only 18%.

5.2.1.2 Slotted ALOHA

The characteristics of pure ALOHA protocol are that there is not constraint on the time of sending data by senders, so the throughput is very low. Therefore a split ALOHA protocol is proposed.

The basic idea is to improve the time by limiting the time of sending information. Stations can only start sending messages at the beginning time of a time slot. It requires user time synchronization; to realize it, one way is to set up a special site, sending a signal like a clock at the start of every time.

The best result is: 37% of the time slot is empty, 37% of the time slot transmission is successful, 26% of the time slot conflict. The utilization of channel with the slotted ALOHA protocol is twice as much as that of the pure ALOHA channel.

5.2.2 Carrier sense multiple access protocols

When we use the split ALOHA protocol, we only control the channel's gap and do not control any process of the node's delivery, so the throughput is only 37%.

Carrier sense multiple access protocol (CSMA) requires network sites to listen to the presence of transmission and execute corresponding actions. Putting forward the method of "listen before talk" is an improved protocol for ALOHA.

Before sending data, each station first listens if any other stations are sending information. If there is none, it sends data; if there are, it will not be sent for a while, and it would try again after a period of retreat. According to the different listening strategies, there are three types of CSMA, Persistent CSMA, p-Persistent CSMA, Non-Persistent CSMA, channel utilization versus load for various random access protocols are shown in Figure 5-6.

5.2.2.1 Persistent CSMA

When listening to the busy channel, we still insist on listening until the channel is idle. Once the channel is idle, it sends data immediately. If there is a conflict (not answered in a specified time), then it is advisable to wait for a random time to intercept.

Persistent CSMA is also known as 1-persistent CSMA, because once the site finds idle channels, the probability of sending data is 1.

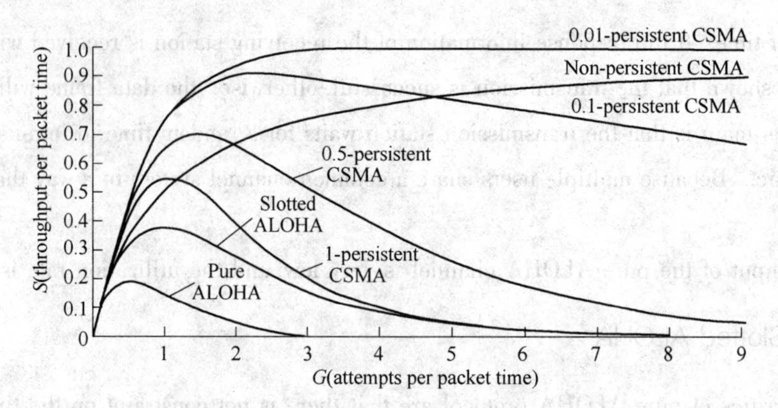

Figure 5-6 Comparison of the channel utilization versus load for various random access protocols

5. 2. 2. 2 P-persistent CSMA: for the gap channel

When listening to the busy channel, we still insist on listening until the channel is idle. When the channel is idle, the data is sent according to the probability of p; in the case of probability $q = 1 - p$, it delays the task of transmitting the data to the next slot and relistens to the channel at the same time. If the next time slot is still idle, it will transmit data again at the probability p or postpone the transmission to the next time slot with probability q. This process has been duplicated and sent or postponed according to the p or q until the transmission is successful or the other station starts sending data.

5. 2. 2. 3 Non-persistent CSMA

Once we hear the busy channel, we will not listen to it any longer, and we will delay the monitoring again after a random time. If we listen to the channel idle, we will send data immediately. If there is a conflict (not answered within a specified time), the site will wait for a long time, and then start again.

There are some problems existing in CSMA, CSMA can effectively reduce the probability of conflict before sending the data, but the conflict cannot be completely eliminated because of the existence of propagation delay, and CSMA cannot detect the conflicts that have occurred in time. In CSMA, even if the conflict has occurred, the workstation will continue to send; only if the response is not received after the arrival of the specified time, it is known that the transmission has been in conflict.

In order to detct the conflict in time, take the way of while listening to the channel while it is trans mitting. Once the conflict is monitored, the two sides of the conflict stop sending immediately, so as not to waste the channel, and the channel quickly enters the idle period.

CSMA with conflict detection (CSMA/CD) is the most commonly used method in bus networks. When a site wants to transmit data, it first listens to the channel to see if other sites are transmitting.

If the channel is idle, the data is sent and monitored. Once the conflict is monitored, the transmission is stopped immediately, and a series of blocking signals (JAM) will be sent to the bus in a short time to strengthen the conflict, and to notify the stations on the bus that there are conflicts, so as to empty the channel early and improve the utilization of the channel.

If the channel is busy, it will wait until it listens to the idle channel, and then sends out the data.

It must be noted that after the site detection conflicts, they do not continue to finish their frames, but stop as soon as possible, quickly end the transmission of the conflict frames to save time and frequency.

Method of conflict detection is that the workstation sends the information to the bus, and receives information from the bus, at the same time then compares the received information with the original information. They will compare same aspects as following:

(1) The size of the comparison signal.

(2) Determine the location of the zero point.

(3) Bit by bit comparison.

5.3 IEEE standard 802 for LANs and MANs

The IEEE802 Committee was founded in 1980.2. Its task is to establish the international standards of LAN. The contents of the study are as follows:

802.1 LAN overview, architecture, network management and performance measurement;

802.2 LLC logic link control protocol;

802.3 Ethernet;

802.4 Token bus;

802.5 Token ring MAC layer;

802.6 MANS DQDB;

802.7 Broadband LAN standard;

802.8 Fiber optic TAG.

The current standards for more mature research are illustrated as Figure 5-7.

802.2LLC					Data link layer
802.3 CSMA/CD	802.4 Token Bus	802.5 Token Ring	802.6 DQDB	…	Physical layer

Figure 5-7 IEEE standard 802

The data link layer has two different protocol data parts: LLC layer and MAC layer. Data from high-level together with LLC's head are LLC. Downwards to the MAC sublayer, plus MAC's head and tail, constitute MAC frame. The physical layer transmits MAC frames as bit streams transparently across data link entities.

5. 3. 1 IEEE standard 802. 3 and ethernet

IEEE 802. 3 is a working group and a collection of Institute of Electrical and Electronics Engineers (IEEE) standards.

In 1972, the first experimental bus LAN system was developed by Xerox Research Center (Bob Metcalfe). The data transmission rate is 2. 94M, and the system was named Ethernet.

In July 1976, Bob Metcalfe published papers:《ethernet: distributed packet switching for local computer networks》the core technology is CSMA/CD.

In 1980, the physical layer and data link level specification of Ethernet was published by DEC, Intel and Xerox.

In 1985, it was named IEEE 802. 3, the LAN standard of CSMA/CD protocol, with a data rate between 1Mbps and 10Mbps, supporting multiple transmission media.

Ethernet refers to the baseband bus LAN, the frame format of Ethernet and IEEE 802. 3 is different.

5. 3. 1. 1 The physical layer standard of IEEE802. 3

IEEE802. 3 defines a variety of physical layer standards, mainly including the following types.

(1) 10Base5—thick Ethernet cable.

(2) 10Base2—thin Ethernet cable.

(3) 10BaseT—twisted pair cable.

(4) 10BaseF—optical fiber cable.

The Ethernet specification is the same as 10Base5.

10Base5 standard: as shown in Figure 5-8 10Base5 (also known as thick Ethernet or thicknet) was the first commercially available variant of Ethernet. As shown in Figure 5-9, 10Base5 uses a thick and stiff coaxial cable up to 500 meters in length. Up to 100 stations can be connected to the cable using vampire taps and they share a single collision domain with 10 Mbit/s of bandwidth shared among them. The system is difficult to install and maintain.

The cable is a thick and stiff coaxial cable. Transceiver can transmit/receive, conflict and is capable of detection, electrical isolation, and ultra-long control. NIC can send data to the network, and receive data from other devices from the network and send it to the workstation.

10Base2 standard: 10Base2 (also known as cheapernet, thin ethernet, thinnet, and thinwire) is a variant of Ethernet. During the mid to late 1980s this was the dominant 10 Mbit/s Ethernet standard, but due to the immense demand for high speed networking, the low cost of Category 5 cable, and the popularity of 802. 11 wireless networks, both 10Base2 and 10Base5 have become increasingly obsolete, though devices still exist in some locations.

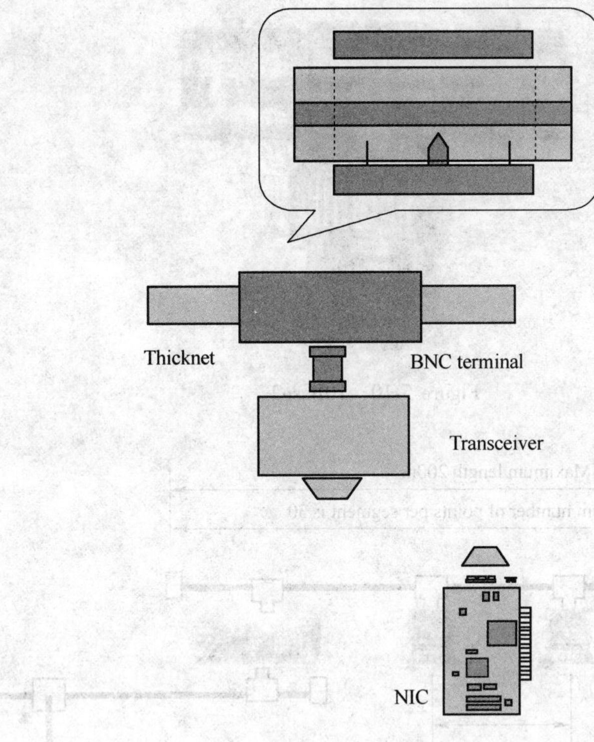

Figure 5-8 10Base5

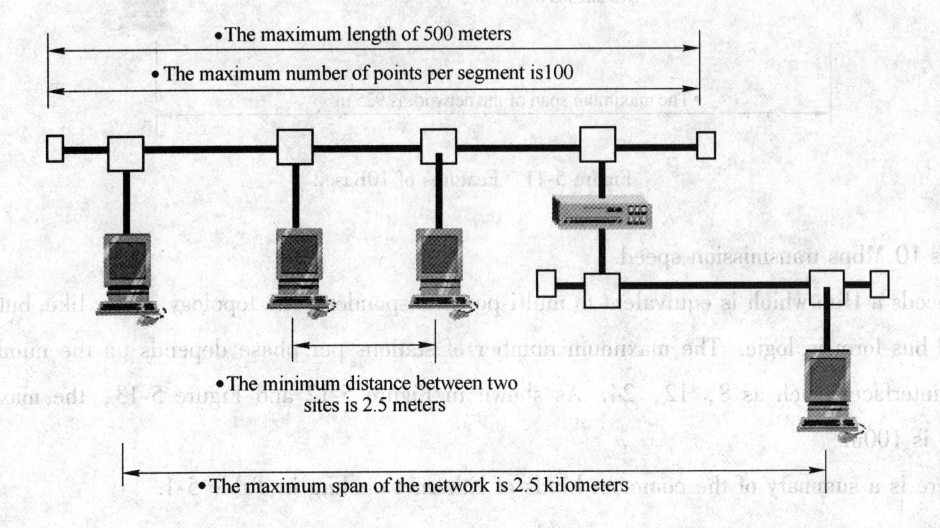

Figure 5-9 Features of 10Base5

As shown in Figure 5-10, the cable is a thin coaxial cable connected through the BNC T joint and its transceiver circuit is on the interface board network card. As shown in Figure 5-11, the maximum of the segment is 200m, with a maximum of 30 sites per segment. The shortest distance between the two sites is 0.5m and the maximum span of the network is 925m.

10Base-T standard: 10Base-T is ordinary telephone twisted pair wire. 10Base-T supports Eth-

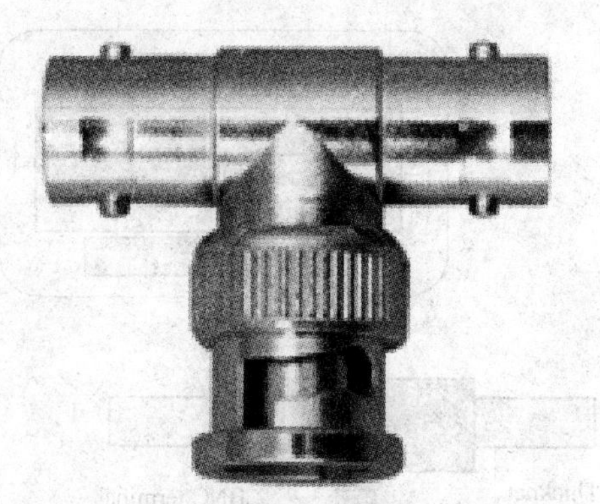

Figure 5-10 10Base2

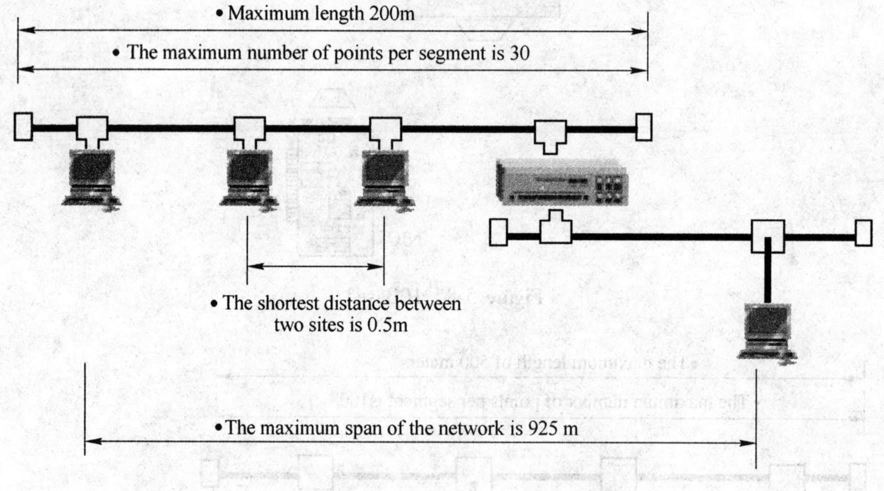

Figure 5-11 Features of 10Base2

ernet's 10 Mbps transmission speed.

It needs a Hub which is equivalent to multi-port transponder. The topology is star like, but logic is still bus form in logic. The maximum number of stations per phase depends on the number of HUB interfaces, such as 8, 12, 24. As shown in Figure 5-12 and Figure 5-13, the maximum length is 100m.

There is a summary of the commom kinds of Ethernet cabling in Table 5-1.

Table 5-1 The most common kinds of Ethernet cabling

Name	Cable	Max. seg.	Nodes/seg	Advantages
10Base5	Thick coax	500m	100	Original cable; now obsolete
10Base2	Thin coax	185m	30	No hub needed
10Base-T	Twisted pair	100m	1024	Cheapest system
10Base-F	Fiber optics	2000m	1024	Best between buildings

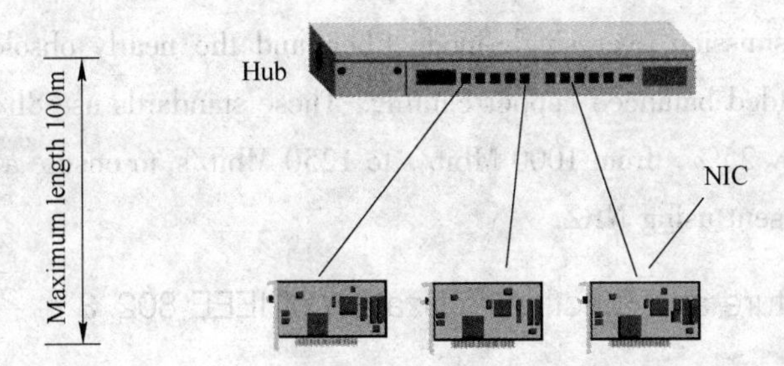

Figure 5-12 Features of 10Base-T

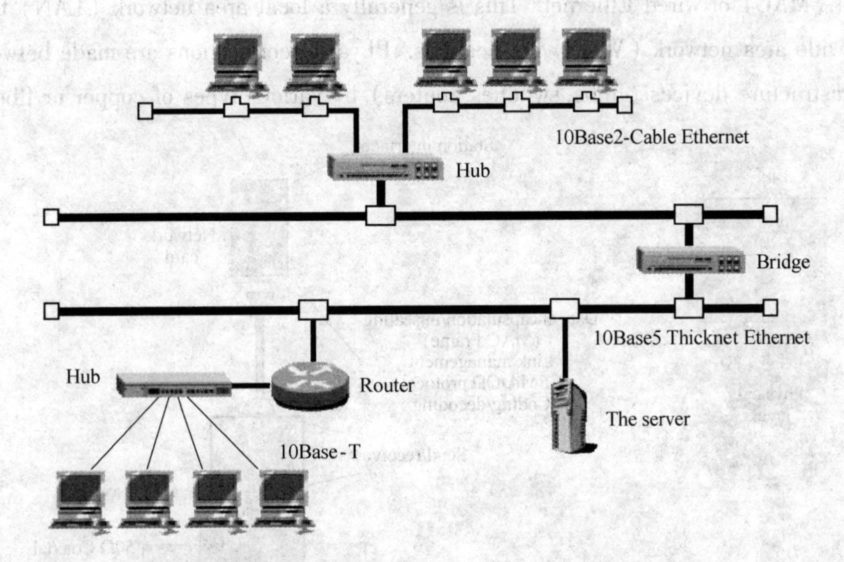

Figure 5-13 Interconnection of multiple network segments

In computer networking, Gigabit Ethernet (GbE or 1 GigE) is a term describing various technologies for transmitting Ethernet frames at a rate of a gigabit per second (1000000000 bits per second), as defined by the IEEE 802.3 standard.

There are five physical layer standards for Gigabit Ethernet using optical fiber (1000Base-X), twisted pair cable (1000Base-T), or shielded balanced copper cable (1000Base-CX), as shown in Table 5-2.

Table 5-2 Gigabit Ethernet cabling

Name	Cable	Max. segment	Advantages
1000Base-SX	Fiber optics	550m	Multimode fiber (50μm, 62.5μm microns)
1000Base-LX	Fiber optics	5000m	Single (10μm) or multimode (50μm, 62.5μm)
1000Base-CX	2 Pairs of STP	25m	Shielded twisted pair
1000Base-T	4 Pairs of UTP	100m	Standard category 5 UTP

The IEEE 802.3z standard includes 1000Base-SX for transmission over multi-mode fiber,

1000Base-LX for transmission over single-mode fiber, and the nearly obsolete 1000Base-CX for transmission over shielded balanced copper cabling. These standards use 8b/10b encoding, which inflates the line rate by 25%, from 1000 Mbit/s to 1250 Mbit/s, to ensure a DC balanced signal. The symbols are then sent using NRZ.

5.3.1.2 Architecture and function realization of IEEE 802.3

As shown in Figure 5-14, IEEE 802.3 defined the physical layer and data link layer's media access control (MAC) of wired Ethernet. This is generally a local area network (LAN) technology with some wide area network (WAN) applications. Physical connections are made between nodes and/or infrastructure devices (hubs, switches, routers) by various types of copper or fiber cable.

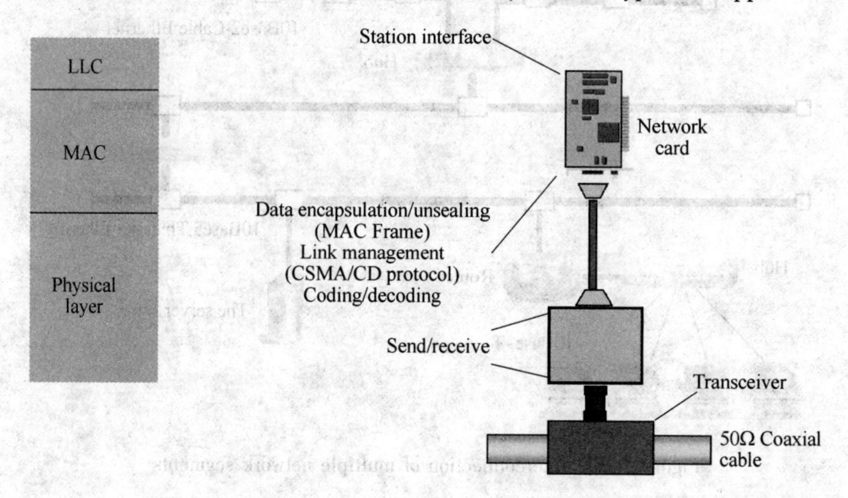

Figure 5-14 Architecture of IEEE 802.3

5.3.1.3 Ethernet/IEEE802.3's operation

Carrier-sense multiple access with collision detection (CSMA/CD) is a media access control method used most notably in early Ethernet technology for local area networking. It uses carrier-sensing to defer transmissions until no other stations are transmitting. There are four steps, listen before talk, listen while talk, collision stop, delayer recurence.

Listen before talk. Before the source site sends the data frame, it first listens to whether the channel is busy. If the channel is busy, that is, there is a carrier signal on the channel, then the site delays the transmission and waits for the channel to recover to the idle state.

Listen while talk. The source site uses the technology of sending and listening on the side to detect whether there is any interference signal indicating the conflict.

Collision stop. If there is an interference signal, it means that the data transmitted on the channel is conflicting, and the source site will stop sending immediately. In order to inform conflicts between other sites involved in the conflict, the source site first sends a series of blocking signals quickly for the line alarm.

Delayed reoccurrence the site that is involved the conflict waiting for a random time, and then repeating the above actions to prepare the retransmitted for the conflicting data frames.

As shown in Figure 5-15, every site can receive all data from other sites. It needs an addressing mechanism to identify the destination site, and only one site copies the frames received, and the other sites will discard frames.

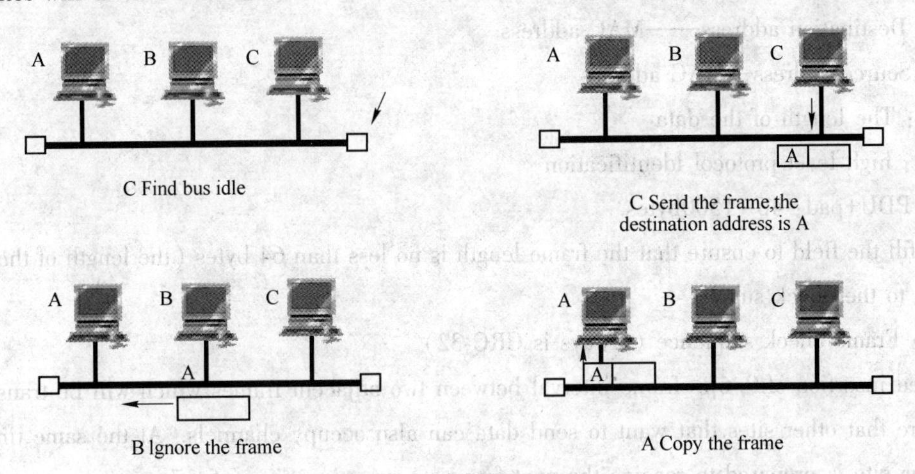

Figure 5-15 Example

5. 3. 1. 4 IEEE802. 3 and Ethernet frame format

As shown in Figure 5-16, there are some differences in IEEE802. 3 and Ethernet with their frame format. In the Ethernet, frames have a type field. This frame format is used on all forms of Ethernet by protocols in the Internet protocol suite. Same as Ethernet except type field is replaced by length, and an 802. 2 LLC header follows the 802. 3 header, it is based on the CSMA/CD process.

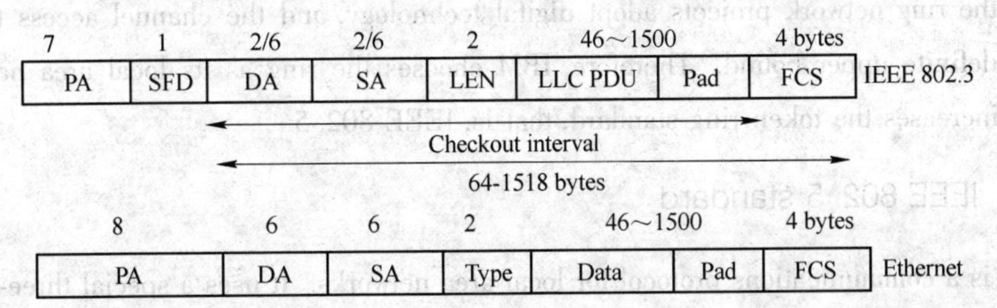

Figure 5-16 Frame of IEEE 802. 3

A data unit on an Ethernet link transports an Ethernet frame as its payload. The internal structure of an Ethernet frame is specified in IEEE 802. 3.

An Ethernet frame is preceded by a preamble and start frame delimiter (SFD), which are both part of the Ethernet packet at the physical layer. Each Ethernet frame starts with an Ethernet header, which contains destination and source MAC addresses as its first two fields. The middle section of the frame is the payload data including any headers for other protocols (for example, internet protocol) carried in the frame. The frame ends with a frame check sequence (FCS), which is a 32-bit cyclic redundancy check used to detect any in-transit corruption of data.

PA: The preamble sequence 10101010 is used to synchronize the receiver with the sender

SFD: Frame header delimitation-10101011 marks the beginning of a frame

DA: Destination address——MAC address

SA: Source address—MAC address

LEN: The length of the data

Type: high level protocol identification

LLC PDU+pad: 46~1500bytes.

Pad fill the field to ensure that the frame length is no less than 64 bytes (the length of the target address to the check sum)

FCS: Frame check sequence (always is CRC-32)

Forced insertion of 9.6μs frame interval between two adjacent frames which will be transmitted to ensure that other sites that want to send data can also occupy channels. At the same time, the receiving site is prepared to receive the next frame as shown in Figure 5-17.

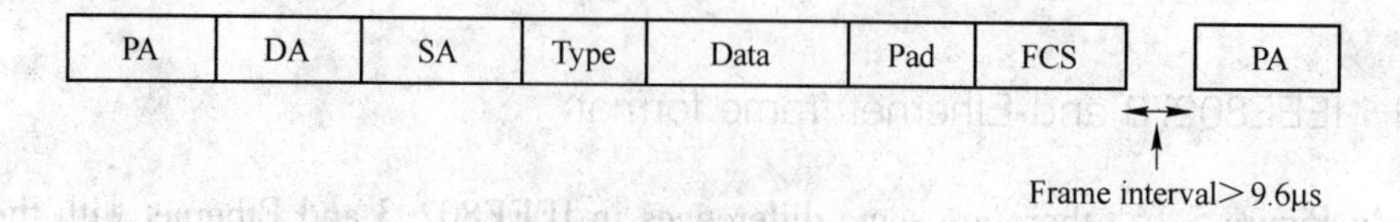

Figure 5-17 Frame

5.3.2 IEEE standard 802.5: token ring

Almost all the ring network projects adopt digital technology, and the channel access time of the ring has a definite upper bound. Therefore, IBM chooses the ring as its local area network, and IEEE also increases the token ring standard, that is, IEEE 802.5.

5.3.2.1 IEEE 802.5 standard

Token ring is a communications protocol for local area networks. It uses a special three-byte frame called a "token" that travels around a logical "ring" of workstations or servers. Introduced by IBM in 1984, the technology was then standardized with protocol IEEE 802.5 and was fairly successful, particularly in corporate environments.

The IEEE 802.5 standard specifies the format and protocol of the data unit used by the media access control sublayer and the physical layer by the token ring. It specifies the services between adjacent entities and the methods to connect the physical media of the token ring.

(1) Transmission media: Shielded twisted pair wire, optical fiber.

(2) Rate: 1M, 4M, 16Mbps.

(3) Maximum number of points: 250.

(4) Signal coding: Manchester code.

(5) Typical signal propagation speed: 200m/μs.

5.3.2.2 Operation mode of ring interface

The structure of ring interface is shown in Figure 5-18, TCU has two operation modes, sending mode (when the site sends data) and listening mode (at other times).

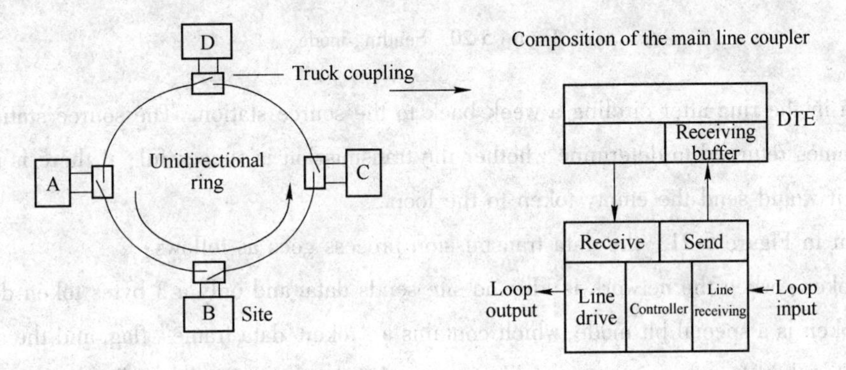

Figure 5-18 The structure of ring interface (TCU—truck coupling unit)

In the listening mode, as shown in Figure 5-19, TCU and DTE disconnect, the input bits are copied with "1bit delay", them be output. And detect whether there is a local address or a token in the frame.

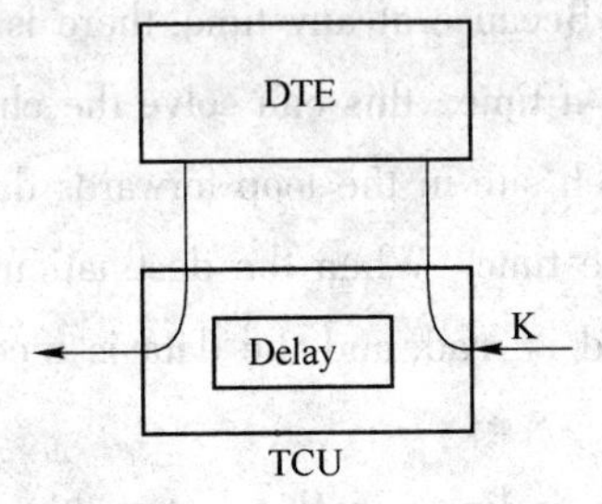

Figure 5-19 Listening method

If there is a local address, the switch K will be disconnected, TCU will connect to DTE, copy the frame to DTE, and continue forwarding to the next node; if a token is sent and the DTE has data to send, the token is captured and transferred to the sending mode, and the data frame is sent.

Capture token: when a site needs to send a data frame, the token's unique logo is transformed into a symbol of the information frame, namely position in a 3 byte token is reversed to make the token the first 3 bytes of a normal data frame.

In the sending mode, as shown in Figure 5-20 the interface connections between the input and output, outputs its data to the ring, and the data is sent from the output of TCU to the input of the next TCU in frame.

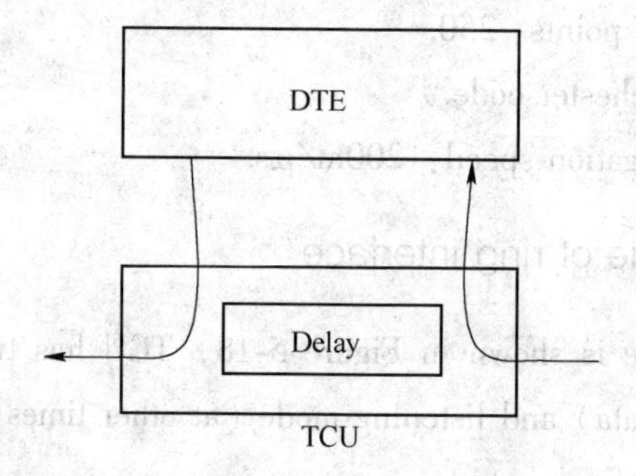

Figure 5-20 Sending mode

The data in the ring after circling a week back to the source station. The source station checks the data frames returned to determine whether the transmission is successful; if there is no data to send, then it would send the empty token to the loop.

As shown in Figure 5-21, the data transmission process goes as follows:

Empty token: when the network is idle, no site sends data, and only a 3 bytes token detours the loop; the token is a special bit mode, which contains a "token/data frame" flag, and the symbol bit "0" indicates that the token is an available empty token, and the flag bit "1" indicates that there is site occupancy. The token is sending the data frame. Note: the ring interface is generally powered by the loop, and closing the site will not affect the way the interface works.

Capture the token, and send data: when a site needs to send data, it must wait for and obtain a token, the token's flag position as "1", plus the header, information, and the end of the message, them change it into a data frame. Because at any time, there is only one token on the token ring and only one site can send data at a time; this can solve the channel access problem.

Forward and receive frames: each site in the loop forwards data and checks the destination address in the data frame of the same time. When the destination address in the frame head is the address of this site, the data carried is read, and the data is received and transmitted at the same time.

Revocation of frame: data frame circling a week to return, the sending station will withdraw from the loop. At the same time, according to the information returned, we confirm whether there is any error in the transmitted data. If there is a mistake, it would repeat the remaining acknowledgement frames in the buffer, or release the confirmed frames in the buffer.

Re-issuing: after sending the data from the sending site, a new empty token is reproduced to the next site to allow other sites to obtain the license to send data frames.

When the traffic is light, the token is mostly in the loop. However when the traffic is heavy, each site has a sending queue, and the transmission right is passed in the ring. It can be seen that the network efficiency is almost 100% when the loop is overloaded.

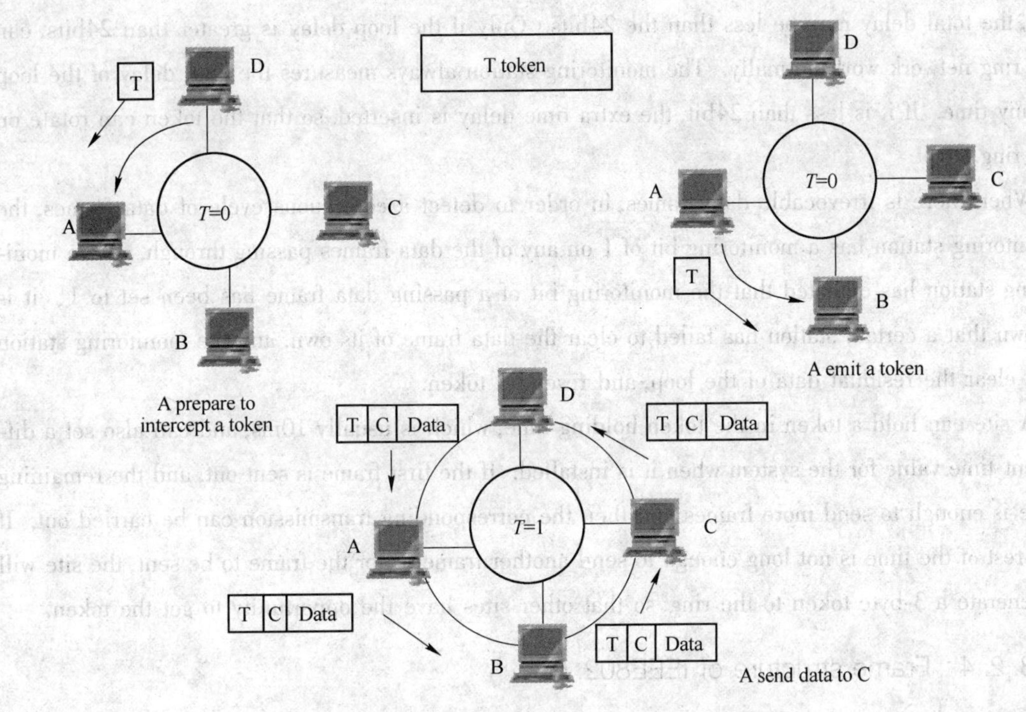

Figure 5-21 The operation of token ring/IEEE802.5

5.3.2.3 Maintenance of token ring

There are two most serious errors in the loop network:

Token loss: the token is a particular format of a bit string, and the token ring transfer process may also be disturbed and make mistakes in the loop without the token.

Data frames are not be revocable: when a station sends a data frame, the data frame cannot be revoked from the network because of the failure; it will cause the error of the continuous cycle of the data frame of the network.

The fault handling function of the token ring is mainly reflected in the maintenance of token and data frame.

The solution is centralized control. Each ring is designated as a monitoring station on the ring road to manage the whole loop operation. The responsibility of the monitoring is to ensure the token is not lost, guaranteeing the minimum time delay of the loop, scavenging frame without site.

When the token is lost, the monitoring station detects token loss using a timeout mechanism. The timeout value is longer than that of the longest frame traversing the loop. If no token is detected during the period, the token is considered to be lost. The management station will clear up the data fragments in the loop and issue a token.

When the loop delay is too short, the total loop delay includes transmission medium delay and 1bit delay at each station. If the length of the ring network is short and the active workstations are few, the total delay may be less than the 24bits. Only if the loop delay is greater than 24bits, can the ring network work normally. The monitoring station always measures the total delay of the loop at any time. If it is less than 24bit, the extra time delay is inserted, so that the token can rotate on the ring.

When there is irrevocable data frames, in order to detect a continuous cycle of data frames, the monitoring station has a monitoring bit of 1 on any of the data frames passing through. If the monitoring station has detected that the monitoring bit of a passing data frame has been set to 1, it is known that a certain station has failed to clear the data frame of its own, and the monitoring station will clear the residual data of the loop, and resend a token.

A site can hold a token in the token holding time, which is usually 10ms, and can also set a different time value for the system when it is installed. If the first frame is sent out, and the remaining time is enough to send more frames and then the corresponding transmission can be carried out. If the rest of the time is not long enough to send another frame or for the frame to be sent, the site will regenerate a 3-byte token to the ring, so that other sites have the opportunity to get the token.

5.3.2.4 Frame structure of IEEE802.5

As shown in Figure 5-22, the MAC frame of IEEE 802.5 token ring has two basic formats: token frame and data frame. The token frame is only 3 bytes long, and the data frame may be very long. The two frames have a start delimiter SD and end delimiter (ED), which are used to determine the boundary of the frame. When no station is sending a frame, a special token frame circles the loop. This special token frame is repeated from station to station until arriving at a station that needs to send data.

	1B	1B	1B
Token frame	SD	AC	ED

	1	1	1	2/6	2/6	≥0	4	1	1B
Data frame	SD	AC	FC	DA	SA	DATA	FCS	ED	FS

Figure 5-22 Frame structure of IEEE802.5

Tokens are 3 bytes in length and consist of a start delimiter, an access control byte, and an end delimiter.

Data frames carry information for upper-layer protocols, while command frames contain control information and have no data for upper-layer protocols. Data/command frames vary in size, depending on the size of the information field.

(1) SD, ED (Start delimiter and end delimiter): used to mark the beginning and end of frames.

(2) AC (Access control field) as shown in Figure 5-23.

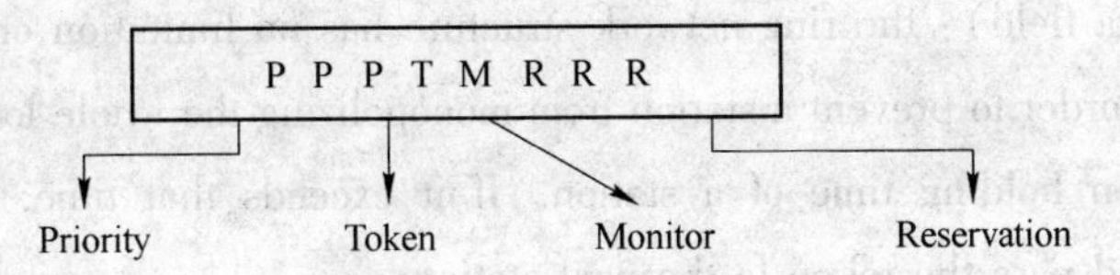

Figure 5-23 Access control field

Priority and reservation: only the sites with priority above the priority within the token are allowed to intercept tokens. A site that sends data can make an appointment when the data frames sent by other sites are passed through this station, and the priority of the station is written to the reservation bit of the frame. A maximum number of priority is 8; in the no priority loop, these two fields do not work and are placed 0.

Token bit: frame type identifier, "0"-token; "1"-information and control frame.

Monitor bit: prevents invalid frames from circulating in the loop indefinitely. When the frame is sent out, $M=0$; when the frame is first monitored, the M is set to 1; if the frame passes through the monitoring station again, the frame will be cleared by the monitoring station.

(3) FC (The frame control field, as shown in Figure 5-24): It is used to separate the general information frame from the control frame in the data frame.

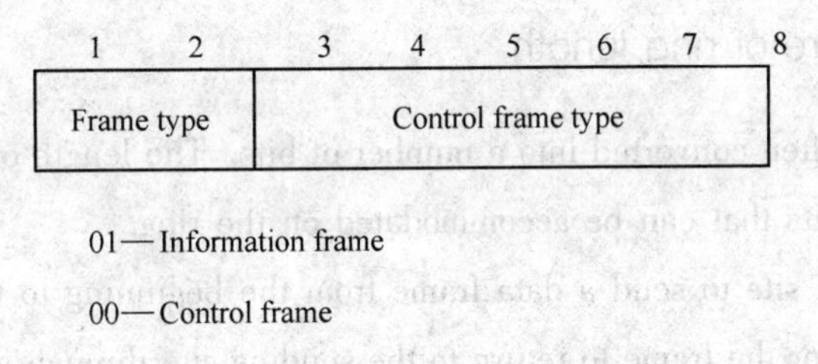

Figure 5-24 The frame control field

Duplicate address test: 00000000, it is used to detect whether the two addresses in the loop are the same;

Standby monitor present: 00000110 to announce the possible monitoring stations;

Active monitor present: 00000101 is sent regularly by the current monitoring station, while other stations monitor this frame.

Claim token: 00000011, when the monitoring station of the current operation fails, the standby monitoring station sends this frame to try to become a new monitoring station.

Purge: 00000100 is cleared. The new monitoring station initializes all other stations to the idle state with this frame.

Alarm beacon: 00000010, when there is a serious failure in the loop, send this frame to notify all stations to stop executing token ring protocol.

(4) DA, SA (Address field): destination address field, and source address field; same as

IEEE 802. 3.

(5) DATA (Data field): the ring network structure has no limitation on the maximum length of the frame. But in order to prevent a station from monopolizing the whole loop, IEEE802. 5 specifies the longest token holding time of a station. If it exceeds that time, the station must stop sending the token and pass the token to the next station.

(6) FCS field (Frame check sequence): any site is checked when it is forwarded, and the check range is from the frame control field to the data field.

(7) ED (End the field): X X X X X X X E, which is similar to the ED in the token frame, but the last E is error bit. When sending, E=0; when the site is forwarded, if it finds a check error, E=1.

(8) FS (Frame status field): it is used for source station to understand data frame transmission situation its format always is A C X X A C X X

If A=0, C=0: destination station does not exist or is not powered;

If A=1, C=0: destination station exists but the frame is not received;

If A=1, C=1: the destination station exists and the frame is copied;

According to the A, C and E bits in the return frame, the source station can understand whether the frame is transmitted correctly.

5. 3. 2. 5 Bit measure of ring length

The length of a ring is often converted into a number of bits. The length of a ring measured by bits reflects the number of bits that can be accommodated on the ring.

If the spent time for a site to send a data frame from the beginning to the end of the frame, equals to the spent time for the frame to return to the sending site through a loop from the beginning of transmission, and then all the bits of the data frame just fill the whole loop. In other words, when the transmission delay of the data frame is equal to the propagation delay of the signal on the loop, the number of bits of the data frame is the loop length measured by bit.

In practice, every interface in the loop will be delayed. The existence of interface delay time is equivalent to increasing the signal propagation delay in the loop, which is equivalent to increasing the bit length of the loop. Therefore, the delay of interfaces can also be measured by bits. Generally, each interface in the loop is equivalent to an increase of 1bit delay.

Thus, the formula of the length of the ring in which the bit measure is measured by bits is as following:

Ring bit length = signal propagation delay×data transfer rate+interface delay

= loop medium length×5(μs/km)×data transfer rate + interface delay digit

5μs/km is the reciprocal of signal propagation speed 200m/μs.

Example: The length of loop medium is 10km, and the data transfer rate is 4Mbps. There are 50 sites on the loop, and the interface of each site is 1 bit delay.

Then the bit length of the ring = 10(km) × 5(μs/km) × 4(Mbps) + 1(bit) × 50

= 10 × 5 × 10 − 6 × 4 × 106 + 1 × 50

= 200 + 50 = 250(bit)

If the length of the loop medium is too short or the number of stations is too small so that the bit length of the loop can not meet the requirement of the length of the data frame, additional delay can be induced at each ring interface, by using a shift register for example.

Example: A token ring network with a media length 10km, a data transfer rate is 16Mbps, a total of 100 sites in the loop, and a 1bit delay for each site, with a signal propagation speed of 200m/μs.

A. How long is the 1bit delay on the ring equivalent to the length of the cable?

B. How many bits are the effective bit length (ring bit length) of the ring?

Answer:

A. equivalent length of cable = signal propagation speed/data transfer speed, so

200m/μs/(16Mbps) = 12.5m/b

B. 10km ÷ 200(m/μs) × 10 − 6 × 16 × 106(bps) + 100 × 1(bit) = 900(bit)

Example: A token ring of 10km, 16Mbps, 100 sites, each site introduces a 1 bit delay bit with a signal propagation speed of 200m/us, There is 1 bit delay are equivalent to __(1)__ meters cables, and the effective length of the loop is __(2)__.

(1) A. 10 B. 12.5 C. 15 D. 20

(2) A. 100 B. 200 C. 500 D. 900

5.3.3 IEEE standard 802.4: token bus

The CSMA/CD media access control by bus contention has the advantages of simple structure and small delay under light load. However, with the increase of load, the collision probability increases and the performance will decrease obviously.

The token ring media access control has the advantages of a high utilization rate, insensitivity to distance and fair access under heavy load, but the structure of the ring network is complex, and there is a problem of error detection and reliability.

The token bus media access control is a media access control method based on the advantages of the two kinds of media access control.

IEEE802.4 puts forward the standard of the token bus media access control method.

5.3.3.1 IEEE 802.4: token bus

Token bus's characteristic is bus network in physical, logic token ring network in logic, an example for token bus is shown in Figure 5-25. Each site knows their addresses of the site left and right site. After the logical ring is initialized, the site with max number can send the first frame, the site sends tokens to its neighboring and transfer the transmission right to it.

The transmission medium is coaxial cable with 75Ω broadband, the topology is the has or tree type, the data rate is 1M, 5M or 10M, and the baseband signal needs to be modulated.

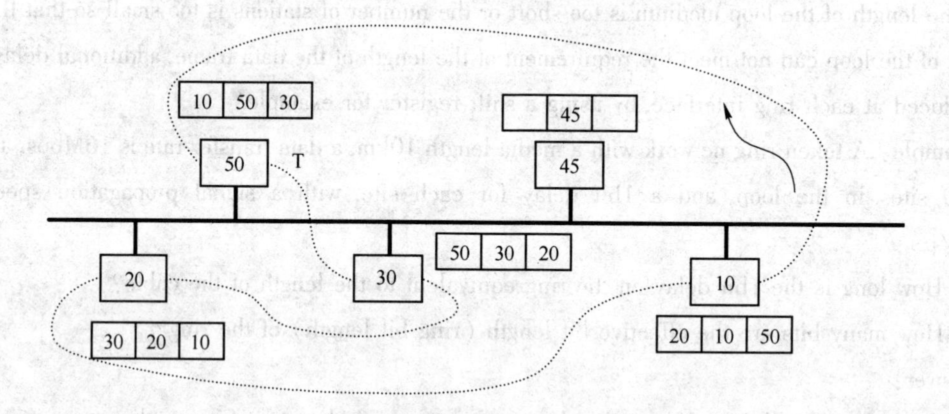

Figure 5-25 Token bus

5.3.3.2 Frame structure of IEEE802.4

As shown in Figure 5-26.

(1) Preamble code: same as IEEE 802.3, used for synchronization of the receiver clock, it is as short as one byte.

(2) Start and end delimiter: same as IEEE 802.5, frame border markers.

(3) Frame control: used to distinguish data frames or control frames; if it is data frames, it includes the priority level of the frame, and may also include an indicator requiring the destination station to acknowledge correct or incorrect receipt of the frame. For control frames, the field specifies the frame type. Detailed description of frame control is shown in Table 5-3.

≥1	1	1	2/6	2/6	0-8182	4	1B
Forward synchronization code	Start	Frame control	Destination address	Source address	Data	FCS	End

Figure 5-26 Frame structure of IEEE802.4

Table 5-3 Frame control

Detection control field	Name	Meaning
00000000	Claim_ token	Request to publish a token during initialization
00000001	Solicit_ successor_ 1	Allowable inlet ring

Continued Table 5-3

Detection control field	Name	Meaning
00000010	Solicit_ successor_ 2	Allowable inlet ring
00000011	Who_ follows	Recover from a lost token
00000100	Resolve_ contention	When multiple stations want to enter the ring
00001000	Token	PT pass token
00001100	Ser_ successor	Set up a new post station

(4) Destination address, source address: the same as IEEE 802.3.

(5) Data part: there are two kinds of maximum length. When using 2B address, it is 8182B, when 6B address is 8174B, 5 times larger than 802.3 frame.

(6) FCS: the same as IEEE 802.3.

5.3.3.3 The management and maintenance of logical rings and Tokens

A Build ring

As shown in Figure 5-27. The order of the logical ring: the arrangement order of each station is determined from big to small according to the address of the site, the arrangement order of each station, and has nothing to do with its physical location. Each site maintains its address, forward address and subsequent station address.

Loop initialization: if any site detects that the network channel is idle for more than a certain amount of time, the Claim-token frame request token is sent; if no other site is competing, the token is generated, and a ring that only includes its own is created; and if there is a competition, one of the sites is left by the arbitration algorithm. Other sites are added as the new station entry method.

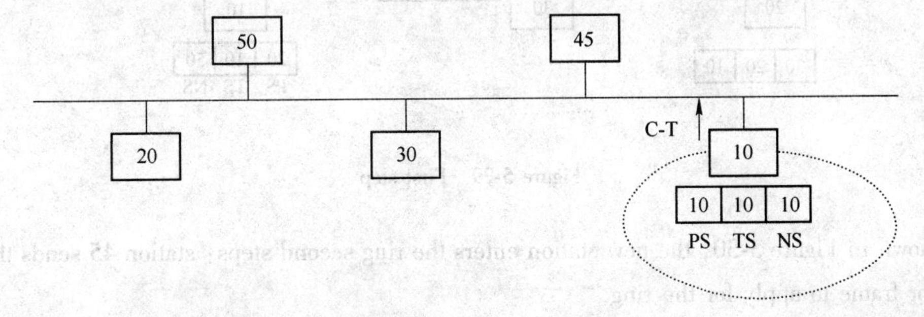

Figure 5-27 Build ring

B Ring in

The new station enters the ring: the token holder asks periodically whether there is an address to join the website at between it and its subsequent stations. If there is a station application, the token holder changes its successor address to the new station address and passes the token to the new station; then the new station sets up the front and rear stations, and the successor station of the new station also modifies its predecessor address. As shown in Figure 5-28, it is a example of the sation 45 enters the ring.

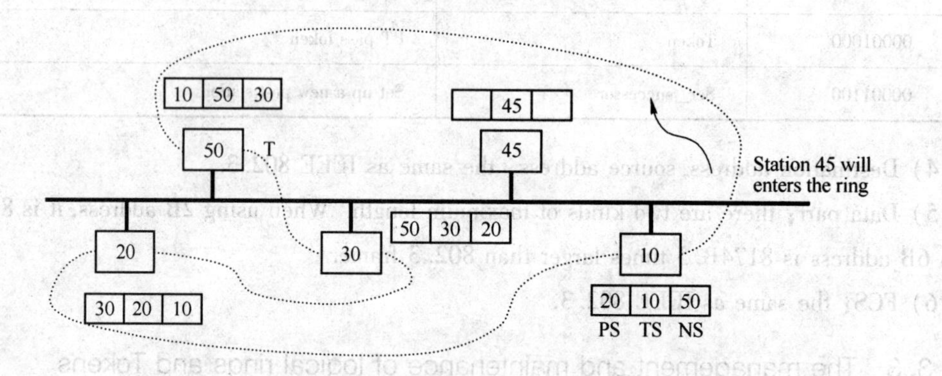

Figure 5-28 Ring in

Example: As shown in Figure 5-29, the first step of the new station entry: the token holder 50 station periodically sends the Solicit-successor-1 frame (allowing the station to enter the ring) to ask if there is an application to apply for the site between 50 and 30.

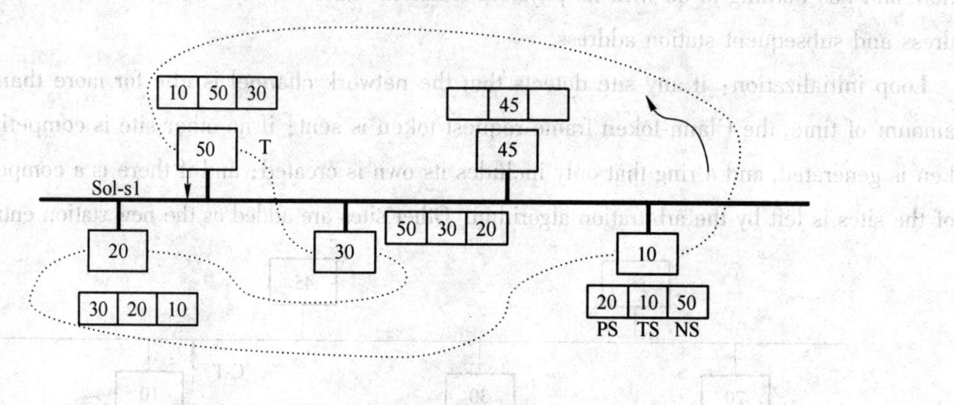

Figure 5-29 First step

As shown in Figure 5-30, the new station enters the ring second steps: station 45 sends the Set-successor frame to apply for the ring.

As shown in Figure 5-31, the third step of the new station entry: token holder 50 will change its saved address 30 to the new station address 45 and pass the token to station 45; then the 45 station sets its front and post station addresses to 50 and 30 respectively.

As shown in Figure 5-32, the new station enters the ring: the 45 station will give the token to the post station 30, and the 30 station will modify its predecessor address to 45.

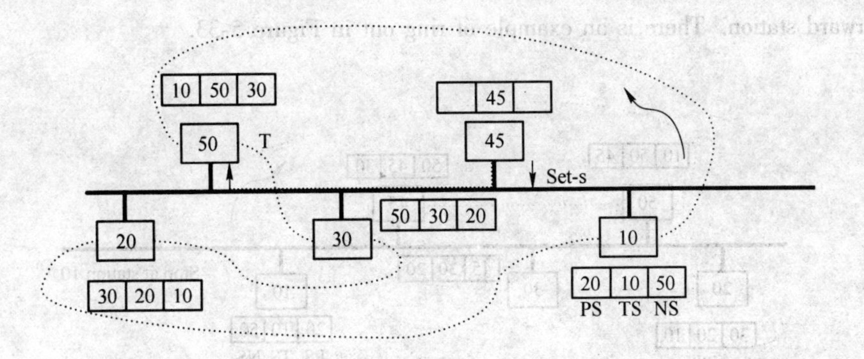

Figure 5-30 Second step

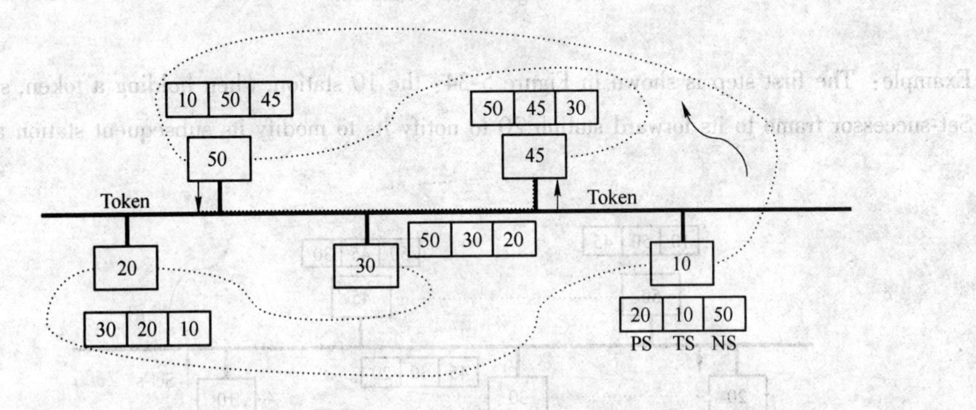

Figure 5-31 Third step

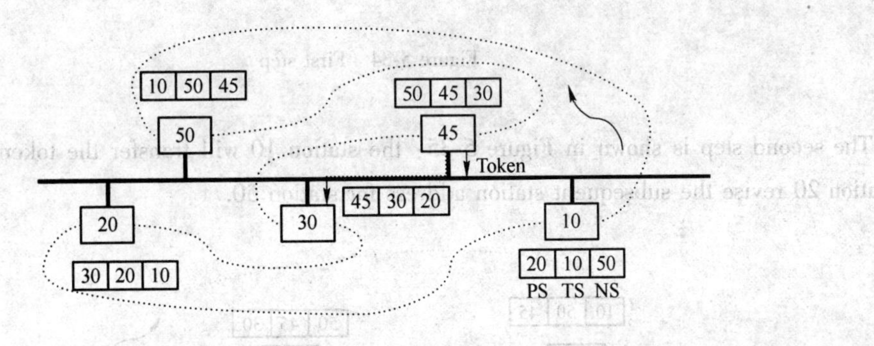

Figure 5-32 Fourth step

Only one station is allowed to enter the ring at a time. If there are multiple simultaneous applications, the arbitration decision will be made through broadcasting to resolve contention.

C Ring out

Ring out: When the holding token station wants to withdraw the ring, it sends Set-success frame to its forward station to inform it to modify the address of the subsequent station and transfer the token to the forward station. There is an example of ring out in Figure 5-33.

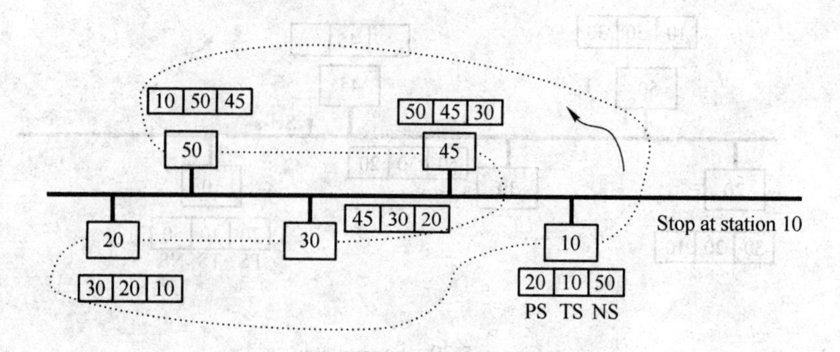

Figure 5-33 Ring out

Example: The first step is shown in Figure 5-34: the 10 station, when holding a token, sends the Set-successor frame to its forward station 20 to notify its to modify its subsequent station address.

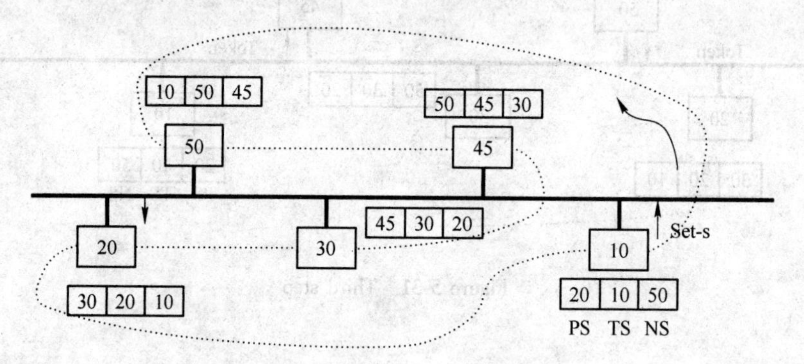

Figure 5-34 First step

The second step is shown in Figure 5-35: the station 10 will transfer the token to station 20. Station 20 revise the subsequent station address for station 50.

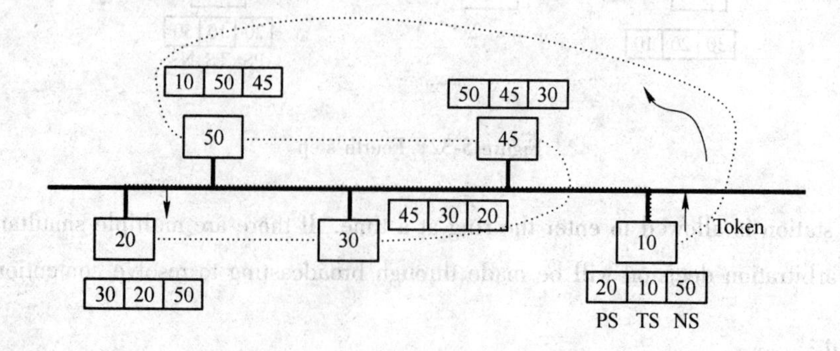

Figure 5-35 Second step

The third step is shown in Figure 5-36: the station 20 transfers the token to its new successor station 50, notifies the modification of its address of predecessor station with station 20.

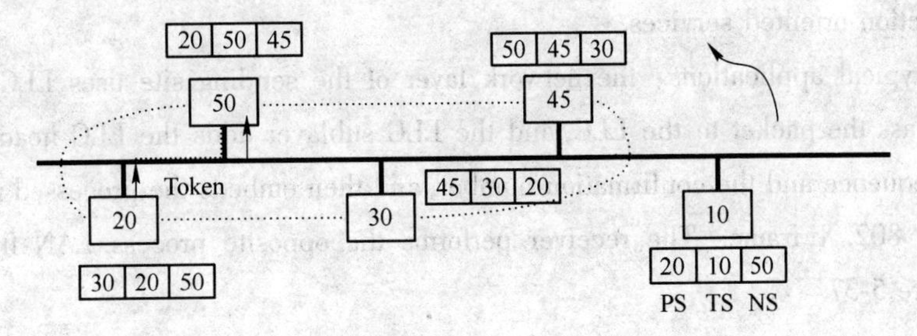

Figure 5-36 Third step

There is a comparison of 802.3, 802.4 and 802.5 in Table 5-4.

Table 5-4 Comparison of 802.3, 802.4 and 802.5

LAN	MAC protocol	Join the site	Data field length	Transmission delay	Performance
IEEE802.3	Simple	The easiest	46~1500B	Uncertain	Light load, time delay=0; Heavy load, worst performance
IEEE802.4	Most complex	Convenient	0~8174B	Certain	Light load, time delay; Heavy load, similar to TDM
IEEE802.5	Complex	Hard	≥0	Certain	Light load, time delay; Heavy load, similar to TDM

LAN	priority	conflict	Administration	Information transmission	Voice/TV	Optical fiber	Analog components	Application
IEEE802.3	Not	Not	Distribution	Broadcast	Not support	Inconvenient	Few	Office
IEEE802.4	Have	Not			Support		Many	Factory real time control system
IEEE802.5	Have	Not	Focus	Point to point forwarding		Support	None	

5.3.4 IEEE standard 802.2: logical link control

The logical link control (LLC) data communication protocol layer is the upper sublayer of the data link layer of the seven-layer OSI model. The LLC sublayer provides multiplexing mechanisms that make it possible for several network protocols (e.g. IP, IPX, Decnet and Appletalk) to coexist within a multipoint network and to be transported over the same network medium. It can also provide flow control and automatic repeat request (ARQ) error management mechanisms.

It hides the differences between the 802 networks and provides a unified format and interface to the network layer. These interfaces, formats and protocols are entirely based on OSI. LLC constitutes the upper part of the data link layer, and MAC constitutes the lower part of the data link layer.

LLC provides three services, unreliable datagram service, confirmed datagram service, and reliable connection oriented services.

There is a typical applications: the network layer of the sending site uses LLC to access the primitives to pass the packet to the LLC, and the LLC sublayer adds the LLC header to it, which contains the sequence and the confirmation number, and then embeds the processed results into the payload of the 802. X frame. The receiver performs the opposite process. LAN frame format is shown in Figure 5-37.

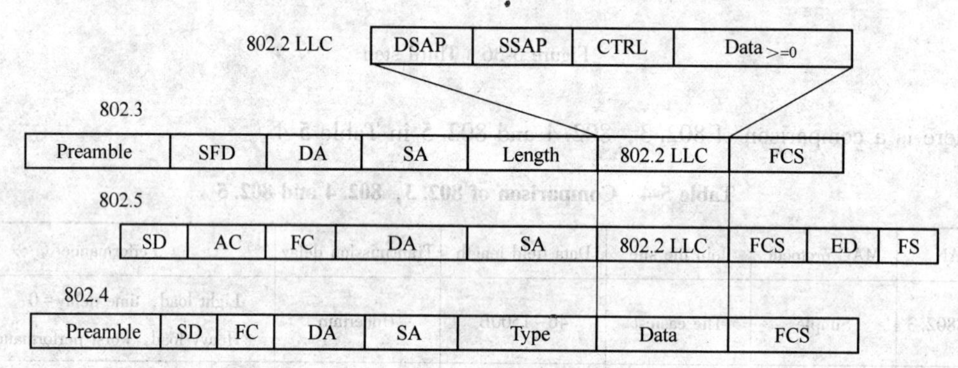

Figure 5-37 LAN frame format

5. 4 Bridge

The role of network interconnection equipment is to connect different networks and achieve cross-network communication. Here are common network interconnection devices working at different levels:

(1) Physical layer: repeater, hub;

(2) Data link layer: bridge;

(3) Network layer: router;

(4) Protocol layer above network layer: gateway.

5. 4. 1 Repeater

In the early stage of the development of the LAN, the transmission medium of the network is mainly coaxial cable, the digital signal attenuates in the transmission process of the coaxial cable, and the waveform will distort. The length of the transmission medium is related to the attenuation of the signal and the delay of the signal transmission. Therefore, in the physical layer protocol using a coaxial cable as a transmission medium, the maximum length of a single coaxial cable and the number of nodes to be connected must be limited.

In order to meet the needs of users, LAN repeater devices are studied, it can regenerate and send the received signal. Repeaters are widely used in early LAN networking.

As shown in Figure 5-38, repeaters can only play the role of receiving, amplifying, rectifying and sending the signal waveform on the transmission medium—the function of the physical layer.

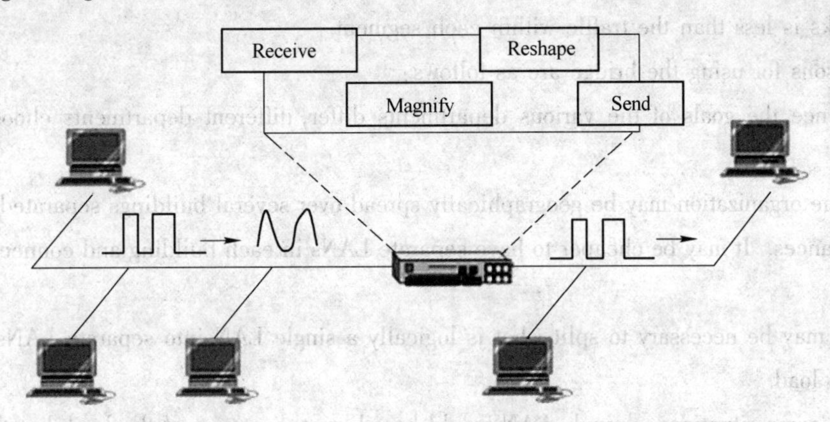

Figure 5-38 Basic principle of work

The duties of repeater are not involve the structure of the frame and not handle any content of the frame. It can only play an important role in increasing the length of the transmission medium.

All nodes are connected to different cable segments. As long as there is one that sent data, all nodes can receive the data, so several network segments connected by the repeater still belong to a LAN.

A general rule using a repeater to interconnect two segments: no repeater can be used to connect a network of two different units or departments with a large number of users; otherwise, the increase in traffic between the two segments of the cable will lead to a sharp decline in the performance of the two networks.

5. 4. 2 Hub

The early Ethernet network mainly uses the 10Base-2 protocol for the thick coaxial cable and 10Base-5 for the thin coaxial cable, using the repeater more. With the emergence of the 10Base-T protocol, the use of unshielded twisted pair lines can achieve the data signal transmission of 10Mb/s. This technology greatly promotes the wide application of Ethernet; in the use of 10BASE-T the function of the hub is very important when the protocol is set up.

All nodes are connected to a hub by twisted pair lines, and the CSMA/CD media access control method is still performed. When a node sends data, all nodes receive.

5. 4. 3 Bridges

Many organizations have multiple LANs and wish to connect them. LANs can be connected by devices called bridges, which operate in the data link layer. Bridges connect two LANs which are independent and related into one LAN; they can also split a single LAN into separate LANs to accommodate the load.

When using the bridge to segment the network, two aspects must be considered: first, reducing the amount of communication on each LAN segment as much as possible; that the traffic between the networks is less than the traffic within each segment.

The reasons for using the bridge are as follows:

(1) Since the goals of the various departments differ, different departments choose different LANs.

(2) The organization may be geographically spread over several buildings separated by considerable distances. It may be cheaper to have separate LANs in each building and connect them with bridges.

(3) It may be necessary to split what is logically a single LAN into separate LANs to accommodate the load.

(4) In some situations, a single LAN would be adequate in terms of the load, but the physical distance between the most distant machines is too great.

(5) The matter of reliability. The bridge can be programmed to decide whether to copy and forward the accepted content.

(6) Contribute to the organization's security. By inserting bridges at various places and being careful not to forward sensitive traffic, it is possible to isolate parts of the network so that its traffic cannot escape and fall into the wrong hands.

5. 4. 4 Bridge from 802. x to 802. y

Bridges work at the data link layer, enabling multiple LAN interconnections.

There are some problems existing in LAN interconnection:

(1) Each of the LANs uses a different frame format.

Any copying between different LANs requires reformatting, which takes CPU time, requires a new checksum calculation, and introduces the possibility of undetected errors due to bad bits in the bridge's memory.

(2) The interconnected LANs do not necessarily run at the same data rate.

When forwarding a long run of back-to-back frames from a fast LAN to a slower one, the bridge will not be able to get rid of the frames as fast as they came in.

(3) All three 802 LANs have a different maximum frame length as shown in Figure 5-39.

For 802. 3, the maximum frame length depends on the configuration parameters, but for the standard 10Mb/s system, the maximum payload is 1500 bytes.

For 802. 4, the maximum frame length is fixed to 8191 bytes.

For 802. 5, no upper limit is provided, as long as the transmission time of the site does not exceed the token holding time. If the token time is lost, the default is 10ms, and the maximum frame length is 5000 bytes.

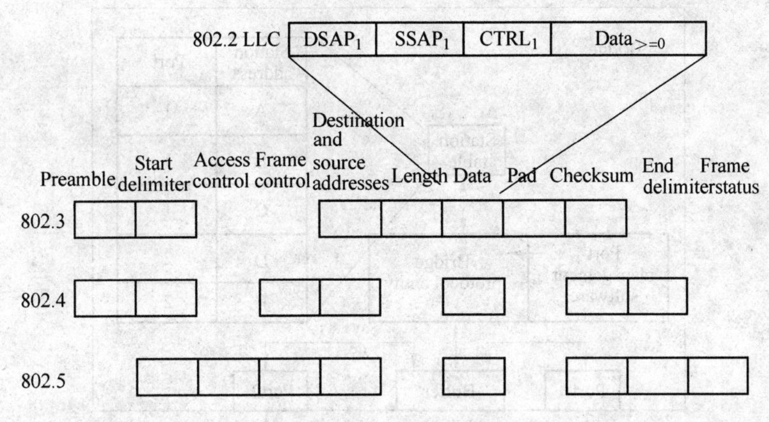

Figure 5-39 LAN frame format

There is a summary of Bridge from 802. x to 802. y as shown in Figure 5-40.

Source LAN \ Objective LAN	802.3	802.4	802.5
802.3		1, 4	1, 2, 4, 8
802.4	1, 5, 8, 9, 10	9	1, 2, 3, 8, 9, 10
802.5	1, 2, 5, 6, 7, 10	1, 2, 3, 6, 7	6, 7

Figure 5-40 Bridge from 802. x to 802. y

1—reframe, recalculate FCS; 2—inversion sequence; 3—replication priority; 4—produce a false priority;
5—discarding priority; 6—outflow ring; 7— set A, C bit; 8—worry about congestion;
9—worry about early surrender of token due to ACK delay or inability to provide ACK;
10—if the frame is too long for the destination LAN, it is discarded

The bridge realizes the connection of multiple LAN systems by sending data frames, filtering address, storing and forwarding data frames, receiving data frames. According to the source address and destination address of the data frame, we decide whether to receive and forward the data frames.

According to different forwarding strategies, bridges can be divided into transparent bridges and source routing bridges.

5.4.5 Transparent bridge

Transparent bridge routing decisions are determined by bridges, and bridges and routed sites are completely transparent to each site.

The working process of the bridge are shown in Figure 5-41, the bridge receives the frame from LANx, and checks whether the destination address of the frame is in the checklist.

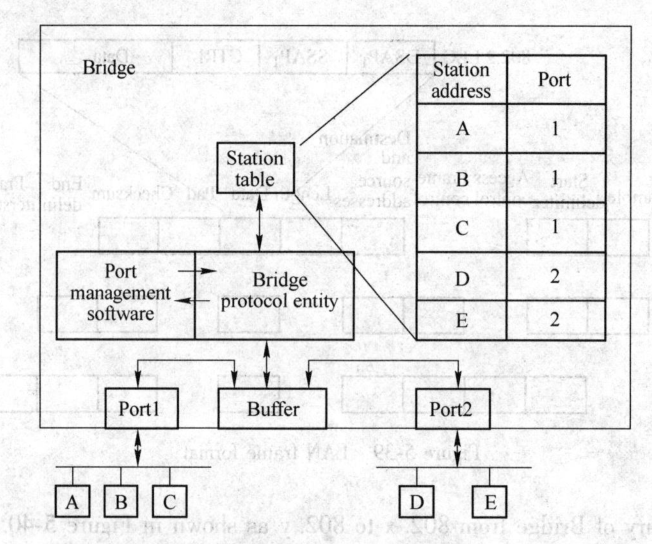

Figure 5-41 The working process of the bridge

If the destination LAN = LANx, the frame is discarded (because the source is the same as the destination).

If the destination LAN is in the station table, the destination LAN is forwarded to the destination.

If the destination LAN is not in the station table, the flooding method is used to forward all ports other than x.

Transparent bridging uses a table called the forwarding information base to control the forwarding of frames between network segments. The table starts empty and entries are added as the bridge receives frames. If a destination address entry is not found in the table, the frame is flooded to all other ports of the bridge, flooding the frame to all segments except the one from which it was received. By means of these flooded frames, a host on the destination network will respond and a forwarding database entry will be created. Both source and destination addresses are used in this process: source addresses are recorded in entries in the table, while destination addresses are looked up in the table and matched to the proper segment to send the frame to.

As shown in Figure 5-42, if the bridge B1 receives frames from E from Port2, it is known that frames sent for E can be forwarded to Port2 later. This is the station table building method: reverse learning in the forwarding process. It is easy to configure, and install, without management. But it cannot make full use of all resources; we cannot guarantee the best routing.

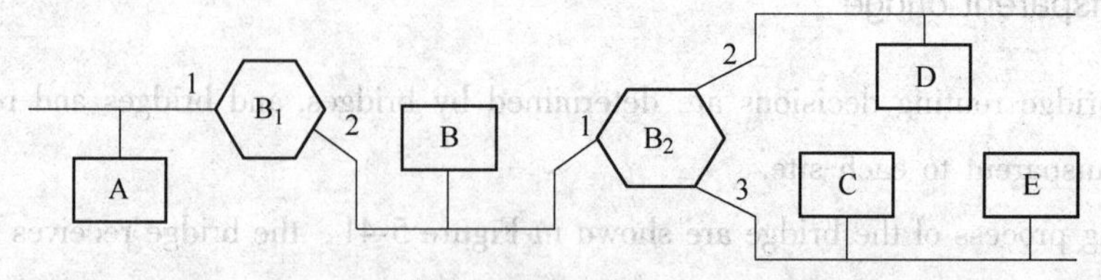

Figure 5-42 Table of forwarding information

5. 4. 6 Source routing bridge

In source routing bridge, the source station is responsible for routing, and the routing information is placed at the beginning of the frame.

Each station source station is configured with a routing table. In the table, establish a list of all workstations and bridges that can be reached by the station, listing the locations of all workstations and bridges along the way from this station to the destination station.

The route selection table establishment method involves sending the probe frame to the destination station, which is transmitted along all possible routes in the extended LAN; each detection frame records the route it passes; these probe frames return to the source station along their respective routes after arriving at the destination; selecting the best route among the source stations and enter the route selection table of the sending station.

Its advantages are the function of the bridge is simple, and the best route can be transmitted.

However its disadvantages are the computation time of the source station is long, and the host computer is heavily loaded.

In a word, the use of bridges to interconnect local area networks overcomes the physical limitations of local area networks. The bridges select filter function, properly isolate some specific frames, improve network performance, and improve network speed and reliability. The working mechanism is simple and easy to realize.

However, transparent bridges have a simple routing function, the source routing bridge does not have the routing function. There is an example for the operation of a bridge to the MAC frame shown in Figure 5-43.

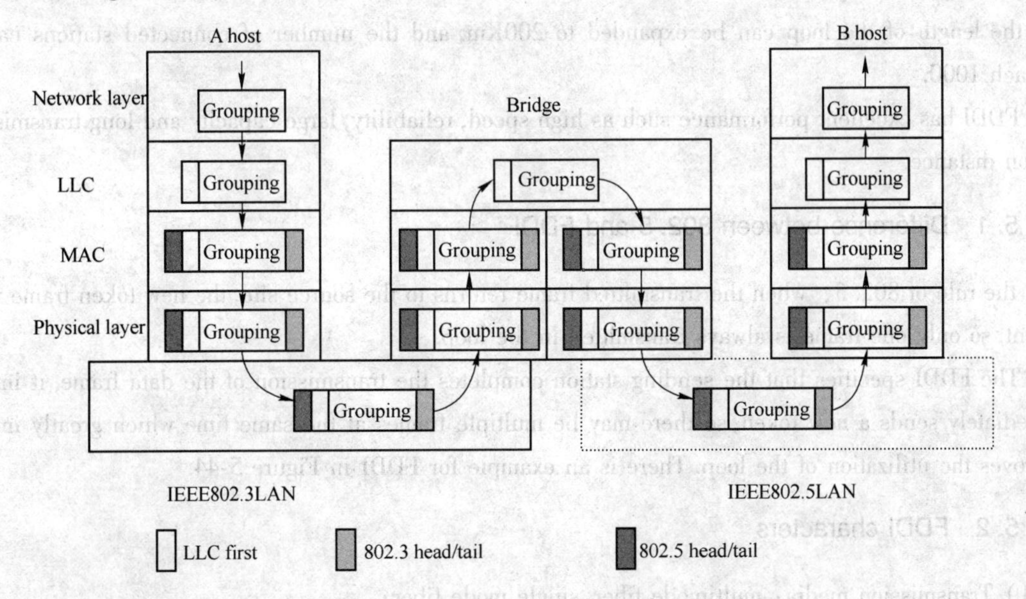

Figure 5-43 The operation of a bridge to the MAC frame

It is a comparison of 802 bridge shown in Table 5-5.

Table 5-5 Comparison of 802 bridge

Class	Transparent bridge	Source station selection network bridge
Service type	Connectionless	Connection oriented
Transparency to the source station	Completely transparent	Opaque
Configuration, management	Automatic configuration, easy to manage	Artificial method
Selected routing	Second best	Optimum
Method of destination determination	Reverse learning	Detection frame
Fault processing and topology change	Bridge is responsible for the bridge	Host is responsible
Complexity and overhead	Bridge burdens	Host burden

5.5 High-speed LANs

Ethernet, token ring and token bus are all networks which are based on copper media. They are suitable for low speed or short distance LAN data transmission and communication. For high speed or a long distance, LAN must be based on optical fiber or highly parallel copper dielectric network. Usually, fast LAN uses optical fiber as the network transmission medium.

Fiber Distributed Data Interface (FDDI) is a standard of optical fiber ring network, developed by American National Standard Association (ANSIA). A token ring protocol similar to 802.5 is used for FDDI. Because fiber is used as the transmission medium, the data rate can reach 100Mb/s, the length of the loop can be expanded to 200Km, and the number of connected stations can reach 1000.

FDDI has excellent performance such as high speed, reliability, large capacity and long transmission distance.

5.5.1 Difference between 802.5 and FDDI

In the rule of 802.5, when the transmitted frame returns to the source site, the new token frame is sent, so only one frame is always transmitted in the loop.

The FDDI specifies that the sending station completes the transmission of the data frame, it immediately sends a new token, so there may be multiple frames at the same time which greatly improves the utilization of the loop. There is an example for FDDI in Figure 5-44.

5.5.2 FDDI characters

(1) Transmission media: multimode fiber, single mode fiber;

(2) Topology: double ring;

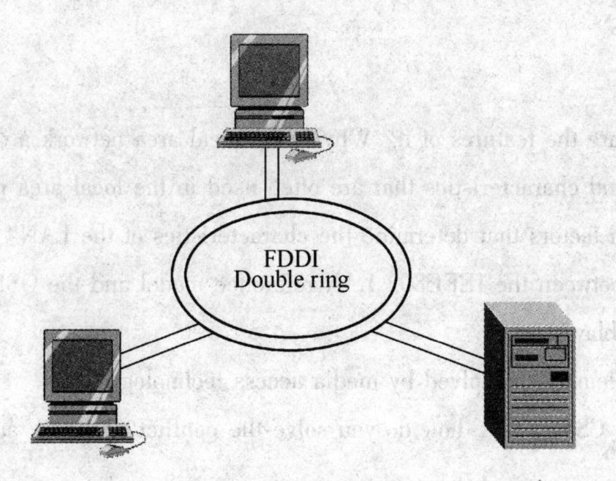

Figure 5-44 FDDI

(3) Data rate 100Mbps;

(4) There are up to 1000 sites on the ring, with a loop length of 100km; The maximum distance between two adjacent stations is 2km;

(5) MAC protocol: IEEE802.5, the maximum frame length is 4500B;

(6) LLC protocol: IEEE802.2;

(7) Dynamically allocate bandwidth, the priority is similar to IEEE802.4.

5.5.3 FDDI frame format

Where PA is the preamble, SD is a start delimiter, FC is frame control, DA is the destination address, SA is the source address, PDU is the protocol data unit (or packet data unit), FCS is the frame check Sequence (or checksum), and ED/FS are the end delimiter and frame status. The Internet Engineering Task Force defined a standard for transmission of the Internet Protocol (which would be the protocol data unit in this case) over FDDI.

PA	SD	FC	DA	SA	INFO	FCS	ED	FS

(1) PA: the preamble, the 16 idle symbol I = 11111, it provides the clock synchronization signal;

(2) SD: the frame header delimiter is composed of two symbols J = 11000 and K = 10001, which indicates the beginning of the frame;

(3) ED: the frame end delimiter is composed of a T symbol, and T=01101 indicates the end of the frame;

(4) FC: frame control field;

(5) DA, SA: destination address, source address; same as IEEE 802.5;

(6) INFO: information part contains LLC data or control related information;

(7) FCS: the same as IEEE 802.5;

(8) FS: frame state field, the same as IEEE 802.5.

Problems

5-1 What is a LAN? What are the features of it? Why is the local area network a communication subnet?

5-2 What are the topology and characteristics that are often used in the local area network?

5-3 What are the three main factors that determine the characteristics of the LAN?

5-4 What is the difference between the IEEE802 LAN reference model and the OSI reference model? Why is the media access control sublayer set up?

5-5 What are the main problems to be solved by media access technology?

5-6 What is the conflict? In CSMA/CD, how do you solve the conflict? Is there any conflict between token ring and token bus network?

5-7 Briefly describe the transmission and reception process of data frames in token ring networks.

5-8 Ten thousand airline reservation stations are competing for the use of a single slotted ALOHA channel. The average station makes 18 requests/hour. A slot is 125μs. What is the approximate total channel load?

5-9 Consider building a CSMA/CD network running at 1 Gbps over a 1km cable with no repeaters. The signal speed in the cable is 200000 km/s. What is the minimum frame size?

5-10 What kinds of network topologies are commonly used in local area networks? What kind of structure is the most popular nowadays? Why did early Ethernet choose bus topology instead of star topology, but now switch to star topology?

5-11 What is called traditional ethernet? What are the two main standards of ethernet?

5-12 Explain the meanings of "10", "Base" and "T" in 10Base-T.

5-13 What is FDDI, and explain the difference between 802. 5 and FDDI.

5-14 How many types ALOHA protocols in its development?

5-15 What is transparent bridge? What is source routing bridge? and explain the difference between transparent bridge and source routing bridge.

5-16 In Figure 5-45, the Ethernet switch of a college has three interfaces which are connected to the Ethernet of the three departments of the college, and the other three interfaces are connected to the E-mail server, the Website server and a router connecting to the Internet. A, B and C in the figure are all 100Mbit/s, and any of the nine hosts in the figure can communicate with any server or host. Try to calculate the maximum total throughput generated by these nine hosts and two servers. Why?

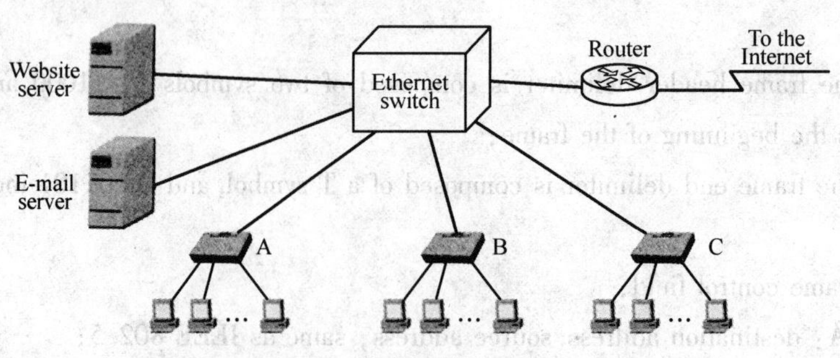

Figure 5-45

6 The Network Layer

Goal:

(1) Understand design issues of the network layer.

(2) Master the IP address rule.

(3) Be able to design IP networks and allocate sublayer.

The network layer is the third layer of the OSI model. It is the most complicated layer in the OSI reference model and the highest layer of the communication subnet. It provides services to resource subnets on the basis of the next two layers. Its main task is to select the most appropriate path for the message or packet through the communication subnet or through the routing algorithm. This layer controls the forwarding of information between the data link layer and the transport layer, establishing, maintaining, and terminating the connection to the network. Specifically, data at the data link layer aims converted into data packets at this layer, and then transmitted from one network device to another through control of path selection, segment combination, sequence, and in/out routing.

Generally, the data link layer aims to solve the communication between nodes in the same network, and the network layer mainly solves the communication between different subnets. For example, during communication between wide area networks, there is bound to be a problem of routing (i. e., there may be multiple paths between two nodes).

The network layer mainly implements two major functions: the first function is to realize datagram transmission between source nodes and destination nodes located in different networks; the second function is to determine which technology is used in the network to select a path for the source node, so that the datagram is finally delivered to the destination node through the intermediate node.

The main issues that need to be addressed when implementing network layer functions are as follows:

(1) Addressing: The physical address (such as MAC address) used in the data link layer only addresses the addressing problem within the network. In order to identify and find devices in

the network when communicating between different subnets, devices in each subnet are assigned a unique address. Since the physical technology used by each subnet may be different, this address should be a logical address (such as an IP address).

(2) Exchange: Specifying different ways of information exchange. Common switching technologies are: circuit switching technology and store-and-forward technology, which in turn includes message exchange technology and packet switching technology.

(3) Routing algorithm: When there are multiple paths between the source node and the destination node, the layer can select the best path for the data packet through the network according to the routing algorithm, and transmit the information from receiving to the end through the most suitable path.

(4) Connection service: Unlike the data link layer flow control, the former controls the traffic between adjacent nodes of the network, and the latter controls the traffic from the source node to the destination node. Its purpose is to prevent clogging and to perform error detection.

6.1 Design points of network layer

The network layer can provide two ideas of services to the transport layer at the network layer/transport layer interface: connection service and connectionless service.

6.1.1 Connection-oriented service

A connection-oriented service is a process in which a communication party establishes a communication line in advance. The process has three processes: establishing a connection, using a connection, and releasing a connection. The TCP protocol is a connection-oriented service protocol, and the telephone system is a connection-oriented model. The working mode for connection-oriented services and telephone systems is similar. The feature is that the data transmission process must go through three processes of establishing a connection, maintaining a connection, and releasing the connection; during the data transmission process, each packet does not need to carry the address of the destination node. A connection-oriented transport connection is similar to a communication pipeline where the sender puts data at one end and the receiver takes the data from the other end.

For a connection-oriented service, we need a virtual-circuit subset. The virtual circuit is the terminal device of the two users, which needs to establish logical connections through the communication network before starting sending and receiving data to each other. Before data transmission, a virtual call establishment stage is needed; after data transmission, a demolition stage is needed. All groups must carry out data transmission along this pre-established virtual circuit, which is essentially different from circuit switching. The characteristics of virtual circuits are as follows:

(1) Similar to circuit switching, including the 3 stages of virtual circuit establishment, data transmission and dismantling. A specific path is selected for transmission, and all nodes in the packet are stored and forwarded to these packets, while circuit switching has no such function.

(2) Message packets do not need to carry auxiliary information such as destination address and source address, but only need to carry virtual circuit identification number. When packets arrive at the destination node, there will be no loss, duplication or disorder.

(3) When packets pass through nodes on each virtual circuit, nodes need to do error detection instead of routing.

The connected network interconnection of a connection-oriented service is linked virtual circuits, as shown in Figure 6-1. The subnet finds its destination at the far end and establishes a virtual circuit of the nearest router to the destination network, and then sets up a virtual circuit from the router to an external "gateway" (multi protocol router). The gateway records the virtual circuit in its table and continues to create another virtual circuit to the router in the next subnet. This process continues until the destination host is reached. Once the data packet is sent along this path, each gateway is responsible for forwarding the input packet and changing the grouping format and virtual circuit number as required. And all data packets must go through the gateways in the same order, and finally arrive in this order. The key is to build a series of virtual circuits from the far end to reach the destination by one or more gateways. Each gateway maintains some record tables, which record which virtual circuits go through, where to go, and how much the new virtual circuit is.

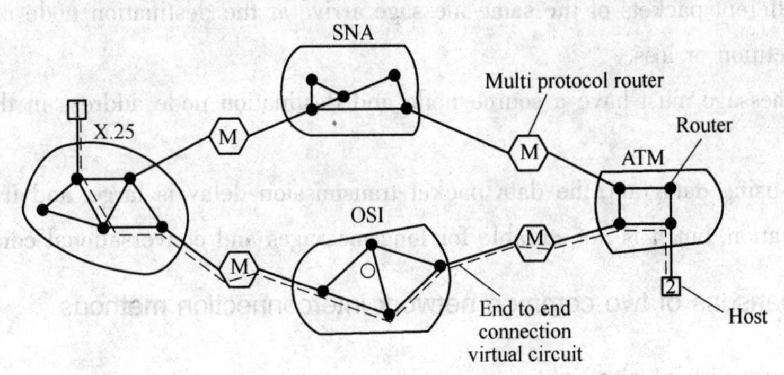

Figure 6-1 Connected network interconnection

6.1.2 Connectionless service

The connectionless network service means that the communication parties do not establish a communication line in advance, but send each packet with the destination address to the line, then the system selects the route for transmission. A switching network treats each packet of a network as a separate "small message", regardless of which packet it belongs to, just as a message is handled separately in a datagram exchange. This packet switching mode is referred to as the datagram transmission mode, and the "small message" as the basic transmission unit is called a datagram.

The only service provided by the network layer for the transport layer is to send data to the subnet without the concept of virtual circuit. All groups are not required to reach the gateway in order. The datagram from Host 1 to Host 2 uses different Internet routes, and each packet selects routing individually, as shown in Figure 6-2; if all the packets arrive at the destination, they do not guarantee that they are in order.

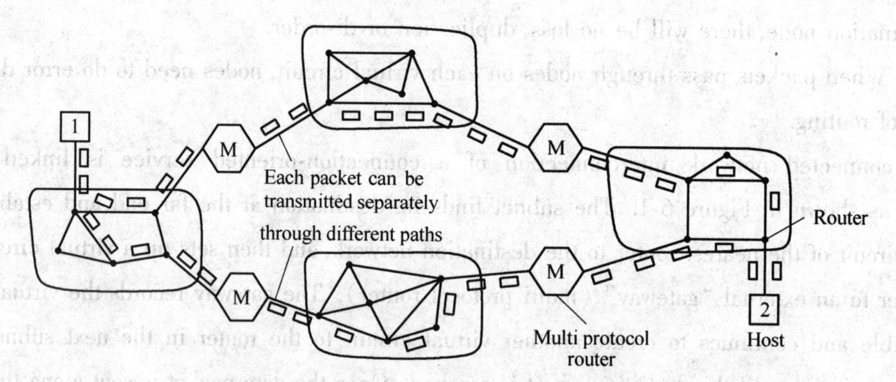

Figure 6-2 Connectionless network interconnection

The IP and UDP protocols are a connectionless protocol, and the postal system is a connectionless mode. The characteristics of the datagram are as follows:

(1) Different packets of the same message can be transmitted by different transmission paths through communication subnets.

(2) The different packets of the same message arrive at the destination node and may appear in disorder, repetition or loss.

(3) Each message must have a source node and destination node address in the transmission process.

(4) When using datagram, the data packet transmission delay is large, and it is suitable for burst communication, but it is not suitable for long messages and conversational communication.

6.1.3 Comparisons of two common network interconnection methods

6.1.3.1 Chain virtual circuit

Advantages: the buffer can be reserved in advance, and can be sent sequentially. Shorter information headers can be used to avoid errors caused by delayed retransmission packets.

Disadvantages: The method need to allocate table space to each open connection, and there is no backup route to bypass the congestion area. Therefore, there is the vulnerability of routers crashing along the route. If there is an unreliable datagram network in the network involved, it is hard or even impossible to achieve it.

6.1.3.2 Datagram method

Advantages: If features with better adaptability to congestion and robustness when routers collapse.

Disadvantages: it is more likely to cause congestion; longer information heads are needed.

Comparison between virtual circuit subnet and datagram subnet is shown in Table 6-1.

Table 6-1 Comparison between virtual circuit subnet and datagram subnet

Project type	Virtual circuit subnet	Datagram subnet
Circuit setting	Needed	Unneeded
Address	Each packet contains a short virtual circuit number	Each group has the active address and the destination address
Status information	Every virtual circuit is required to occupy subnet table space	Subnet does not store state information
Routing selection	When the virtual circuit is built, the routing is determined, and all packets are routed through this route.	Independent selection for each group
The effect of router failure	All virtual circuits of failed routers will be terminated	There is no other influence except for the total loss of packets during the crash
Congestion control	It is easy to control if enough buffers are allocated to each virtual circuit that has been established	Hard

6.2 Routing algorithm

The main task of the network layer is to transfer data packets from the source node to the target node, and to select the appropriate path for the data packets to be transmitted, which is related to the utilization of network resources and the performance of the network. Routing aims to decide which node to output from the output line, and involves generating the output line selection table of nodes.

The routing algorithm is part of the network layer software, and is responsible for determining the outgoing routes through which the packets received should be transmitted. If a datagram is used within the subnet, a routing choice is made for each of the packets received, because the best route to each packet may have been changed for each packet; if a virtual circuit is used within the subnet, then only one routing decision is needed when a new virtual circuit is built, and later, the data is transmitted on this previously established route. What is more, the routing algorithm should have the following characteristics, such as: correctness, simplicity, robustness, stability, fairness and optimality.

The routing algorithm is classified to non-adaptive algorithm and adaptive algorithm. A non-adaptive algorithm is a kind of static routing selection, where routing is not based on the current traffic and topology of the measured or estimated network. The routing from I to J is calculated beforehand and downloaded to the router when the network is started. An adaptive algorithm is a kind of dynamic routing selection. Its routing is changed according to topology and traffic volume. Adaptive algorithms can be divided into different categories according to their ways of obtaining information, changing the routing conditions, and optimizing the parameters.

6.2.1 Non-adaptive algorithm

6.2.1.1 Shortest route selection

There are several algorithms for computing the shortest path between two nodes of a graph. One is by Dijkstra (1959), where each node is labeled (in parentheses) with its distance from the source node along the best known path. Initially, no paths are known, so all nodes are labeled with infinity. As the algorithm proceeds and paths are found, the labels may change, reflecting better paths. A label may be either tentative or permanent. Initially, all labels are tentative. When it is discovered that a label represents the shortest possible path from the source to that node, it is made permanent and never changed thereafter.

6.2.1.2 Diffusion method

Any node should copy the packets received by it to the adjacent nodes, but it cannot send back to the node it just left. Selective diffusion is that the router does not send every incoming packet from every output line, but only to those lines that are close to the right direction-the selected flood method. The advantage is high reliability and good robustness, and the shortcoming is congestion in the network. It is suitable for occasions with small communication and high reliability.

6.2.1.3 Fixed routing method

A pre-defined routing table is stored at every node in the network. This table gives the shortest path from this node to the destination node. When a packet needs to be sent from this node, it can find out its output line from the routing table according to the destination address. The advantage is this method is simple. The disadvantage is that it does not adapt to topological changes; when the selected route fails, it affects normal transmission. It is suitable to the network with good reliability and stable load, and it also provide suboptimal routing for each node related to other methods.

6.2.2 Adaptive algorithm

6.2.2.1 Isolated routing strategy

Each node determines routing only according to the state of this node, and does not need to exchange state information with other nodes. The choice of routing is to calculate the length of each link queue and send the packets to the shortest queue waiting to be sent.

The characteristic of this method is that the queuing time is reduced, but it is obvious that the best path is not chosen.

6.2.2.2 Centralized routing strategy

Like fixed routing in static routing algorithms, centralized routing stores a routing table on each node. However, unlike fixed routing, the node routing table in the fixed routing algorithm is manually set. In the centralized routing algorithm, the node routing table is calculated by the Network Control Center (NCC) according to the network state, generated and distributed to each corresponding node. Since the RCC utilizes the information of the entire network, the obtained routing is relatively complete, and the burden of calculating the routing of each node is also alleviated.

6.2.2.3 Distributed routing strategy

A distributed routing algorithm means that all nodes periodically exchange routing information with each of their neighbors. Each node stores a routing table, and this routing table maintains synchronization with routing information of other nodes. Each node in the network has an entry in the table. Each item is further divided into two parts: one is the output link to the destination node that is desired to be used; the other is the estimated delay or distance required to reach the destination node.

6.3 Congestion control algorithm

When too many packets are present in the subnet, its performance degrades. This situation is called congestion. Figure 6-3 depicts the phenomenon. When the number of packets stored in a communication subnet is within its transmission capacity, they are all delivered to the destination. However, when traffic increases too fast, routers can no longer cope and they, start losing packets, causing the situation to deteriorate. In the case of high traffic volume, the network is completely paralyzed, and few packets can be served.

When the number of packets stored in a communication subnet is within its transmission capacity, they are all delivered to the destination, and the number of packets sent is proportional to the number transmitted. However, when traffic increases too fast, routers can no longer cope, and they start losing packets, causing the situation to deteriorate. In the case of high traffic volume, the network is completely paralyzed, and few packets can be served.

Congestion can be caused by several factors. If there are too many groups when there is insufficient memory to hold all of them, packets will be lost. Slow processors can also cause congestion. If the router's CPUs are slow at performing the bookkeeping tasks required of them, queues can build up, even though there is excess line capacity. Similarity, low-bandwidth lines can also cause

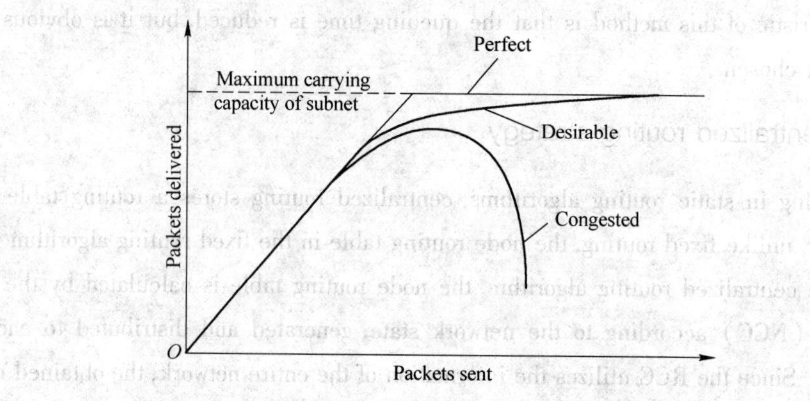

Figure 6-3 Because the amount of information in the communication subnet is too large, the performance degrades

congestion. Upgrading only part of the system rather than the whole is often equivalent to shifting the bottleneck to other parts of the system. All components can be balanced to solve this problem.

It is worth pointing out clearly the difference between congestion control and traffic control, as the relationship between then is subtle. The problem of congestion control is the problem of resource congestion in the network. That is, the demand for resources is greater than the available resources. It must be ensured that the communication subnet can transport the data to be transmitted, which is a global problem. Traffic control is to solve the problem that the data sender's ability is greater than the receiver's ability. It only relates to the point-to-point communication between a sender and a receiver. Its task is to ensure that a fast sender cannot transmit data at a higher rate than the receiver can withstand.

6. 3. 1 The basic principle of congestion control

Many problems in complex systems, such as computer networks, can be viewed from a control theory point of view. All solutions are divided into two types: open loop and closed loop. Open-loop solutions aim to avoid problems by good design and ensure that problems will not happen at the very beginning. Once the system is installed and running, no intermediate stage corrections will be made.

Functions for doing open-loop control include deciding when to receive new communications and when to discard packets as well as which packets to discard, open-loop control also creats schedules at different points of the network. When deciding, the current network condition is not considered.

In contrast, closed-loop solutions are built on the feedback loop. When used for congestion control, there are 3 parts:

(1) The monitoring system detects when and where congestion occurs;

(2) Transmit this information to where action may take place;

(3) Adjust the operation of the system to correct the problem.

6.3.2 Load shedding

When other methods can not eliminate congestion, routers can bring out the ultimate weapon: load shedding. Load shedding is a fancy way of saying that when the router is overwhelmed by the packets it cannot control, all these packets have to be thrown away.

A router drowning in packets can just pick packets at random to drop, but usually it can do better than that. Which packet to discard may depend on the application running. For filter transfer, older packets are more valuable than new ones, this policy is often called wine strategy; for multimedia transmission, new packets are much more important than older ones, this policy is often milk strategy.

To implement an intelligent discard policy, applications must mark their packets in priority classes to indicate how important they are. If they do this, then when packets have to be discarded, routers can first drop packets from the lowest class, then the next lowest class, and so on. Of course, unless there is some significant incentive to mark packets as anything other than very important, are should never ever discard.

6.4 Internet interconnection

Network interconnection refers to connecting different networks to form a larger-scale network system, realizing data communication, resource sharing and collaborative work between networks. The purpose of network interconnection is to enable users on one network to access resources on other networks, so that users on different networks can communicate and exchange information with each other. This not only facilitates resource sharing, but also improves the reliability of the network as a whole.

Network interconnection mainly includes LAN-LAN, LAN-WAN, WAN-WAN and LAN-WAN-LAN. Figure 6-4 depicts the system of network interconnection, and Figure 6-5 depicts the different levels network interconnection.

A variety of networks and protocols will coexist for a long time:

Almost all personal computers-run TCP/IP;

Mainframe of large commercial organization—SNA running IBM;

A considerable number of telephone companies-running ATM networks;

The rising wireless network runs many different protocols.

As the price of computers and networks is getting cheaper, the decision-making power of purchase gradually drops. Different networks adopt completely different technologies, so when new hardware technology advances, new software will come into being.

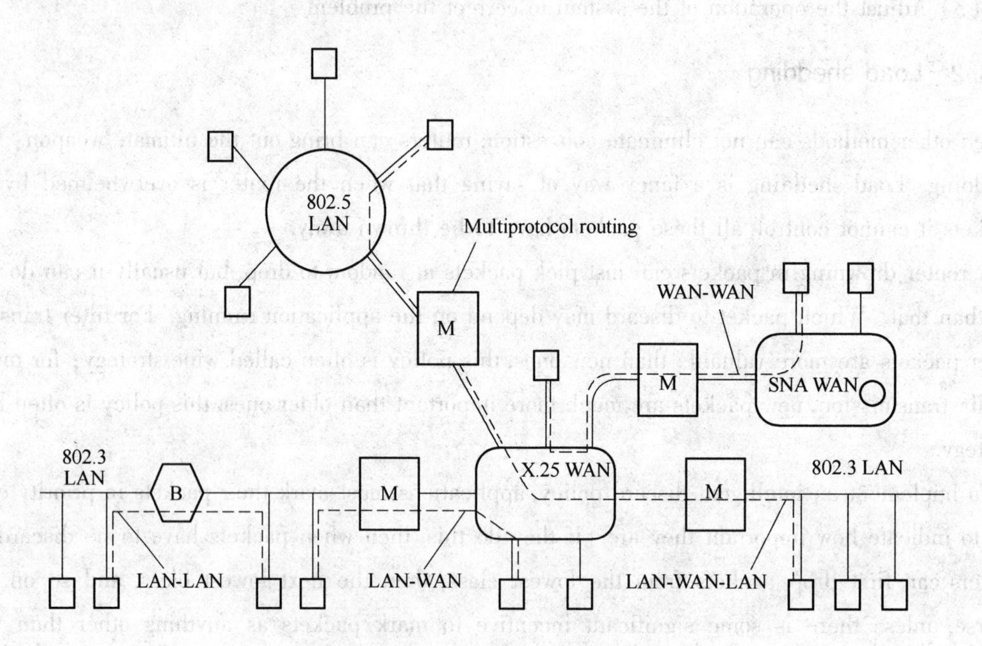

Figure 6-4 Network interconnection

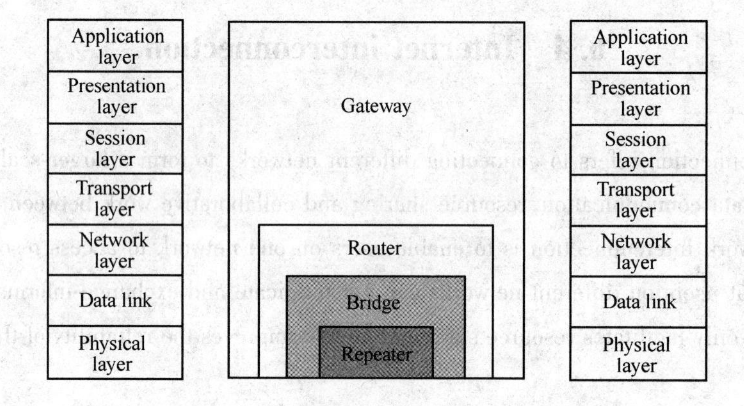

Figure 6-5 The levels of network interconnection

6.4.1 Network interconnected equipment

The main network interconnection devices are listed as follows:

The first level: repeaters, which copy every bit between the two cable segments;

The second layer: bridges, which store and forward data link frames between LAN;

The third layer: routers, which forward packets between heterogeneous networks;

The fourth level: gateways, which allow more than four layers of network interconnection. We will describe these devices in more detail.

6.4.1.1 Repeater-physical layer connection equipment

The repeater is used to connect several network segments in the LAN, and it only uses a simple signal amplification function to extend the length of the LAN. Figure 6-6 depicts the role of repeaters in the network.

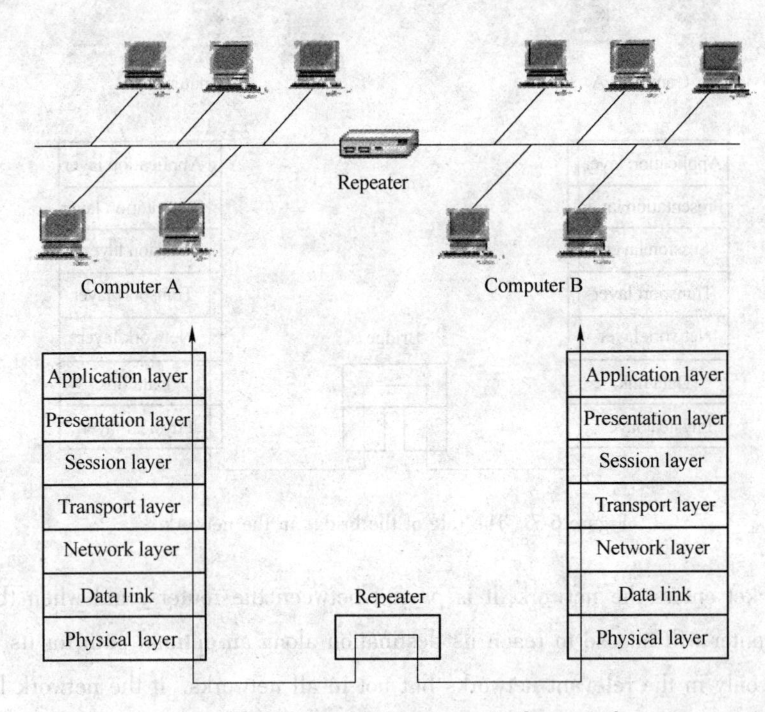

Figure 6-6 The role of repeaters in the network

6.4.1.2 Bridge-data link layer connection equipment

The bridge works on the data link layer of the OSI reference model, and is used to realize the interconnection of different LANs. A bridge can connect 2 or more LAN segments to receive, store and forward the data frames of each segment, provide data flow control and error control, and connect two physical networks (segments) into a logical network.

When the bridge network is interconnected, the data link layer of the Internet is allowed to be the same or different from the physical layer protocol. Figure 6-7 depicts the role of the bridge in the network.

6.4.1.3 Router-network layer connection equipment

The router works on the network layer of the OSI reference model, and the most basic function is to forward packets. In an interconnected network realized by routers, routers detect data packets, determine their destination addresses, and decide whether to receive or forward packets.

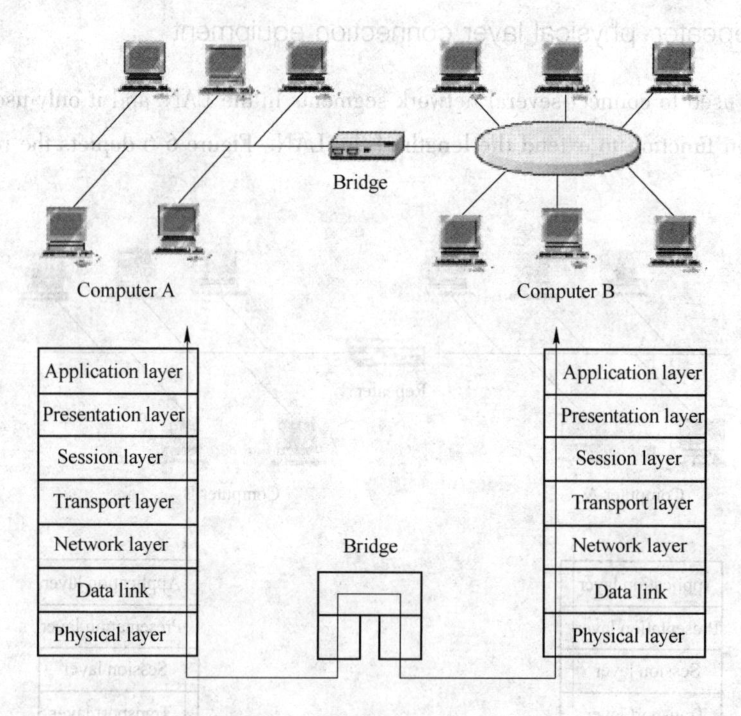

Figure 6-7 The role of the bridge in the network

After the packet enters the network, it is passed between the routers, and when the packet arrives, and the router will choose to reach its destination along an optimal path for its choice. Data are transmitted only in the relevant networks but not in all networks. if the network layer protocol is the same, interconnection will mainly solve the routing problem; if network layer protocols are different, multiprotocol routers are needed. When the router interconnects with the Internet, the network layer and the following protocols can be the same. Figure 6-8 depicts the role of the router in the network.

6.4.1.4 Gateway-high level connection equipment

The gateway, which works at the top of the OSI reference model, is the most complex network device; it can interconnect between different protocol networks, between networks of different network operating systems, and between LAN and remote networks. The main function of a gateway is to address translation, protocol conversion and data format transformation between heterogeneous networks. In order to realize the interconnection between different protocols, the gateway must realize the transformation between different network protocols. It is commonly used for interconnection between different types of network systems, and can also be used for interconnection between different mainframe and different databases. Figure 6-9 depicts the role of a gateway in the network.

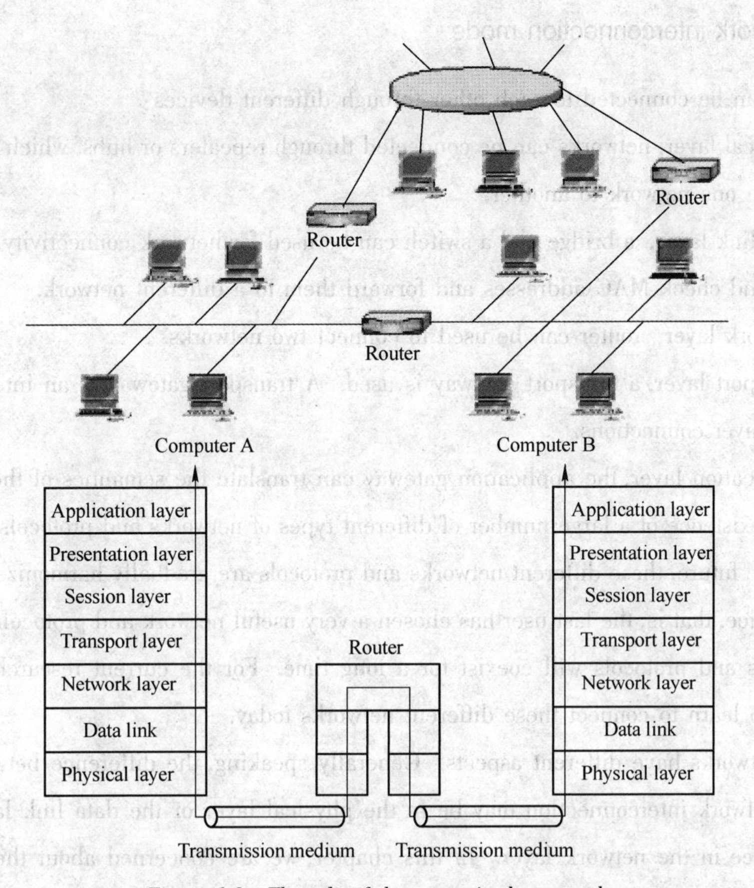

Figure 6-8 The role of the router in the network

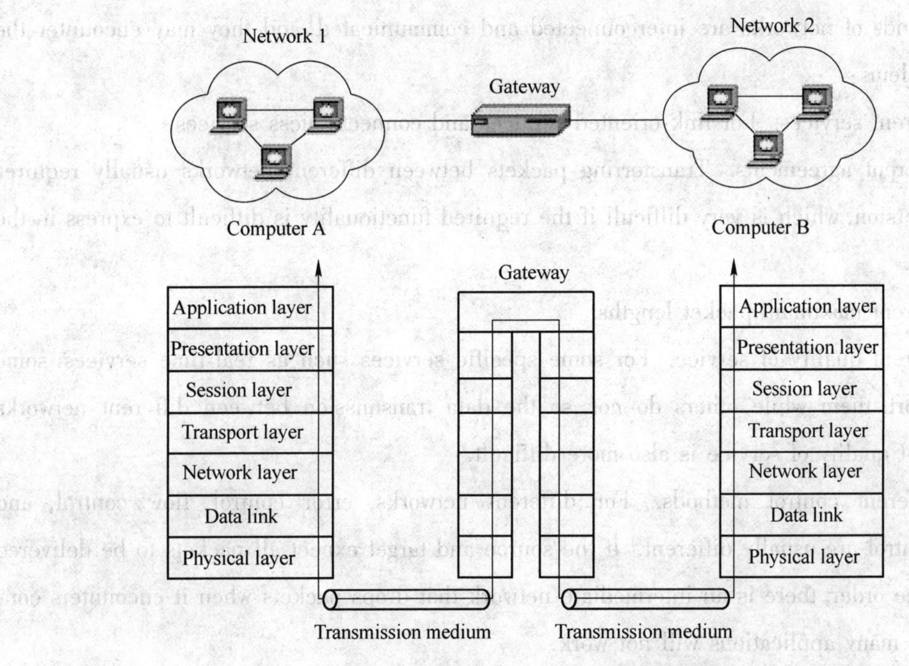

Figure 6-9 The role of a gateway in the network

6. 4. 2 Network interconnection mode

The network can be connected to each other through different devices:

At the physical layer, networks can be connected through repeaters or hubs, which usually simply move data from one network to another.

At the data link layer, a bridge and a switch can be used for network connectivity. They can receive frames and check MAC addresses and forward them to a different network.

At the network layer, router can be used to connect two networks.

At the transport layer, a transport gateway is used. A transport gateway is an interface between two transport layer connections.

At the application layer, the application gateway can translate the semantics of the message.

Due to the existence of a large number of different types of networks and protocols in today's reality and in the future, these different networks and protocols are gradually harmonized by the user's personal choice, that is; the last user has chosen a very useful network and protocol. Or these different networks and protocols will coexist for a long time. For the current research and study, it makes sense to learn to connect these different networks today.

Different networks have different aspects. Generally speaking, the difference between computer networks in network interconnection may be in the physical layer or the data link layer, or it may be the difference in the network layer. In this chapter, we are concerned about the difference in the network layer.

Different kinds of networks are interconnected and communicated, and they may encounter the following problems:

(1) Different services. For link-oriented services and connectionless services.

(2) Different agreements. Transferring packets between different networks usually requires protocol conversion, which is very difficult if the required functionality is difficult to express in the protocol.

(3) Different maximum packet lengths.

(4) Different quality of service. For some specific services such as real-time services, some network support, them while others do not, so the data transmission between different networks which different quality of service is also more difficult.

(5) Different control methods. For different networks, error control, flow control, and congestion control are usually different. If the source and target expect all packets to be delivered in an error-free order, there is an intermediate network that drops packets when it encounters congestion. Then many applications will not work.

6.5 Internet network layer

At the network level, the Internet can be regarded as a set of interconnected subnets or autonomous systems. Figure 6-10 shows that there are many computer networks interconnected by some routers. Since the computer networks participating in the interconnection use the same Internet Protocol (IP), the computer networks can be regarded as a virtual interconnection network when interconnected. The so-called virtual interconnected network is also a logical interconnected network. It means that the heterogeneity of the interconnected physical networks is inherently objective, but we can use the IP protocol to make these networks with different performances in the network. The layer looks like a unified network. Such a virtual interconnection network using the IP protocol may be simply referred to as an IP network (the IP network is virtual, but it is not necessary to emphasize the word "virtual" every time). The advantage of using an IP network is that when hosts on an IP network communicate, as if they are communicating on a single network, they do not see the specific heterogeneous details of the interconnected networks (such as specific addressing schemes, routing protocols) and many more. If the TCP protocol is used in the upper layer of such a global IP network, then it is the Internet.

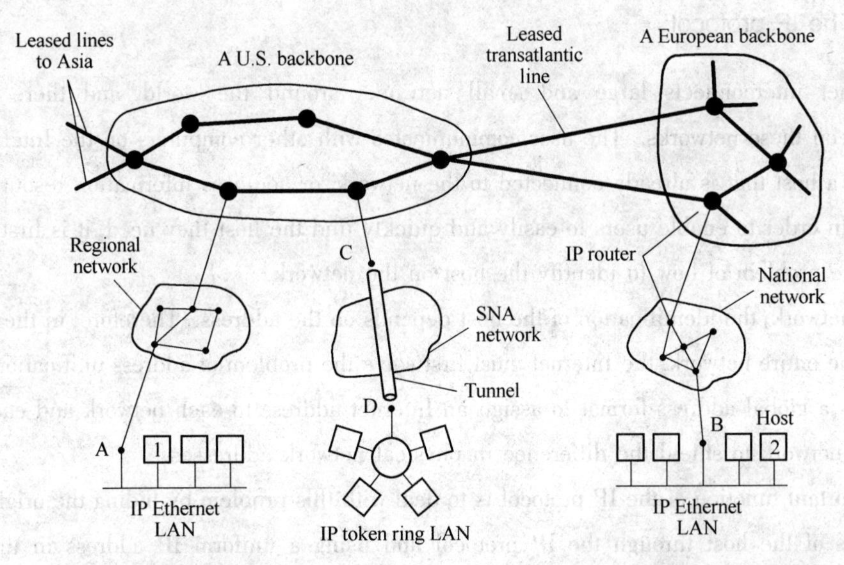

Figure 6-10 The internet is an interconnected collection of many networks

There are four steps of communication an the Internet, as shown in Figure 6-11:

(1) The transport layer obtains data streams and divides them into datagrams.

(2) Each datagram is transmitted over the Internet and may be subdivided into smaller units.

(3) When all packets are finally transmitted to the destination, they are recombined by the

network layer to the original datagram.

(4) The datagram is then transmitted up to the transport layer.

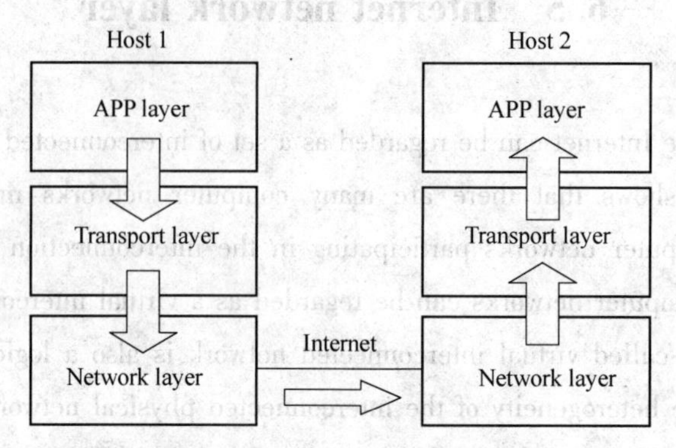

Figure 6-11 The steps of communication in the Internet

The virtual Internet is also a logical Internet. It means that the heterogeneity of various physical networks is inherently objective, but using IP protocols can make these networks with different performances appear to users as if they were a unified network. A virtual internetwork using the IP protocol may be referred to simply as an IP network. When hosts on the Internet communicate, it is as if they are communicating on a network, and the specific network heterogeneous details of the interconnection are not visible.

6.5.1 The IP protocol

The internet interconnects large and small networks around the world, and there are many computers on these networks. The user communicates with other computers on the Internet by operating on a host that is already connected to the network, or acquires information resources on the network. In order to enable users to easily and quickly find the host they need, it is first necessary to solve the problem of how to identify the host on the network.

In the network, the identification of the host depends on the address. Therefore, in the process of unifying the entire network, the Internet must first solve the problem of address unification. The Internet uses a global address format to assign an Internet address to each network and each host on the entire network to shield the difference in physical network addresses.

An important function of the IP protocol is to deal with this problem by hiding the original physical address of the host through the IP protocol and using a uniform IP address in the network layer. Figure 6-12 depicts the IP protocol.

The IP protocol has the following functions:

(1) Define datagram, the most basic transmission unit of the internet.

(2) Define the addressing mode of the Internet, the IP address.

(3) Transfer data between the network access layer and the host.

(4) Send data to the remote host, routing.

(5) Decompose and reorganize datagram; for example, data transmission between IP and X. 25.

(6) Connectionless protocol; no virtual circuit is established before data is exchanged; IP does not perform error detection during transmission, and thus is not reliable. The datagram may be lost, the order of delivery may be incorrect, and the IP does not check whether the destination host actually receives the data.

(7) Send: If the destination address specified by the IP is the local address, that is, the source and destination addresses are on the same network, the IP will directly transfer the data to the destination address; if the IP address is a remote address, the IP will check the local routing table and be remote. The host provides the route; if found, the IP will use the path to send the datagram; otherwise, the datagram will be sent to the host's default route.

(8) Receive: When the IP receives the datagram from the network access layer, if the destination address matches the local machine, the data of the header will be stripped and forwarded to the protocol specified by the header for further processing.

The characteristics of the IP protocol:

(1) IP protocol is an unreliable, connectionless datagram delivery service protocol;

(2) IP protocol is a point-to-point network layer communication protocol;

(3) The IP protocol shields the transport layer from differences in physical networks.

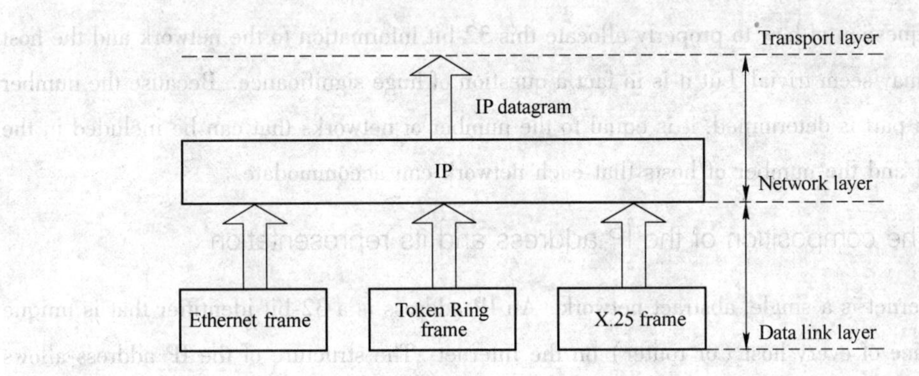

Figure 6-12 The IP protocol

6. 5. 2 The IP address

On the internet, each computer and each user has its own network address. However, the network address on the Internet must be unique and must be found in a certain way. The Internet uses a globally common address format to assign an Internet address to each grid and each computer across the network, thereby masking differences in physical network addresses. How to identify an address is a problem that every network has to face. The address is used to identify a resource in the network system.

The internet is a virtual network that interconnects physical networks through routers or gateways. In any physical network, the device at each node must have an identifiable address in order for the information to be exchanged. This address is called the physical address. Since the physical address is embodied in the data link layer, the physical address is also referred to as a "hardware address" or a "media access control address MAC". The physical address of the network brings some problems to the Internet's unified global network address:

(1) The physical address is the embodiment of physical network technology. Different physical networks have different physical addresses and formats. This creates an obstacle to cross-network communication.

(2) The physical address is solidified in the network device and usually cannot be modified. A physical address is a non-hierarchical address that represents only a single device and does not identify the network to which the device is connected.

(3) The Internet addresses the problem of physical network addresses, using the addressing scheme of the network layer IP address.

The IP protocol provides a uniform network address format. Address assignment is performed under unified management to ensure that each address corresponds to one host (including routers and gateways). The difference in physical addresses is masked by the IP layer.

(4) According to the TCP/IP protocol, the IP address consists of 32 bits, which consists of three parts: address class, network number, and host number.

(5) The question of how to properly allocate this 32-bit information to the network and the host as a number, may seem trivial, but it is in fact a question of huge significance. Because the number of bits in each part is determined, it is equal to the number of networks that can be included in the entire Internet and the number of hosts that each network can accommodate.

6.5.2.1 The composition of the IP address and its representation

The entire Internet is a single, abstract network. An IP address is a 32-bit identifier that is unique to each interface of every host (or router) on the Internet. The structure of the IP address allows us to easily address it on the Internet. IP addresses are now assigned by the Internet Corporation for Assigned Names and Numbers.

The IP address addressing method has gone through three historical stages.

(1) The classified IP address. This is the most basic method of addressing. In 1981, the corresponding standard protocol was adopted.

(2) Division of subnets. This is an improvement to the most basic addressing method, and its standard RFC 950 was adopted in 1985.

(3) The super network. This is a relatively new method of unclassified addressing. It was quickly promoted and applied after it was introduced in 1993.

Only the most basic classified IP addresses are discussed here. The so-called "classified IP address" means dividing the IP address into several fixed classes, and each type of address is composed of two fixed-length fields. First field is the network number (net-ID), which marks the host (or the network to which the router is connected). A network number must be unique across the Internet. The second field is the host number (host-ID), which identifies the host (or router). A host number must be unique within the network range indicated by the network number preceding it. Thus, an IP address is unique across the Internet.

The IP address consists of two parts: the network ID and the host ID.

(1) Network ID: Used to identify the network where the computer is located, or the number of the network.

(2) Host ID: Used to identify the different computers within the network, i. e. the number of the computer.

The IP address specifies that the network number cannot start with 127. The first byte cannot be all 0s, and it cannot be all 1s. The host number cannot be all 0s, and it cannot be all 1s. Each host (router or gateway) on the Internet has a globally unique IP address consisting of 32-bit binary digits, representing the address class, network number, and host number, Figure 6-13 depicts the composition of the IP address.

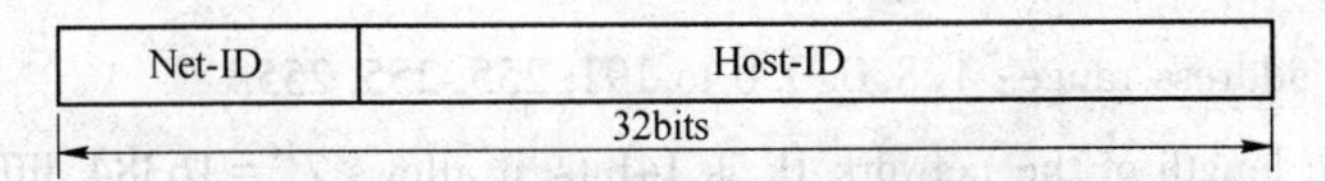

Figure 6-13 The composition of the IP address

6.5.2.2 The format and classification of IP addresses

Since the IP address is a limited resource, in order to better manage and use the IP address, INTERNIC divides the IP address into five categories (ABCDE) according to the size of the network as shown in Figure 6-14.

Class A IP Addresses:

The network number of a class IP address is 7bits, and the host number is 24bits.

(1) Class A IP address range: 1.0.0.0 to 127.255.255.255.

(2) The length of the network number: 7bits. In theory a class can have $2^7 = 128$ networks.

(3) The network number is reserved for special purposes by two IP addresses of all 0s and 1s (0 and 127 in decimal system). There are 126 different Class A networks.

(4) Because the host number is 24bits, the number of hosts of each Class A network is theoretically $2^{24} = 16777216$. The host number is reserved for all 0s and 1s of the two IP addresses for special purposes, so a class network actually allows 16777214 hosts to connect.

(5) Class A IP address structure is suitable for large networks with large hosts.

(6) The effective range of Class A IP address is: 1.0.0.1-126.255.255.254.

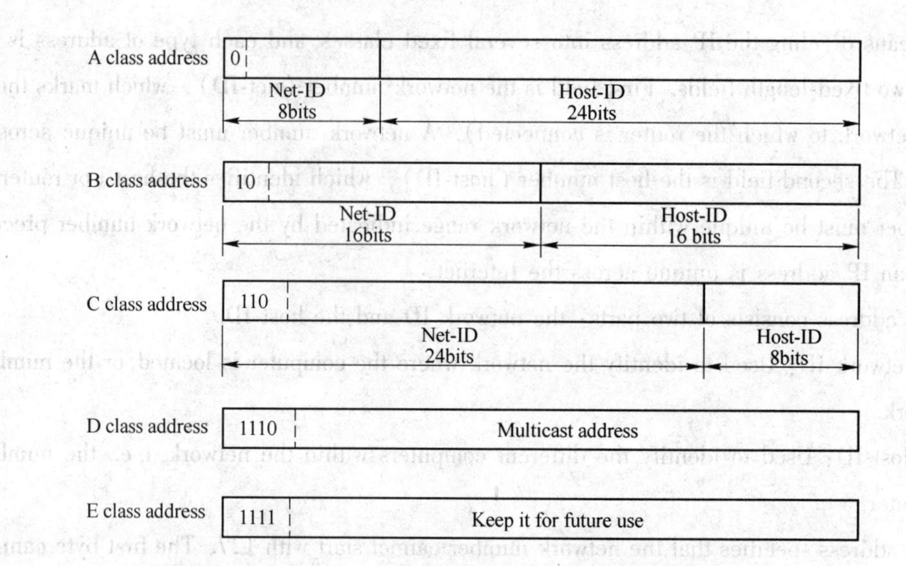

Figure 6-14 The format and classification of IP addresses

Class B IP addresses:

The length of the network IP of the Class B IP address is 14bits, and the host IP length is 16 bits.

(1) Class B IP address range: 128. 0. 0. 0 to 191. 255. 255. 255.

(2) Because the length of the network IP is 14bits, it allows $2^{14}=16384$ different Class B networks, and actually allows the connection of 16382 networks.

(3) Because the length of the Host IP is 16bits, each Class B network can have $2^{16}=65536$ hosts or routers, and a class B IP address is allowed to connect to 65534 hosts or routers.

(4) The B IP addresses are applicable to medium sized organizations such as some large international companies and government agencies.

(5) The valid range of Class B IP address is: 128. 1. 0. 1-191. 254. 255. 254.

Class C IP addresses:

The length of the network address of the IP address is 21 bits, and the host number is 8 bits.

(1) Class C IP address range: 192. 0. 0. 0 to 223. 255. 255. 255.

(2) The network number is 21bits long, so there are $2^{21}=2097152$ different Class C networks.

(3) The host number is 8bits, and the number of hosts per Class C network is $2^{8}=256$, which allows 254 hosts or routers to be connected.

(4) Class C IP addresses are applicable to small companies and ordinary research institutions.

(5) The valid range of Class C IP address is: 192. 0. 1. 1-223. 255. 254. 254.

Class D IP Addresses:

A Class D IP address does not identify the network.

(1) Class D address range: 224. 0. 0. 0 to 239. 255. 255. 255.

(2) For other special purposes, such as multicast address Multicasting.

Class E IP Addresses:

The Class E IP address is temporarily reserved.

(1) Class E address range: 240.0.0.0 to 255.255.255.255.

(2) Used for certain experiments and reserved for future use.

In addition, there are some important features of IP address.

(1) Hierarchical address structure. The benefits of two grades are:

1) The IP address management only assigns the net-ID, while the remaining host-ID is allocated by the affiliate that gets the net-ID. This facilitates the management of IP addresses.

2) The router forwards packets only according to the net-ID connected by the destination host (instead of the destination host-ID), greatly reduce the number of items in the routing table, thus reducing the storage space occupied by the routing table.

(2) In fact, the IP address is the interface to mark a host (or router) and a link.

1) When a host is connected to 2 different networks at the same time, the host must have two corresponding IP addresses, and the net-ID have to be different.

2) Since a router should at least connect to two networks (so it can forward the IP datagram from one network to another), a router should have at least two different IP addresses.

(3) Several LANs connected by switches or bridges are still a network, so these LANs all have the same net-ID.

(4) All networks assigned to the same net-ID, whether they are small LANs or WANs that may cover large areas are all equal, as shown in Figure 6-15.

Special IP address direct broadcast address, as shown in Figure 6-16.

(1) In Class A, B and C, the address with host-ID "all 1" is the direct broadcast address.

(2) Make A router send a packet to all hosts on a specific network by broadcasting.

(3) Only used as the destination address.

Special IP address limited broadcast address, as shown in Figure 6-17.

(1) The 32bits IP address with "all 1" is a limited address.

(2) It is used to send a packet to all hosts in the same physical network broadcast.

(3) The destination packet with a limited address will be received by all hosts, meanwhile the routers block the packet and restrict its broadcast function to the internal network.

Special IP address the net-ID is all "0", as shown in Figure 6-18.

(1) A host or router sends packets to a specific host on the network.

(2) Such packets are restricted to the internal network.

(3) The "0" address: when the host-ID is all "0", it means the local network.

Example. 172.17.0.0 represents a Class B network; "172.17" is the Class B net-ID.

Example. 192.168.1.0 represents a Class C network, "192.168.1" is the Class C net-ID.

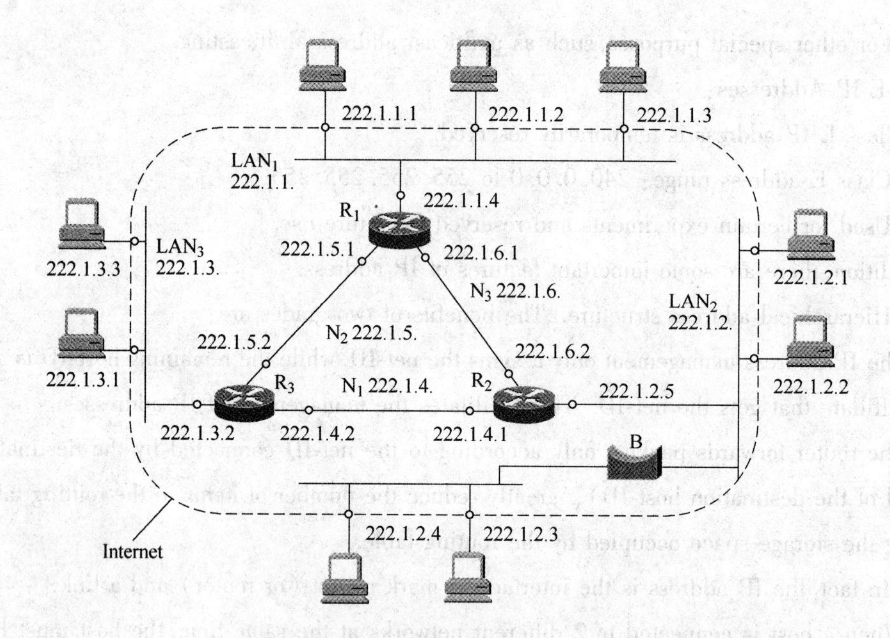

Figure 6-15 IP address in the Internet

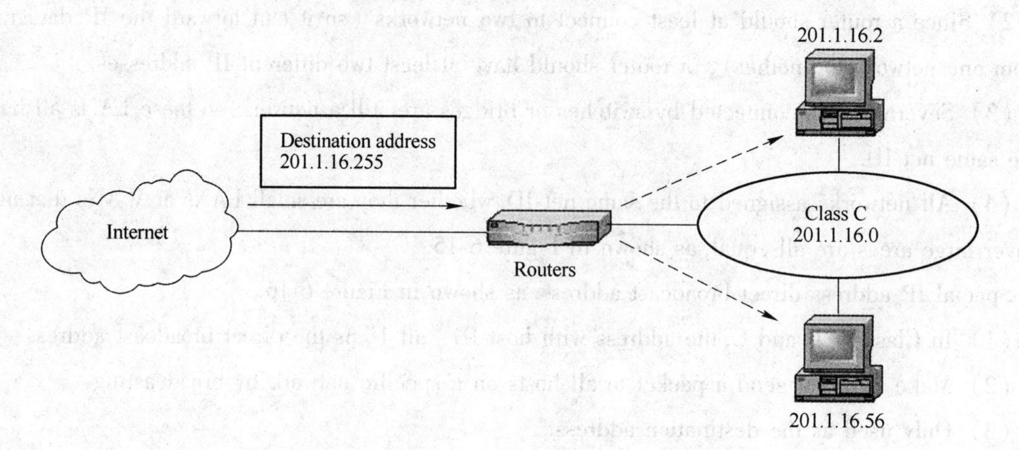

Figure 6-16 Direct broadcast address

Special IP address return address, as shown in Figure 6-19.

(1) Return address: the IP address starting with 127 is the reserved address.

(2) Used for network software testing and local host inter-process communication.

(3) The TCP/IP protocol specifies that packets with net-ID 127 cannot appear on any network; hosts and routers cannot broadcast any addressing information.

6.5.2.3 Management of IP address

Internet's IP address is globally valid, so the allocation and recovery of IP addresses need to be managed uniformly. The top management of the IP address is the internet network information cen-

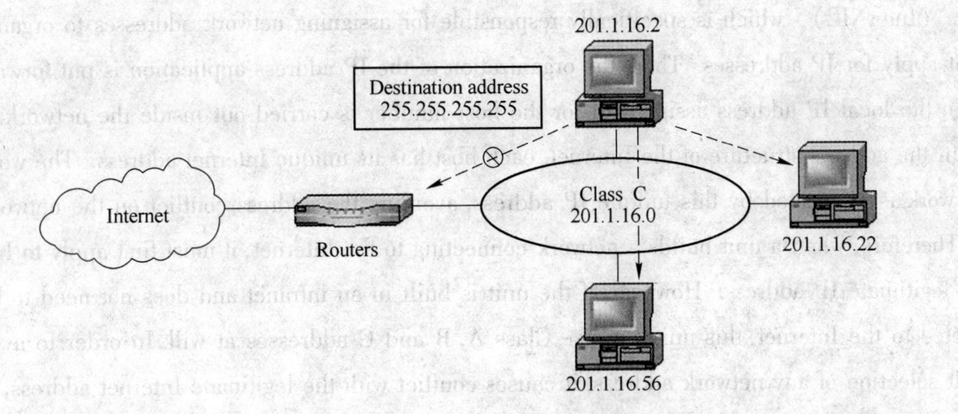

Figure 6-17 Limited broadcast address

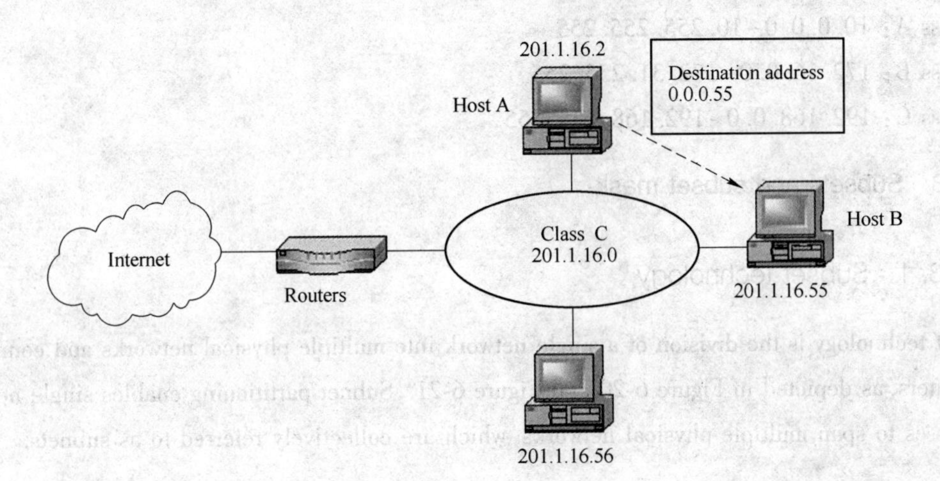

Figure 6-18 The net-ID is all "0"

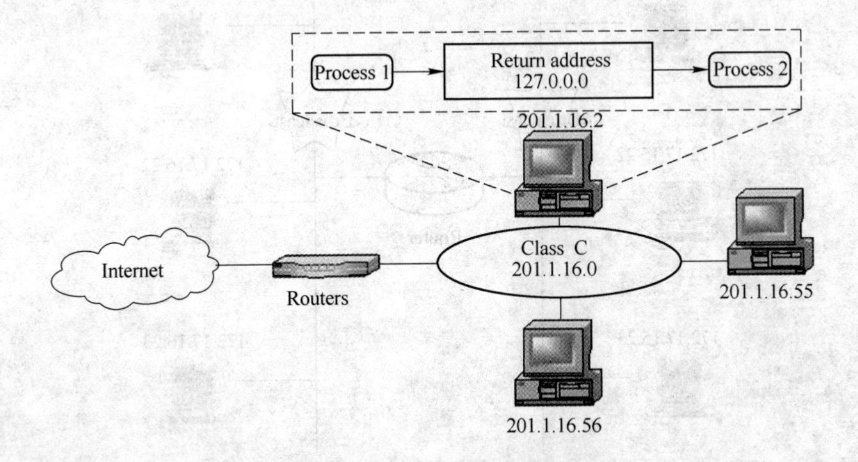

Figure 6-19 Return address

ter, (InterNIC), which is specifically responsible for assigning network addresses to organizations that apply for IP addresses. Then, the organization of the IP address application is put forward, and then the local IP address assignment for the host number is carried out inside the network.

In the address structure of the Internet, each host has its unique Internet address. The worldwide network is connected by this unique IP address, avoiding the address conflict on the network.

Therefore, when a unit builds a network connecting to the Internet, it must first apply to InterNIC for legitimate IP address. However, if the unit is built in an intranet and does not need to be connected to the Internet, this unit can use Class A, B and C addresses at will. In order to avoid any unit selection of any network address, it causes conflict with the legitimate Internet address, assigns specific Class A, B, and C address range for those unit intranets that do not need to be connected to Internet. These address ranges are:

Class A: 10. 0. 0. 0 ~ 10. 255. 255. 255

Class B: 172. 16. 0. 0 ~ 172. 31. 255. 255

Class C: 192. 168. 0. 0 ~ 192. 168. 255. 255

6. 5. 3 Subsets and subset mask

6. 5. 3. 1 Subset technology

Subset technology is the division of a single network into multiple physical networks and connected by routers, as depicted in Figure 6-20 and Figure 6-21. Subnet partitioning enables single network addresses to span multiple physical networks, which are collectively referred to as subnets.

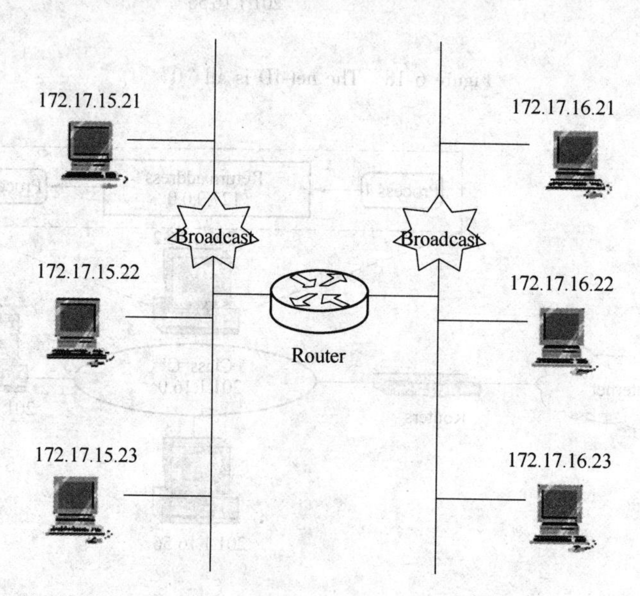

Figure 6-20 Subnet division

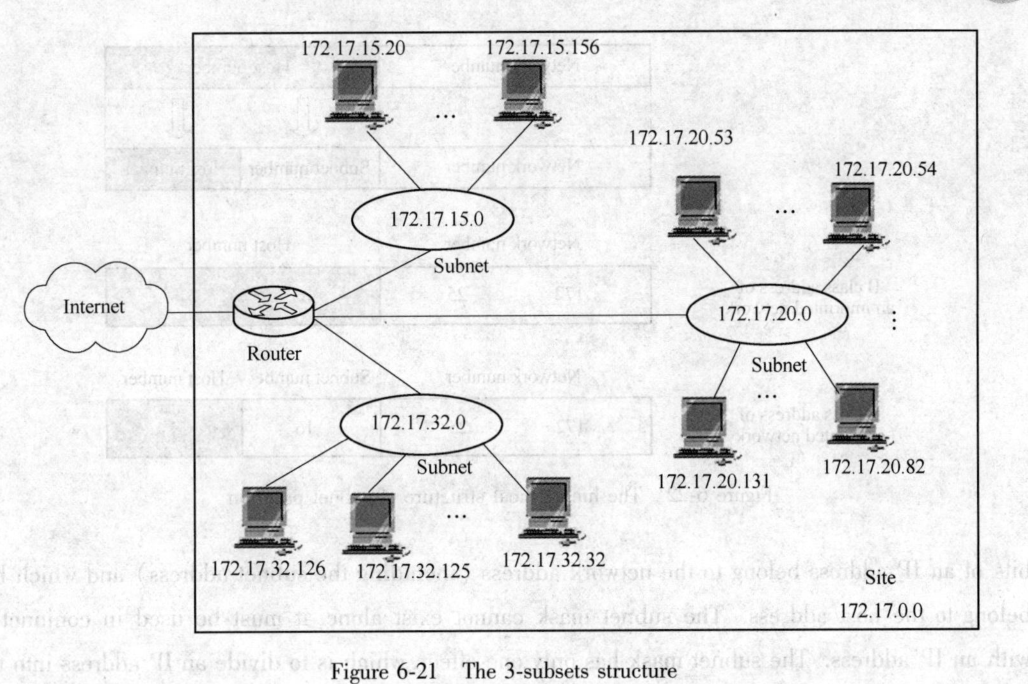

Figure 6-21 The 3-subsets structure

Subnet division has the following advantages:

(1) The full use of the address.

(2) Division of management functions.

(3) Improvement in the network performance.

6. 5. 3. 2 Hierarchical structure and method of subnet partition

Subset partition has two-layer structure and three-layer structure. The following is a detailed description, and Figure 6-22 depicts the hierarchical structure of subnet partition.

(1) The two-layer structure: includes the network ID and the host ID:

Standard class A, B , and C address.

(2) The three-layer structure includes:

Network ID, subnet ID and host ID.

1) The network ID: a site;

2) The subnet ID: a physical subnet;

3) The host ID: the host address connected to the subnet.

(3) The routing process of an IP packet experiences 3 steps:

1) 1 sent to the site;

2) 2 transmitted to the subnet;

3) 3 sent to the host.

6. 5. 3. 3 Subset mask

The subnet mask is also called the network mask and address mask. It is used to indicate which

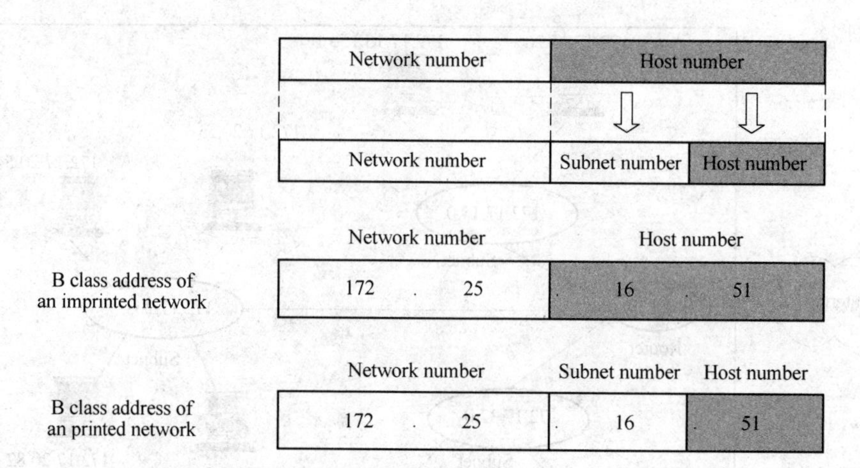

Figure 6-22 The hierarchical structure of subnet partition

bits of an IP address belong to the network address (including the subnet address) and which bits belong to the host address. The subnet mask cannot exist alone, it must be used in conjunction with an IP address. The subnet mask has only one effect, which is to divide an IP address into two parts: the network address and the host address.

The subnet mask is a 32-bit address that is used to mask a portion of the IP address to distinguish between the network ID and the host ID, and whether the IP address is on the local area network or on the remote network. All locations of the network address (network number and subnet number) in the IP address are "1", corresponding to the host address (host number) at all positions of "0". The 32 bit binary number represented by dotted decimals. Table 6-2 shows the default subnet mask for Class A, B, and C addresses.

Table 6-2 Default subnet mask for A, B, C class addresses

Address type	Dotted decimal notation	The binary bit of a subnet mask			
A	255. 0. 0. 0	11111111	00000000	00000000	00000000
B	255. 255. 0. 0	11111111	11111111	00000000	00000000
C	255. 255. 255. 0	11111111	11111111	11111111	00000000

After subnet division, the number of subnet numbers is not necessarily integer multiples of 8 bits, and it can be chosen according to needs. Same as the standard IP address, in order to reserve the subnet address and the subnet broadcast address, the subnet number and the host number are not allowed to be all 0s or 1s. The host number is all 0s, representing the subnet address. The host number is all 1s, representing the broadcast address of the subnet, and Figure 6-23 and Figure 6-24 depicts this.

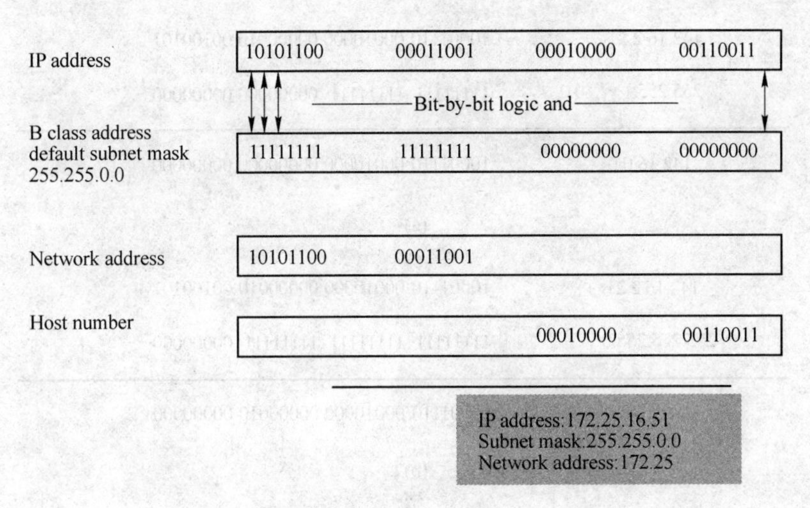

Figure 6-23 Subnet mask and network address

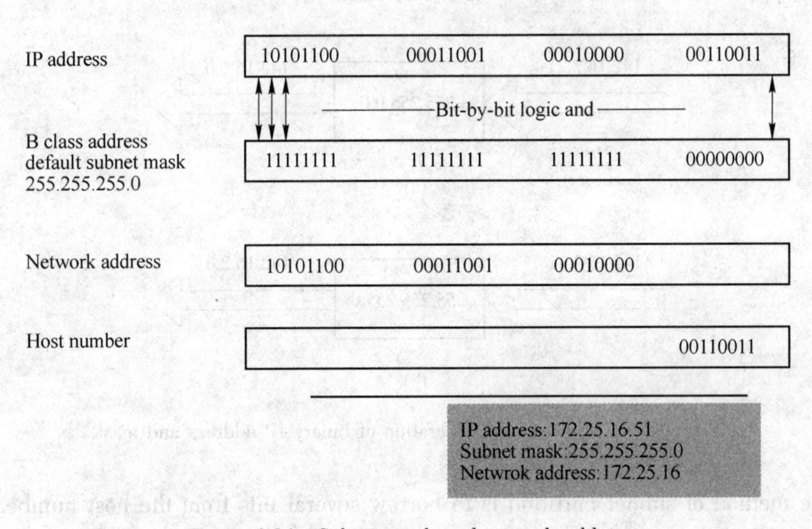

Figure 6-24 Subnet mask and network address

Subnet mask operation is the process of extracting the network number from an IP address, as depicted in Figure 6-25 and Figure 6-26.

6.5.3.4 Subset partition examples

Since 1985, a "subnet number field" has been added to the IP address to make the two-level IP address a three-level one. This is called subnet partition, or subnet addressing subnet partition has become the official standard protocol for the Internet. The basic idea of subnet partition is as follows:

(1) A unit with many physical networks can divide its own physical network into several subnets. Subnet partition is purely a matter within a unit. The network outside this unit is not able to see how many subnets this network consists of, because this unit still acts as a network;

142.16.2.21	10001110 00010000 00000010 00100101
255.255.0.0	11111111 11111111 00000000 00000000
142.16.0.0	10001110 00010000 00000000 00000000

(a)

142.16.2.21	10001110 00010000 00000010 00100101
255.255.255.0	11111111 11111111 11111111 00000000
142.16.2.0	10001110 00010000 00000010 00000000

(b)

Figure 6-25 Bit-wise and operation of binary IP address and mask

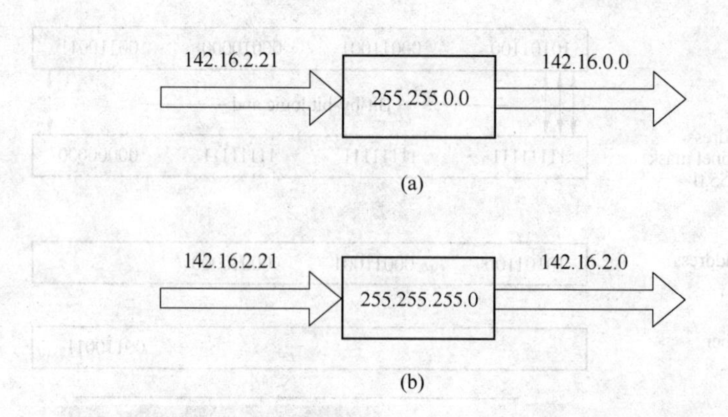

Figure 6-26 Bit-wise and operation of binary IP address and mask

(2) The method of subnet partition is to borrow several bits from the host number of the network as the subnet number (subnet-ID). Of course, the host number is reduced by the same number of bits. Therefore, the two-level IP address becomes a three-level IP address within the unit: the network number, the subnet number, and the host number. It can also be expressed in the following notation:

IP address: <network number>, <subnet number>, <host number>

(3) Any IP datagram sent from another network to a host of the unit is still found on the network connected to the unit according to the destination network number of the IP datagram. However, after receiving the IP datagram, the router finds the destination subnet according to the destination network number and the subnet number, and delivers the IP datagram to the destination host.

(4) Each subnet needs a unique subnet number for identification. It requires two subnet numbers; (all 0s and all 1s as the subnet numbers cannot be allocated).

(5) For each sub-network host and two ports of the router, a unique host number is assigned

(all 0s and all 1s as the host number cannot be allocated) There are 100 hosts in the network. Considering the two ports of the router, the number of hosts that need to be identified is 102. It is assumed that each subnet accounts for half and 51 each.

The following uses examples to illustrate the concept of subnet partition.

Example: Figure 6-27 ~ Figure 6-30 shows that a unit has a class C IP address with a network address of 192. 168. 1. 0 (the network number is 192. 168. 1). Datagrams with a destination address of 192. 168. 1. x are sent to the router on this network. There are 3 steps in subnet partition.

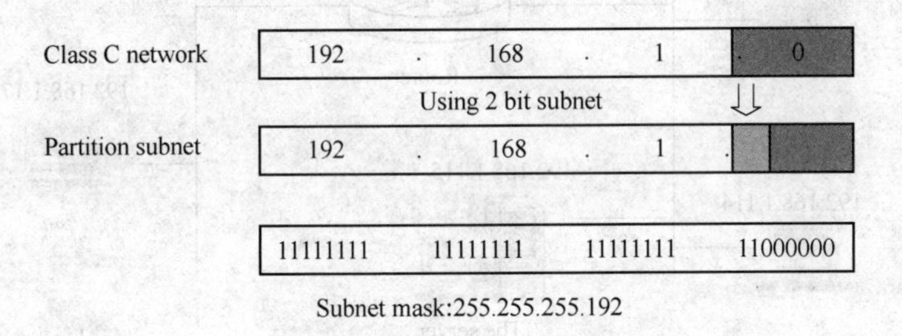

Figure 6-27 Step 1 in subnet partition

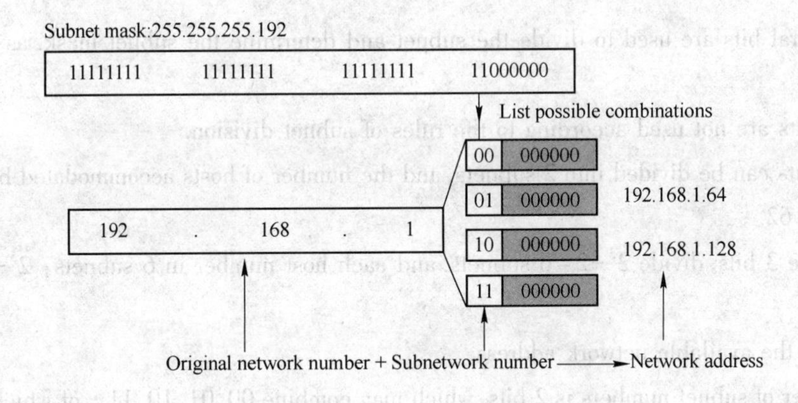

Figure 6-28 Step 2 in subnet partition

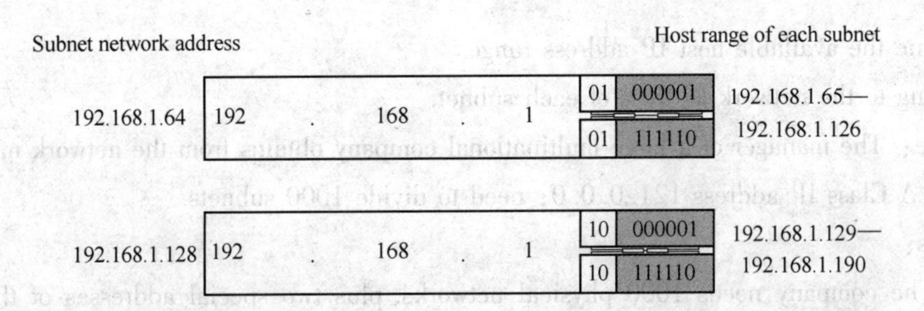

Figure 6-29 Step 3 in subnet partition

Step 1:

The Class C address is divided into two subnets. From the fourth byte representing the host

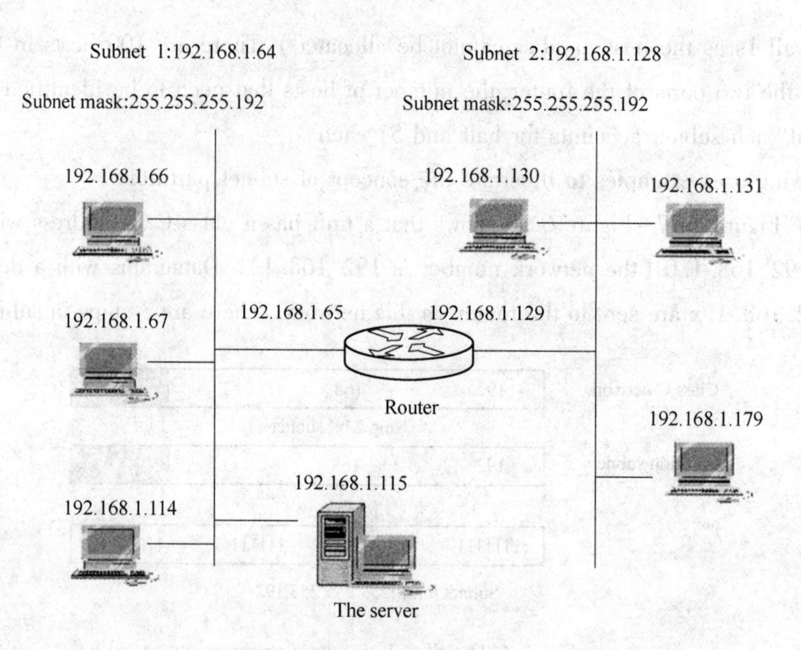

Figure 6-30 Address allocation for each host in each subnet

number, several bits are used to divide the subnet and determine the subnet mask, as depicted in Figure 6-24.

(1) 1 bits are not used according to the rules of subnet division.

(2) 2 bits can be divided into 2 subnets, and the number of hosts accommodated by each subnet is $2^6-2=62$.

(3) Take 3 bits, divide $2^3-2=6$ subnets, and each host number in 6 subnets: $2^5-2=30$.

Step 2:

Determine the available network address.

The number of subnet numbers is 2 bits, which may combine 00,01,10,11, of which 01 and 10 are available subnet numbers.

Step 3:

Determine the available host IP address range.

According to the network address of each subnet.

Example: The manager of a large multinational company obtains from the network management center. A A Class IP address 121.0.0.0; need to divide 1000 subnets.

Analysis:

(1) The company needs 1000 physical networks, plus two special addresses of the network numbers 0 and 1, and the number of subnets is at least 1002.

(2) The bit length of the selected subnet number is 10, and the subnet can be used to allocate up to $2^{10}=1024$. Except the subnet numbers all 0 and all 1 as two reserved addresses, the number of subnets can be allocated to 1022, which can meet the requirements of the user.

We can get the answer from the descriptions in Figure 6-31.

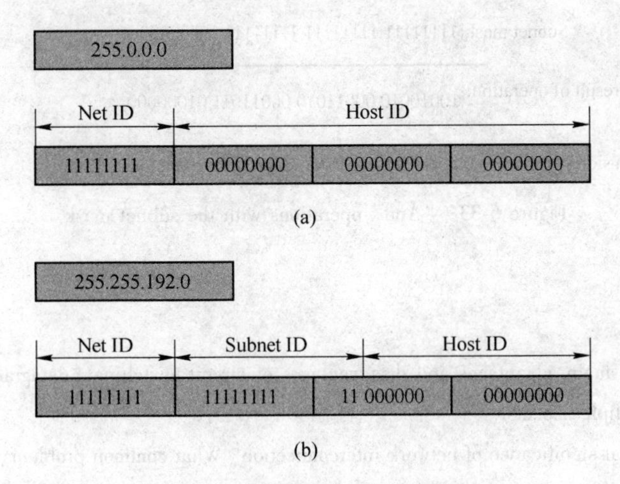

Figure 6-31 Structure of Class A sub-network subnet

In the view of external internet users, the 1000 sub nets appear as a whole, and its network address is: 121.0.0.0, and the address range after subnet is shown in Figure 6-32.

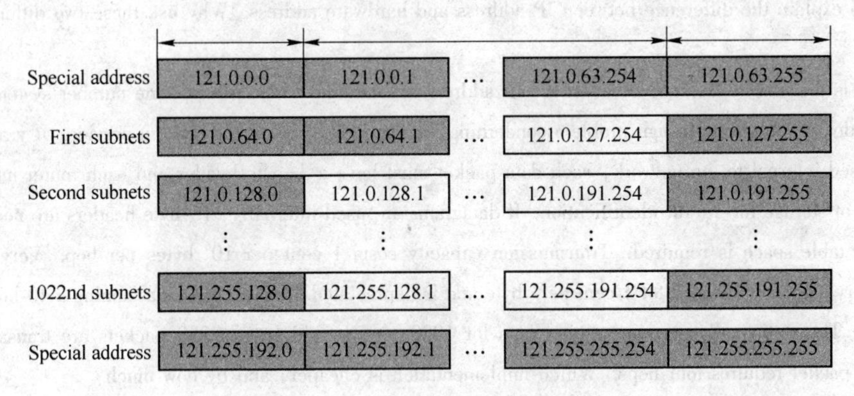

Figure 6-32 Address range after subnet

Last but not least, how can we decide whether the two hosts belong to the same subnet according to the host's IP address? There are usually two methods to achieve it.

(1) In the case of the subnet, we determine whether the two hosts are in the same subnet and see if their network number is the same as that of the subnet mask.

Example:

1) The IP address of the Host 1 is 156.26.27.71;

2) The IP address of the Host 2 is 156.26.27.110;

3) Their subnet masking is 255.255.255.192.

(2) The IP address of Host 1 performs "and" operations with the subnet mask, as depicted in Figure 6-33.

The IP address of the host 1:10010010.00011010.00011011.01101110

Subnet mask:11111111.11111111.11111111.11000000

And the result of operation: 10010010.00011010.00011011.01000000

Conclusion:the subnet numbers are 0001101101,so they belong to a subnet.

Figure 6-33 "And" operations with the subnet mask

Problems

6-1 Try to compare the main advantages and disadvantages of circuit switching, datagram switching and packet switching from multiple aspects.

6-2 What is the practical significance of network interconnection? What common problems need to be solved when interconnecting networks?

6-3 As an intermediate device, what is the difference between a repeater, a bridge, a router, and a gateway?

6-4 Try to explain the role of the following protocols: IP, ARP, RARP and ICMP.

6-5 What are the IP addresses? How to express each? What are the main characteristics of the IP address?

6-6 Try to explain the difference between IP address and hardware address. Why use these two different addresses?

6-7 What is the main difference between the IP address scheme and China's telephone number system?

6-8 Consider the following design problem concerning implementation of virtual-circuit service. If virtual circuits are used internal to the network, each data packet must have a 3-byte header and each router must tie up 8 bytes of storage for circuit identification. If datagrams are used internally, 15-byte headers are needed but no router table space is required. Transmission capacity costs 1 cent per 10^6 bytes per hop. Very fast router memory can be purchased for 1 cent per. byte and is depreciated over two years, assuming a 40-hour business week. The statistically average session runs for 1000 sec, in which time 200 packets are transmitted. The mean packet requires four hops. Which implementation is cheaper, and by how much?

6-9 Suppose that instead of using 16 bits for the network part of a Class B address originally, 20 bits had been used. How many Class B networks would there have been?

6-10 Convert the IP address whose hexadecimal representation is C22F1582 to dotted decimal notation.

6-11 (1) What does the subnet mask 255. 255. 255. 0 mean?

(2) The current mask of a network is 255. 255. 255. 248, how many hosts can the network connect to?

(3) The subnet number of a Class A network and a Class B network is 16 1s and 8 1s respectively. What is the difference between the subnet masks of the two networks?

(4) The subnet mask of a Class B address is 255. 255. 240. 0. What is the maximum number of hosts on each subnet?

(5) The subnet mask of a Class A network is 255. 255. 0. 255. Is it a valid subnet mask?

(6) The hexadecimal representation of an IP address is C2. 2F. 14. 81. Try converting it to dotted decimal notation. What kind of IP address is this address?

(7) Does the Class C network use a subnet mask for practical significance? Why?

6-12 Try to identify the network category of the following IP addresses:

(1) 128. 36. 199. 3

(2) 21. 12. 240. 17

(3) 183. 194. 76. 253

(4) 192. 12. 69. 248

(5) 89. 3. 0. 1

(6) 200. 3. 6. 2

6-13 A router has the following (CIDR) entries in its routing table:

Address/mask	Next hop
135. 46. 56. 0/22	Interface 0
135. 46. 60. 0/22	Interface 1
192. 53. 40. 0/23	Router 1
Default	Router 2

For each of the following IP addresses, what does the router do if a packet with that address arrives?

(1) 135. 46. 63. 10

(2) 135. 46. 57. 14

(3) 135. 46. 52. 2

(4) 192. 53. 40. 7

(5) 192. 53. 56. 7

6-14 Use the traceroute (UNIX) or tracert (Windows) programs to trace the route from your computer to various universities on other continents. Make a list of transoceanic links you have discovered. Some sites to try are

www. berkeley. edu (California)

www. mit. edu (Massachusetts)

www. ustb. edu. cn (Beijing)

www. tsinghua. edu. cn (Beijing)

www. fudan. edu. cn (Shanghai)

www. ucl. ac. uk (London)

www. usyd. edu. au (Sydney)

www. u-tokyo. ac. jp (Tokyo)

www. uct. ac. za (Cape Town)

6-15 Try to find the subnet mask (using a continuous mask) that produces the following number of Class A subnets:

(1) 2 (2) 6 (3) 30 (4) 62 (5) 122 (6) 250

6-16 There are four subnet masks below, which ones are not recommended? Why?

(1) 176. 0. 0. 0 (2) 96. 0. 0. 0 (3) 127. 192. 0. 0 (4) 255. 128. 0. 0

6-17 An autonomous system has five local area networks, and its connection diagram is shown in Figure 6-34. The number of hosts from LAN_2 to LAN_5 are: 91, 150, 3, and 15, respectively. The autonomous system has an instantaneous IP address block of 30. 138. 118/23. Try to give each LAN address block (including the prefix).

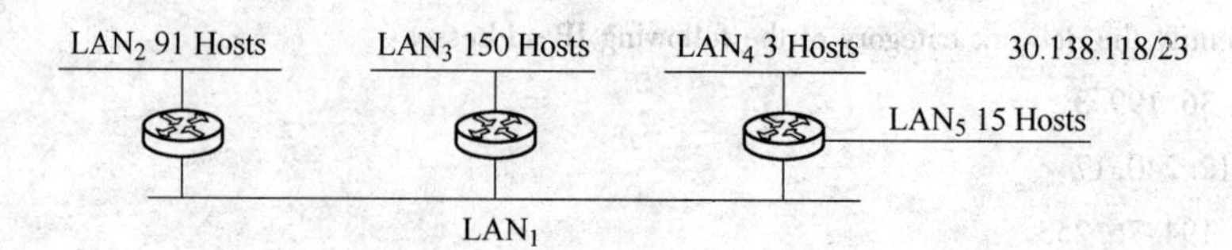

Figure 6-34

6-18 A large company has one headquarters and three subordinate departments. The network prefix assigned to the company is 192. 77. 73/24. The company's network layout is shown in Figure 6-35. There are five LANs in the headquarters, of which $LAN_1 \sim LAN_4$ are connected to router R_1, and R_1 is connected to router R_2 through LAN_5. R_2 are connected to $LAN_6 \sim LAN_8$ of the three departments through the WAN. The number indicated next to each LAN is the number of hosts on the LAN. Try to assign a suitable network prefix to each LAN.

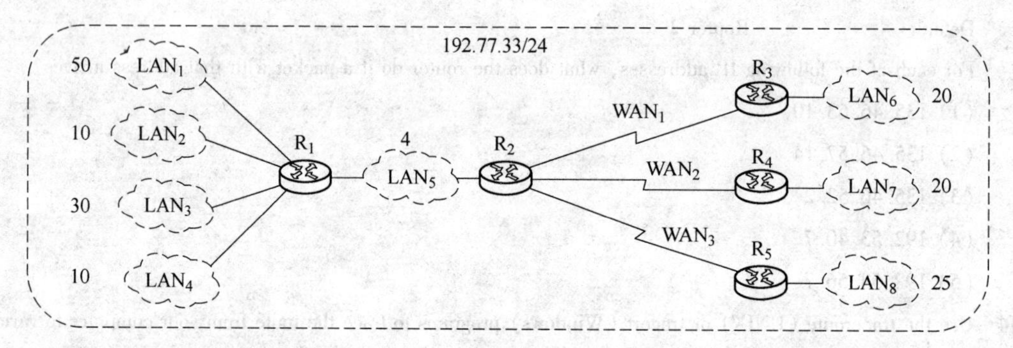

Figure 6-35

6-19 One of the known address blocks is 140. 120. 84. 24/20. Try to find the minimum and maximum addresses in this address block. What is its address mask? How many addresses are there in the address block? How many Class C addresses is it equivalent?

6-20 One of the known address blocks is 190. 87. 140. 202/29. Recalculate the previous question.

6-21 A unit is assigned to an address block 136. 22. 3. 64/26. It now needs to be further divided into four equally large subnets. Ask:

(1) How long is the network prefix for each subnet?

(2) How many addresses are there in each subnet?

(3) What is the address block for each subnet?

(4) What is the minimum and maximum address that each subnet can assign to a host?

6-22 Try to convert the following IPv4 addresses from binary notation to dotted decimal notation.

(1) 10000001 00001011 00001011 11101111

(2) 11000001 10000011 00011011 11111111

(3) 11100111 11011011 10001011 01101111

(4) 11111001 10011011 1111011 00001111

6-23 Is there an error with the following IPv4 address? If yes, please indicate.

(1) 111. 56. 045. 78

(2) 221. 3478. 20

(3) 75. 45301. 14

(4) 11100010. 23. 14. 67

6-24 Find the categories for each of the following addresses.

(1) 00000001 00001011 00001011 11101111

(2) 11000001 10000011 00011011 11111111

(3) 10100111 11011011 10001011 01101111

(4) 11110011 10011011 11111011 00001111

6-25 Find the categories for each of the following addresses.

(1) 227. 12. 14. 87

(2) 193. 14. 56. 22

(3) 14. 23. 120. 8

(4) 252. 5. 15. 111

6-26 Given that a network has an address of 167. 199. 70. 82/27, what is the network mask, network prefix length, and network suffix length of this network?

6-27 A unit is assigned an address block with a starting address of 14. 24. 74. 0/24. The unit needs to use three subnets. The specific requirements of their three sub-address blocks are: subnet N_1 requires 120 addresses, subnet N_2 requires 60 addresses, and subnet N_3 requires 10 addresses. Please give the allocation scheme of the address block.

6-28 As shown in Figure 6-36, the network 145. 13. 0. 0/16 is divided into four subnets N_1, N_2, N_3, and N_4. The interfaces connected to the router R are m0, m1, m2, and m3. The fifth interface of router R (m4) is connected to the internet. Ask:

(1) Try to give the routing table of router R.

(2) Router R receives a packet whose destination address is 145. 13. 160. 78. Try to give an indication of how this packet was forwarded.

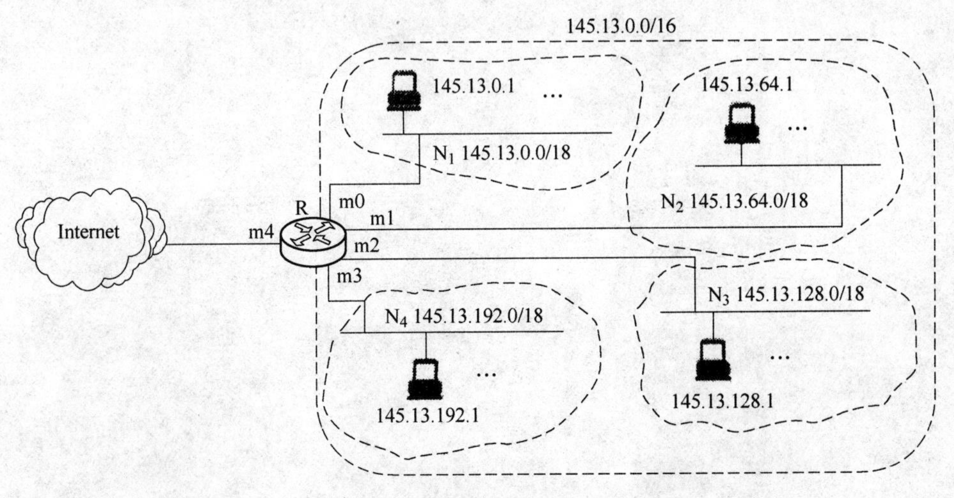

Figure 6-36

6-29 Try to find the subnet mask that can produce the following number of Class A subnets.

(1) 2 (2) 6 (3) 30 (4) 62 (5) 122 (6) 250

6-30 Which of the following prefixes matches the addresses 152. 77. 7. 159 and 152. 31. 47. 252, please explain the reason.

A. 152. 40/13 B. 153. 40/9 C. 152. 64/12 D. 152. 0/11

6-31 A network is divided into several subnets, one of which has been assigned a subnet mask of 74. 178. 247. 96/29. Which of the following network prefixes can no longer be assigned to other subnets.

A. 74. 178. 247. 120/29 B. 74. 178. 247. 64/29

C. 74. 178. 247. 80/28 D. 74. 178. 247. 104/29

7 The Transport Layer

Goal:

(1) Understand the rale of the transport layer.

(2) Understand the UDP protocols in the transport layer.

(3) Master the TCP protocols in the transport layer.

7.1 An overview of the transport layer

The transport layer (as shown in Figure 7-1) is the core of the entire protocol level. Its task is to provide a reliable, cost-effective data transmission function between the source host and the destination host. It is completely independent of the physical network currently used. The network layer and the following layers have realized data communication between hosts in the network, but data communication is not the ultimate goal of building a computer network. The essential activity of the computer network is to realize the process communication between hosts distributed in different geographic locations.

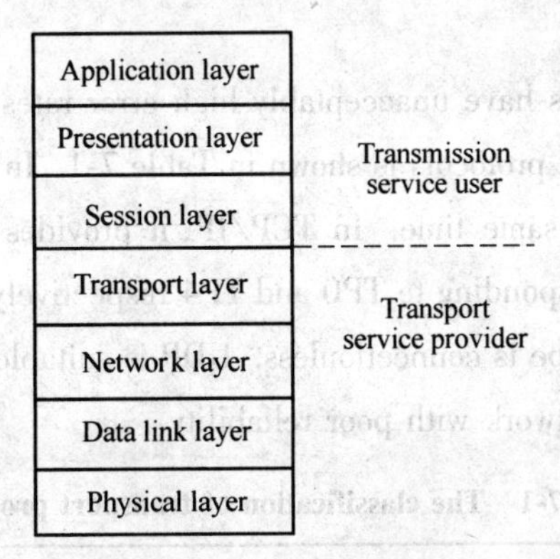

Figure 7-1 The transport layer

The transport layer is set up separately due to following reasons. The network layer is the highest level of the communication subnet, but the wide area network service is provided by the telecommunication company. Users cannot control the subnet, and cannot solve the service quality problem of the network layer. So adding a layer on the network layer to improve the service quality is necessary. From the point of data communication, the transport layer service is similar to the network layer service. First, the transport layer is also connection-oriented and connectionless transport services. Then, they have similar addressing and traffic control methods. The transport layer provides effective, reliable and reasonably priced services to the user application layer process.

Bottom three layers (1~3) of OSI are for data communication. Top three layers (5~7) OSI are for information processing. The transport layer (4) makes up for the differences and deficiencies of services provided by each sub-network, improve service quality and increase service function. The presence of the transport layer makes transport service much more reliable than its low-level network services. If the network service is complete, the transport layer is easy to work. If the quality of network service is poor, the transport layer must make up for the difference between the requirements of transmission users and the services enhanced by the network layer. The transport layer enhances and complements services provided by communication subnets.

Requirements of the transport layer protocol depend on two factors: requirements of the upper layer service for the transport layer (T) and services of communication subnet for the transport layer (N). The transport layer protocol (=T−N) is the difference set of the above two sets. The quality of service can be described by some specific parameters, such as connection set up delay, probability of setting up a failure, throughput rate, transmission delay, error rate, (safety) protection, priority and recovery function.

The communication subnet is divided into three types according to the quality of service:

Class A: network connections have acceptably low error rates and acceptably low failure notification rates.

Class B: network connections have acceptably low error rates and unacceptably failure notification rates.

Class C: network connections have unacceptably high error rates.

The classification of transport protocols is shown in Table 7-1. In OSI implementations, TP0 and TP4 are often provided at the same time. In TCP/IP, it provides both connectionless UDP and connection-oriented TCP, corresponding to TP0 and TP4 respectively. The TCP/IP communication subnet based on IP datagram type is connectionless. UDP is suitable for high reliability LAN. TCP is suitable for the wide area network with poor reliability.

Table 7-1 The classification of transport protocols

Type	Symbol	Network oriented	Function
0	TP0	A	Provide process communications with piecewise assembly functions
1	TP1	B	Adding the basic error recovery function on the basis of the 0 class

Continued Table 7-1

Type	Symbol	Network oriented	Function
2	TP2	A	On the basis of the 0 categories, the network multiplexing function and the corresponding flow control function are added, but there is no recovery function for network connection failures
3	TP3	B	With the functions of 1 and 2 transport protocol, it has the function of error recovery and multiplexing
4	TP4	C	It has the functions of error detection, recovery and multiplexing, and can guarantee high reliability data transmission when the network quality is poor

7.2 The transport layer in the TCP/IP system

The transport layer is the core of the entire protocol hierarchy. It is located in the third layer of the TCP/IP architecture. TCP provides reliable packet streaming and connection services to upper-layer applications. TCP uses sequential response to deliver messages on demand. The transport protocol provides a communication session between computers, and the choice of transport protocol depends on the mode of the data transmission.

The main function of the transport layer is to separate and reorganize the data flow provided by the upper layer, and provide end-to-end transport services for the data flow. At the sender, the transport layer segments the data flow provided by the application layer and adds the data amount to the identification. These identifiers include which application is issued, which application is processed, transport layer protocol, checksum, message length, and so on. This type of identifier is called a transport layer packet header, such as TCP headers and UDP headers. In the receiver, the transport layer removes the packet header of the transport layer. It checks whether the data is wrong in the transmission process by using the checksum in the packet header, and reassembles the data segments into data flow in a certain order for the application to process.

The transport layer provides logical communication for the application process of mutual communication as shown in Figure 7-2. The communication between the two hosts is actually the communication process between the two host applications. Communication between applications is also known as end-to-end communication. A very important function of the transport layer is reuse and separation. Packets from different application layers are sent down to the transport layer through different ports, and the services provided by the network layer are shared below. "Transport layer provides logical communication between application processes" . Logical communication means that communication between transport layers seems to transmit data horizontally. But in fact, there is no horizontal or physical connection between the two transport layers.

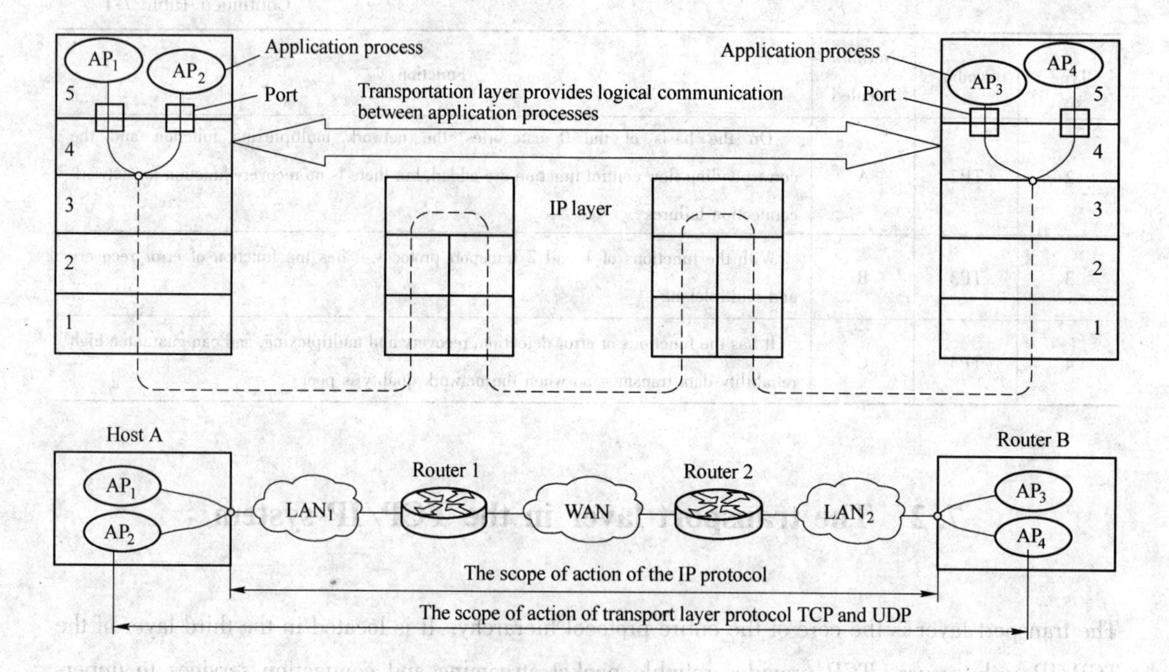

Figure 7-2 The transport layer provides logical communication for the application process of mutual communication

7. 2. 1 Two protocols in the transport layer

The transport layer of the internet TCP/IP protocol stack has two parallel protocols. TCP is connection-oriented, and equivalent to the TP4 of the OSI reference model transport layer. UCP is connectionless, and equivalent to OSI reference model transport layer TP0.

TCP is a protocol designed to provide reliable, end-to-end byte stream communication on unreliable Internet. The data clamp protocol transmitted by TCP is a TCP segment. The TCP entity receives the user's data stream from the local process and divides it into data fragments, which are sent as separate IP datagrams. The receiver is an IP entity containing TCP data, which is sent to the TCP entity of the target machine, and is recombined into the original byte stream. The TCP protocol provides a reliable communication connection for applications, so that the byte stream sent by one computer is sent to other computers on the network without errors. Data communication systems with high reliability requirements often use the TCP protocol to transmit data. However, TCP only provides connection-oriented services, rather than broadcast or multicast services. Since TCP provides reliable, connection-oriented transport services, it inevitably adds a lot of overhead.

UDP is a protocol corresponding to TCP, and it is a non-connected protocol. "Non-connection" means that you do not have to establish a connection with the other party before the formal communication; instead, UDP sends data directly regardless of the other party's status. The data clamping protocol transmitted by UDP is UDP packet or user datagram. On the sending end, the data transfer speed of UDP is limited by the data generation speed of applications, the capabilities of the computer, and the bandwidth of the transmission. At the receiving end, UDP puts each message segment in a queue, and the application reads a message segment from the queue each time. UDP is suitable for application environments that transmit only a small amount of data at a time and do not require high reliability. Since the UDP protocol has no connection process, its communication efficiency is high. However, because of this, its reliability is not as high as TCP protocol.

In general, TCP corresponds to applications with high reliability requirements, while UDP corresponds to applications with low reliability requirements and economical transmission. Application protocols supported by TCP mainly include: Telnet, File transfer protocol (FTP), Simple message transfer protocol (SMTP), etc. Application protocols supported by UDP mainly include: Network file system (NFS), Simple network management protocol (SNMP), Primary domain name system (PDNS), and (Trivial file transfer protocol (TFTP).

7.2.2 The concept of port and socket

The operation and resources of each computer in the network are managed by its own operating system, the state of the activity of other hosts, the state of each process, what time these processes want to participate in the network activity, and what process to communicate with which host of the network, and so on. A major problem that must be solved in the realization of process communication in the network is process naming and addressing methods.

The port number is the access point that TCP and UDP protocols use to connect to the application. It is part of TCP and UDP protocol software. The transport layer protocol of TCP/IP specifies some standard reserved port numbers for service processes; users can apply for non-reserved ports, and the port numbers of these non-reserved ports are unique in each host. Therefore, the port number can be used as the process ID in the network environment.

There are two basic allocations for ports: global allocation and local allocation. Global allocation, a centralized control mode, is carried out by an authoritative body in accordance with the needs of users, and the results are released to the public. Local allocation is a dynamic joint coding mode. When a process needs to access a transport service, a dynamic application is proposed to the local operating system, the operating system returns a local unique port number, and the process connects itself to the corresponding port number through a suitable system call.

The TCP/IP has 2^{16} ports, and these ports are divided into two groups: reservation ports and free ports. The reservation ports are allocated globally, and each standard server has a globally recognized port number. Services of the same nature on different computers has the same port number. The reserved ports only account for a small number, and the port number should be less than 256. The free ports are allocated locally, which account for the vast majority of all ports.

TCP takes the connection as the most basic abstraction. There are two endpoints for each TCP connection. The endpoint of TCP connection is not the host, the host IP address, the application process, or the transport layer protocol port. The endpoint of TCP connection is called socket. The port number is concatenated with the IP address to form the socket.

socket = (IP address : The port number)

Each TCP connection is uniquely determined by the two endpoints (i. e., two sockets) at both ends of the communication. It is expressed as:

TCP connection :: = {socket1, socket2}

= {(IP1: port1), (IP2: port2)}

There are some partially reserved ports (including TCP, UDP) as shown in Table 7-2. The socket address of the telnet service of the web server with IP address 10. 43. 3. 87 is 10. 43. 3. 87: 23.

Table 7-2 Partially reserved port (including TCP, UDP)

Port number	Service name	Port number	Service name
7	ECHO	25	SMTP
13	DAYTIME	69	TFTP
20	FTP (data)	80	HTTP
21	FTP (control)	110	POP3
23	TELNET	161	SNMP

7. 3 UDP protocol

In addition to multiplexing and demultiplexing, UDP adds almost nothing to IP, so it does the least amount of work that the transport protocol can do. UDP gets the data from the application, appends the source port number, destination port number fields, and two other fields, and then transmits the resulting segment to the network layer. UDP has the disadvantage of not providing packet grouping, assembly, and sorting of packets. That is to say, after the message is sent, it is impossible to know whether it arrives safely and completely. UDP is used to support web applications that need to transfer data between computers. The UDP protocol has been in use for many years since its inception. Although its initial brilliance has been overshadowed by some similar protocols, UDP is still a very practical and feasible network transport layer protocol today.

7.3.1 UDP overview

UDP is a simple datagram-oriented transport layer protocol. UDP only adds port function and error detection function on the basis of IP datagram service. Although UDP can only provide unreliable delivery, it has some special advantages in some aspects. UDP is suitable for interactive short messages and is highly efficient. When the communication subnet is quite reliable, its application is higher.

In the TCP/IP model, UDP provides a simple interface above the network layer and below the application layer. It only provides unreliable delivery of data. Once it sends out the data sent to the network layer by the application, it does not retain data backup. UDP only adds multiplexing and data validation (fields) to the header of the IP datagram as shown in Figure 7-3.

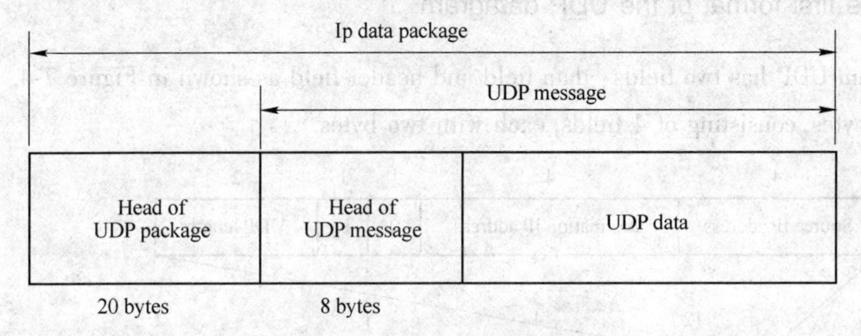

Figure 7-3 The user datagram protocol

Due to the lack of reliability and non-connection-oriented protocols, UDP applications must generally allow a certain amount of packet loss, error. Sometimes, in some applications, such as TFTP, must add a fundamentally reliable mechanism at the application layer if it needed. However, most UDP applications do not require a reliable mechanism; in fact, they may even reduce performance by introducing a reliable mechanism. Streaming media, instant multimedia games and voice over IP (VOIP) must be typical UDP applications. If an application requires high reliability, then TCP can be used instead of UDP.

UDP has the following characteristics:

(1) UDP is a connectionless protocol. Before the data is transmitted, the source and the terminal do not establish a connection. When UDP wants to transmit, it simply grabs the data from the application and throws data to the network as quickly as possible.

(2) Since the transmission data does not establish a connection, there is no need to maintain the connection status. Therefore, a server can transmit the same message to multiple clients simultaneously.

(3) The header of the UDP packet is very short, only 8 bytes, and its overhead is very small relative to the 20 byte information package of TCP.

(4) The throughput is not affected by congestion control algorithm. UDP has no congestion

control and is suitable for multimedia communication. UDP supports one-to-one, one-to-many, many-to-one and many-to-many interactive communication.

(5) UDP uses the best effort to deliver, so the host does not need to maintain a complex link state table (there are not many parameters here).

(6) UDP is message oriented. The sender's UDP packet to the application is delivered to IP layer after adding the header. Instead of splitting or merging, the boundaries of these messages are preserved, so the application needs to choose the appropriate message size.

Compared to TCP protocol, UDP protocol does not provide a guarantee mechanism for data transfer and it cannot guarantee the order in which data is sent and received. Although UDP is an unreliable network protocol, in some cases UDP protocol may become very useful.

7.3.2 The first format of the UDP datagram

User datagram UDP has two fields: data field and header field as shown in Figure 7-4. The header field has 8 bytes, consisting of 4 fields, each with two bytes.

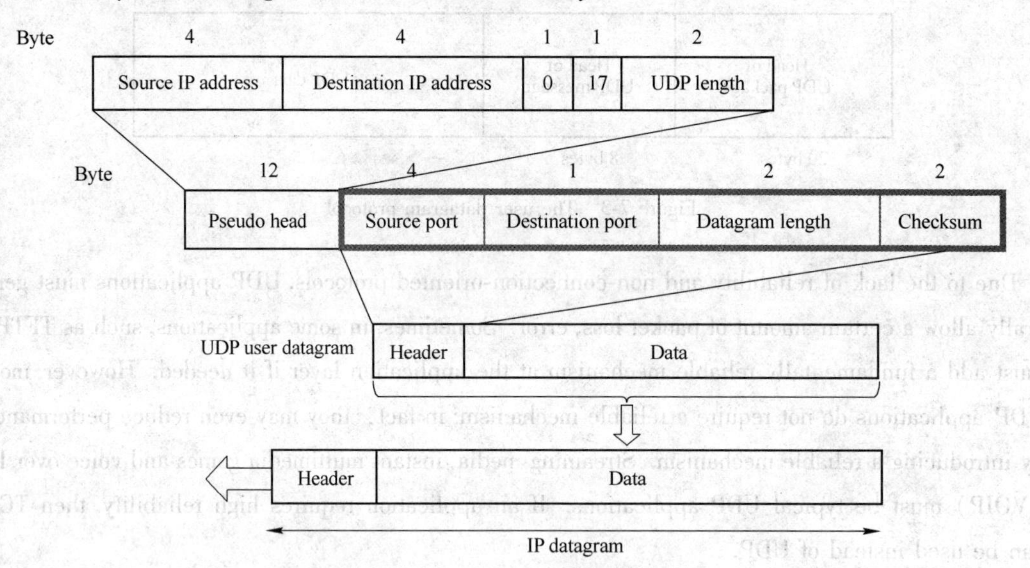

Figure 7-4 The composition of UDP

The UDP header consists of four fields, each of which occupies 2 bytes. The details are as follows: UDP source port, destination port, datagram length, checksum (as shown in Figure 7-5). Two of them are optional. UDP source port and destination port are used to mark the sending and receiving application process. Since UDP does not require an acknowledgment, the source port is optional and is set to zero if the source port is not used. Behind the destination port is a fixed-length length field that specifies the length of the data portion of UDP datagram. The minimum length is 8 bits. The remaining 16 bits of the header are used to checksum between the header and data part. This part is optional, but it is generally used in practical applications.

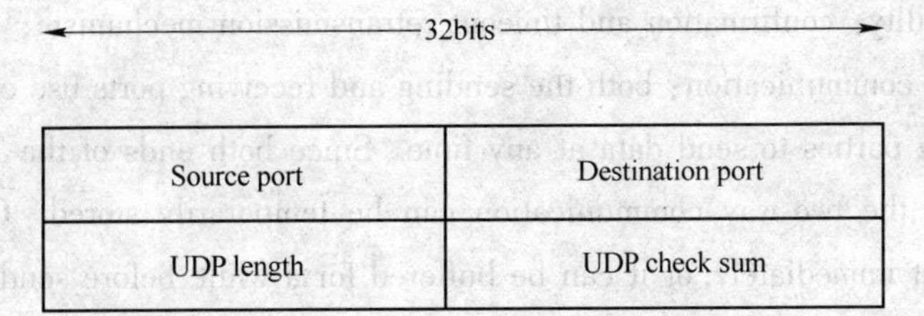

Figure 7-5 The domains of UDP header

UDP protocol uses checksum in the header to ensure data security. The check value is first calculated by the data sender through a special algorithm, and then needs to be recalculated after being passed to the receiver. If a datagram is falsified by a third party during transmission or is damaged due to line noise, etc. , the checksum calculation values of the sender and the receiver will not match, and UDP protocol can detect whether an error has occurred. Although UDP provides error detection, it does not perform error correction when an error is detected. It simply throws away the corrupted message segment or provides a warning message to the application.

7.4 TCP protocol

Transmission control protocol (TCP) is a protocol designed to provide reliable, end-to-end byte stream communication on unreliable Internet defined by IETF's RFC 793. In the simplified computer network OSI model, it performs the functions specified by the fourth layer of the transport layer. In the Internet protocol suite, the TCP layer is an intermediate layer located above the IP layer and below the application layer.

7.4.1 Main features

TCP is a connection-oriented transport layer protocol. Each TCP connection only has two endpoints, each TCP connection can only be point-to-point, it provides full-duplex communication. TCP connection is a virtual connection rather than a real physical connection. It does not care how long the application process sends packets to the TCP cache at a time. It determines how many words should be included in a segment according to the window value given by the peer and the current network congestion. TCP can divide the data blocks that are too long and then transfer them. It can also wait for the accumulation of enough bytes before the message segment is sent out.

The main features of the TCP protocol:

(1) Connection oriented service: the transmission link must be established between the original process and the destination process before the actual datagram transmission is carried out.

(2) High reliability: confirmation and timeout retransmission mechanism;

(3) Full duplex communication: both the sending and receiving ports use caching. TCP allows both communicating parties to send data at any time. Since both ends of the TCP connection are cached, the data of the two-way communication can be temporarily stored. Of course, TCP can send a data segment immediately, or it can be buffered for a while before sending.

(4) Supporting streaming transmission: provides a stream interface, sends continuous data streams, and interprets data streams by both applications.

(5) Reliable establishment and release of transmission links: only uses the 3 handshake.

(6) Providing traffic control and congestion control: uses the sliding window method.

7.4.2 The first part of the TCP message segment

The first part of the TCP message segment is shown in Figure 7-6.

(1) Source port and destination port fields-each 2 bytes. The port is the service interface between the transport layer and the application layer. The multiplexing and demultiplexing functions of the transport layer must be implemented through ports.

(2) Sequence number field-4 bytes. Each byte in the data stream transmitted in the TCP connection is programmed with a sequence number. The value of the sequence field refers to the sequence number of the first byte of the data sent by this segment.

(3) The acknowledgment field, which is 4 bytes, is the sequence number of the first byte of the data expected to receive the next segment of the other party.

(4) Data offset (that is, the length of the header), which accounts for 4 bits, indicates how far away from the beginning of the TCP message segment from the beginning of the TCP message segment. The data offset unit is 32 bit (calculated in 4 bytes).

(5) Reserved fields, which occupy 6 place, are reserved for future use, but should be set at 0.

(6) Emergency URG (E)-when URG is 1, it indicates that the emergency pointer field is valid. It tells the system that there are urgent data in this message segment and should be transmitted as soon as possible (equivalent to high priority data).

(7) Confirm ACK-the confirmation number field is valid only when ACK is 1. When ACK is 0, the confirmation number is invalid.

(8) Push PSH (PuSH) while receiving TCP gets a message segment of PSH = 1, and deliver the application process as quickly as possible, and no longer wait until the entire cache is filled and then delivered up.

(9) Reset RST (ReSeT)-when RST=1 indicates a serious error in a TCP connection (such as a host crash or other cause), a connection must be released and then a transport connection is re-established.

(10) Synchronous SYN-synchronous SYN = 1 indicates that it is a connection request or a connection receiving message.

(11) Terminate FIN-to release a connection. FIN 1 indicates that the data sent to the sender of this message segment has been sent out and requires the release of the transport connection.

(12) The window field, which occupies 2 bytes, is used to set the sending window for the other party. The unit is byte.

(13) Test and-2 bytes. The scope of inspection and field inspection includes the two parts of header and data. When calculating the check and add, 12 bytes pseudo head should be added to the front of the TCP message segment.

(14) The emergency pointer field, which accounts for 16 bits, points out the number of bytes in the emergency data in this newspaper section (emergency data is in the front of the newspaper section).

(15) Option fields-the length is variable. TCP initially specified only one option, that is, the maximum segment size (MSS). MSS tells each TCP: "the maximum length of the data field that my cache can receive is MSS bytes."

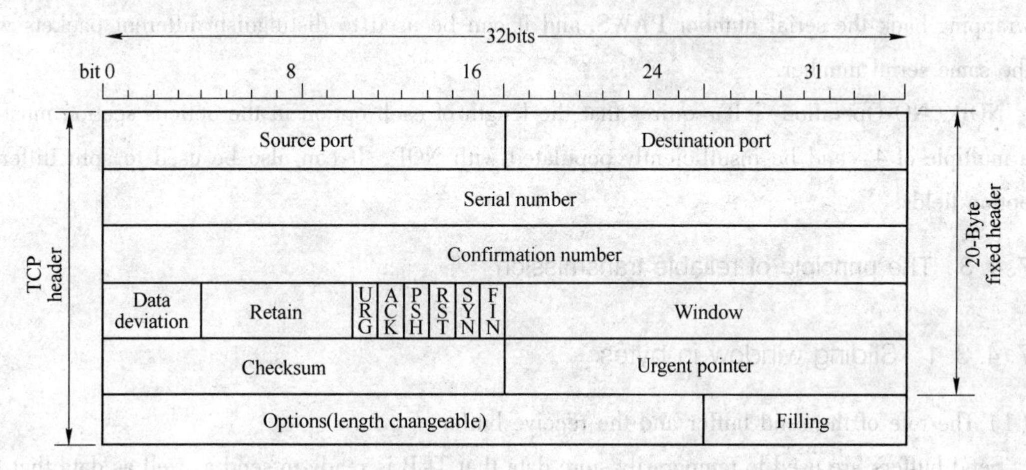

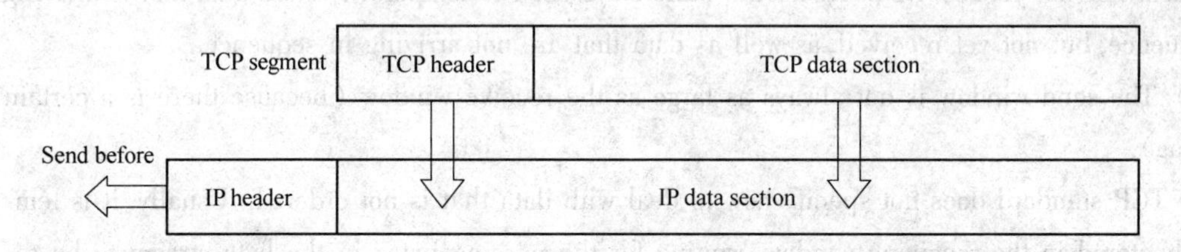

Figure 7-6 The first part of the TCP message segment

TCP sets the source port and destination port to identify the communication process. The final address of network communication is not only the host address, but also can be extended to an identifier describing the computer application process. The port is equivalent to OSI's transport layer service access point TSAP—an abstract software structure (including data structure and I/O buffer).

TCP option fields mainly appears in TCP connection establishment stage, and it actually uses the following options:

The maximum length of the maximum segment size (MSS) segment indicates the maximum length of the data field. The length of the data field plus the TCP header is equal to the length of the entire TCP segment. The MSS value indicates the length of the data field that it expects the other party to send a TCP segment. Both parties to the communication can have different MSS values. If not filled in, the default is 536 bytes. MSS only appears in the SYN message. That is, MSS appears in the segment with SYN=1.

Windows scaling: Since the length of window size field in TCP header is 16 bits, the maximum number it represents is 65535. Larger windows are needed to meet requirements of performance and throughput while communication with large latency and bandwidth is generated, so the window scaling option is created.

Selective acknowledgement: Ensure that only missing message segments are re-transmitted, instead of re-transmitting all message segments.

Timestamps: It can be used to calculate RTT (round trip time). It can also be used to prevent wrapping back the serial number PAWS, and it can be used to distinguish different packets with the same serial number.

NOP (NO-Operation): It requires that the length of each option in the options section must be a multiple of 4, and be insufficiently populated with NOP. It can also be used to split different option fields.

7.4.3 The principle of reliable transmission

7.4.3.1 Sliding window in bytes

(1) The role of the send buffer and the receive buffer.

Send buffers are used to temporarily store data that TCP is ready to send as well as data that has been sent but not yet received. Receiving buffers are used to temporarily store data that is arriving in sequence, but not yet received, as well as data that are not arriving in sequence.

(2) The send window is not always as large as the receive window (because there is a certain time lag).

The TCP standard does not specify how to deal with data that is not ordered. Usually, it is temporarily stored in the receiving window, waiting for the missing bytes in the byte stream to be received, and then the data is delivered to the upper application process sequentially. TCP requires the receiver to have the function of cumulative acknowledgement, which can reduce the transmission overhead.

7.4.3.2 Timeout retransmission time selection

(1) The retransmission mechanism is one of the most important and complex problems in TCP. Each time TCP sends a segment, it sets a timer for this segment. This segment is retransmitted as

long as the retransmission time set by the timer has not yet been acknowledged.

(2) Because the lower level of TCP is an Internet environment, the routing chosen by IP datagram varies greatly. Therefore, the variance of the transport layer's round-trip time is also very large.

(3) Weighted average round trip time.

TCP retains a weighted average round trip time RTTs for RTT. When the RTT sample is first measured, the RTTs value is taken as the measured RTT sample value. Each time a new RTT sample is measured, the RTTs are recalculated as follows:

$$\text{New RTTs} = (1 - a)(\text{old RTTs}) + a \times (\text{new RTT})$$

In the formula, $0<a<1$, if a is close to 0, it means that the RTT value is updated slowly. If a is close to 1, it means that the RTT value is updated faster. The recommended a value of RFC 2988 is 1/8, which is 0.125.

(4) Retransmission timeout.

The retransmission timeout (RTO) should be slightly larger than the weighted average round trip time RTTs obtained above. RFC 2988 recommends using the following formula to calculate RTO:

$$\text{RTO} = \text{RTT} + 4 \times \text{RTTD}$$

The RTTD is a weighted average of the deviations of the RTT. RFC 2988 recommends calculating RTTD in this way: the RTTD value is taken as half of the measured RTT sample value for the first measurement. In subsequent measurements, the weighted average RTTD is calculated using the following equation:

$$\text{New RTTD} = (1 - \beta) \times (\text{old RTTD}) + \beta \times |\text{RTTS} - \text{new RTT}|$$

β is a coefficient less than 1, and its recommended value is 1/4, which is 0.25.

(5) RTT: Karn algorithm.

When calculating the average RTT, as long as the segment is retransmitted, the round trip time samples are not used, so the weighted average round trip time RTTs and the timeout retransmission time RTO are more accurate.

(6) The modified Kam algorithm retransmits the segment once. It increases the RTO by a bit:

$$\text{New RTO} = r \times (\text{old RTO})$$

The typical value of the coefficient r is 2.

When retransmission does not occur again, average RTT and RTO are updated according to the round-trip delay of the segment.

7.4.3.3 SACK

(1) The receiver receives two byte blocks which are discontinuous from the previous word stream. If these byte ordinal numbers are in the receiving window, the receiver will accept the data, but we should tell the sender exactly that information so that the sender will not send the data again.

(2) Regulations of RFC 2018.

If you want to use selection validation, add the "Allow SACK" option in the TCP header when you create a TCP connection. The usage of confirmation number in the original header remains unchanged.

7.4.4 Connection management

There are three stages in transportation connection: connection establishment, data transmission and connection release. The management of transportation connection is to make the establishment and release of transportation connections normal. The following three steps should be taken in the process of connection establishment.

(1) Ensure that each party is able to know the others, existence.

(2) Some parameters (such as maximum message segment length, maximum window size, quality of service, etc.) should be allowed to be negotiated between the two sides.

(3) It can allocate resources to transport entity (such as cache size, items in connection table, etc.).

TCP connections are established using a client server. The application process that initiates the establishment of a connection is called a client. An application process that passively waits for connection establishment is called a server.

7.4.4.1 Establish a TCP connection with a three-way handshake.

In Figure 7-7, A's TCP sends the connection request message section to B, in its header. Synchronous bit SYN = 1, and select serial number seq = x, indicating transmission. The number of the first data bytes at the time of the data is x.

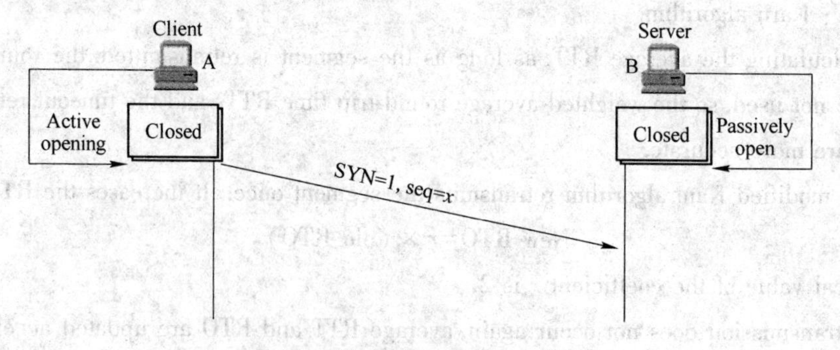

Figure 7-7 TCP connection (1)

In Figure 7-8, if B's TCP receives the connection request segment, and if it agrees, then return confirmation. B should make SYN = 1 in the confirmation message segment, making ACK = 1. Its confirmation number is ACK = x 1, and its serial number seq = y.

In Figure 7-9, a receives the message segment and gives confirmation to B, ACK = 1. Confir-

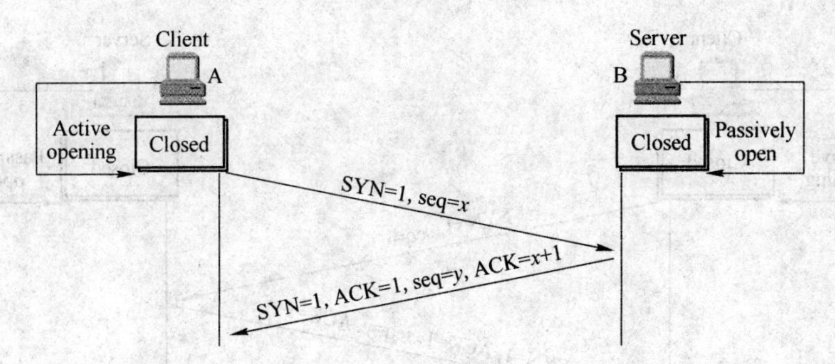

Figure 7-8 TCP connection (2)

mation number ACK = y 1. A's TCP notifies the upper application process, and the connection has been established.

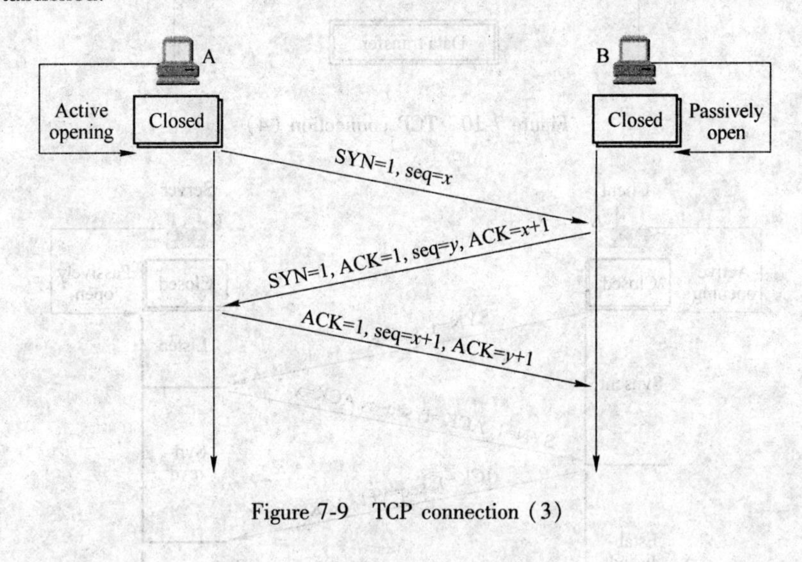

Figure 7-9 TCP connection (3)

In Figure 7-10, B's TCP receives the confirmation of Host A, and the upper layer is notified. Application process: the TCP connection has been established.

In Figure 7-11, establishing the status of a TCP connection with a three-way handshake.

7.4.4.2 The release of TCP's connection

In Figure 7-12, after the data transmission ends, both sides of the communication can release the connection. Now the application process of A sends a connection to its TCP to release the message segment, and stops sending the data again, and closes the TCP connection on its own initiative. A connects the release of the header of the message segment FIN = 1, its serial number seq = u, waiting for the confirmation of B.

In Figure 7-13, B issued a confirmation, confirmation number ACK = u = 1. The number of the message segment is seq = v. The TCP server process notifies the high-level application process. From A to B, the connection in this direction is released, and the TCP connection is in a

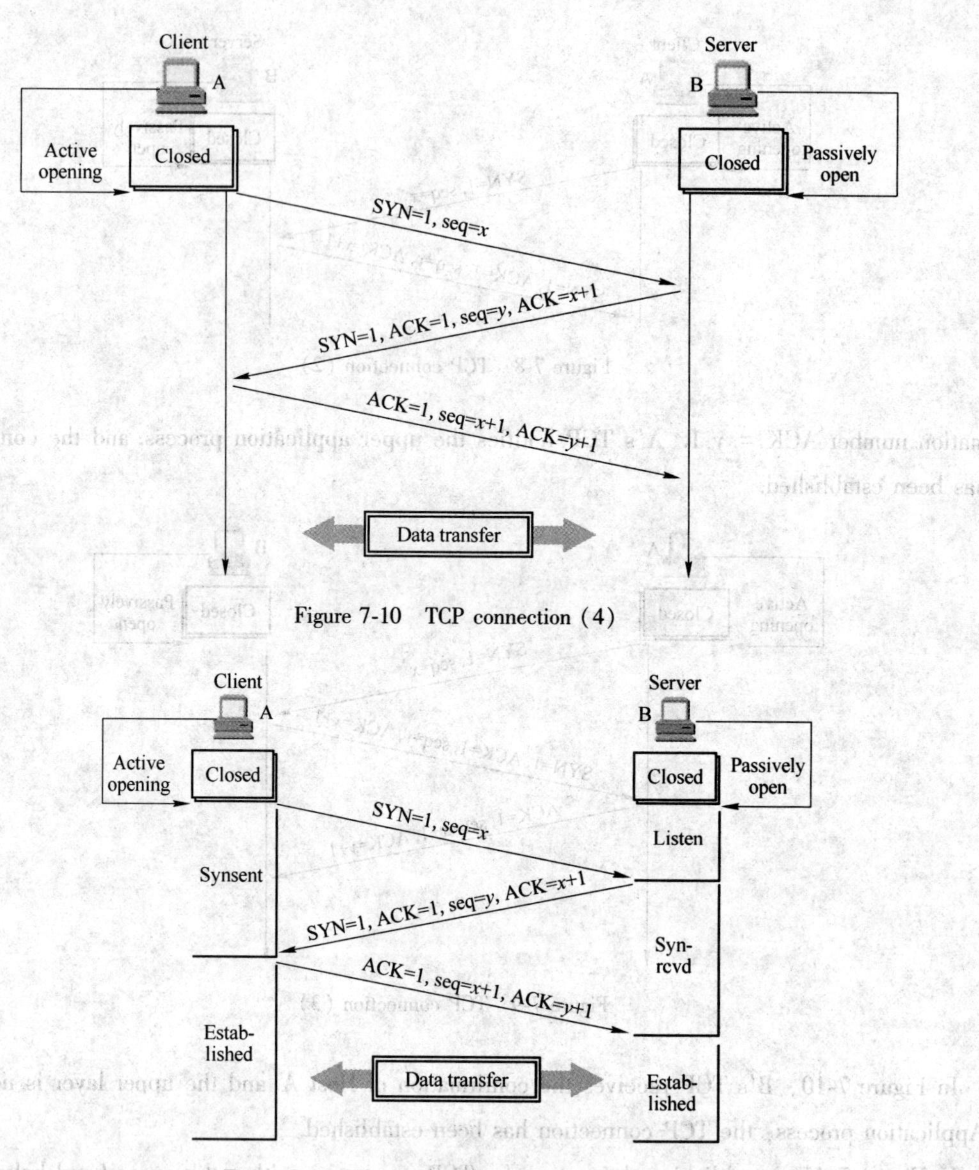

Figure 7-10 TCP connection (4)

Figure 7-11 The last status of a TCP connection

semi-closed state. If B sends the data, A will still receive it.

In Figure 7-14, if B has no data to send to A, its application process will notify TCP to release the connection.

In Figure 7-15, A must receive confirmation after receiving the release-connection message segment.

In Figure 7-16, in the confirmation message segment, ACK = 1, confirmation number ACK = $w + 1$, and its serial number seq = $u + 1$.

In Figure 7-17, the TCP connection must be released after a delay of 2MSL for two reasons. Firstly, this ensures that the last ACK segment sent by A can reach B. Secondly, it prevents the "failed connection request message segment" from appearing in this connection. After A sends the

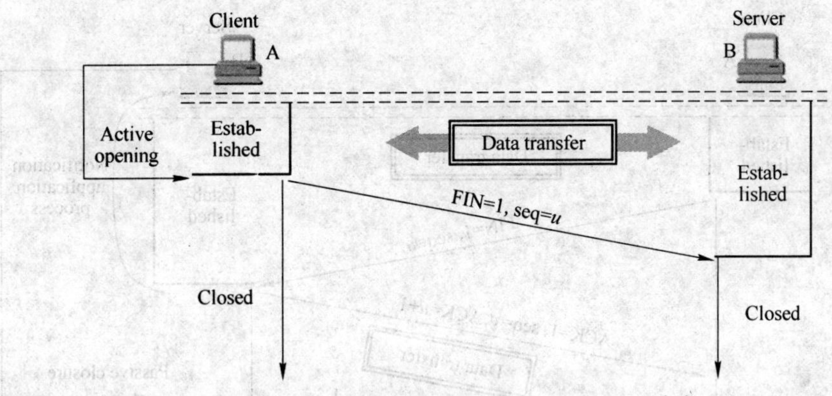

Figure 7-12 The release of TCP's connection (1)

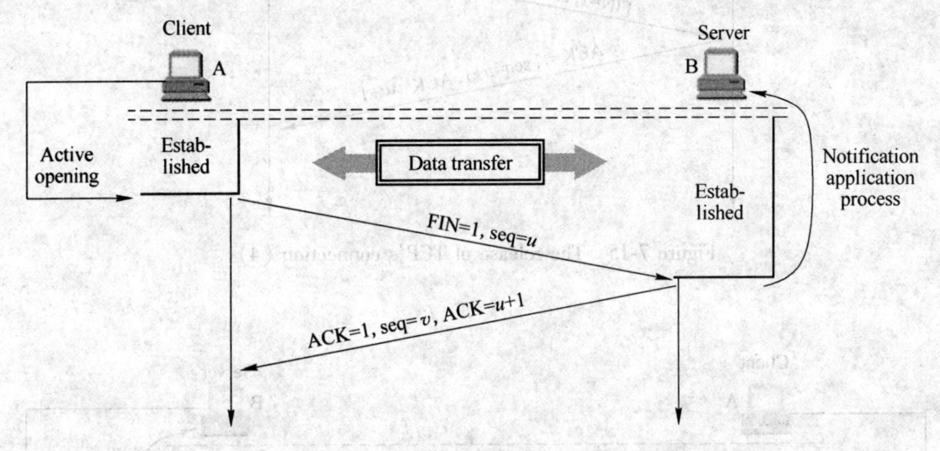

Figure 7-13 The release of TCP's connection (2)

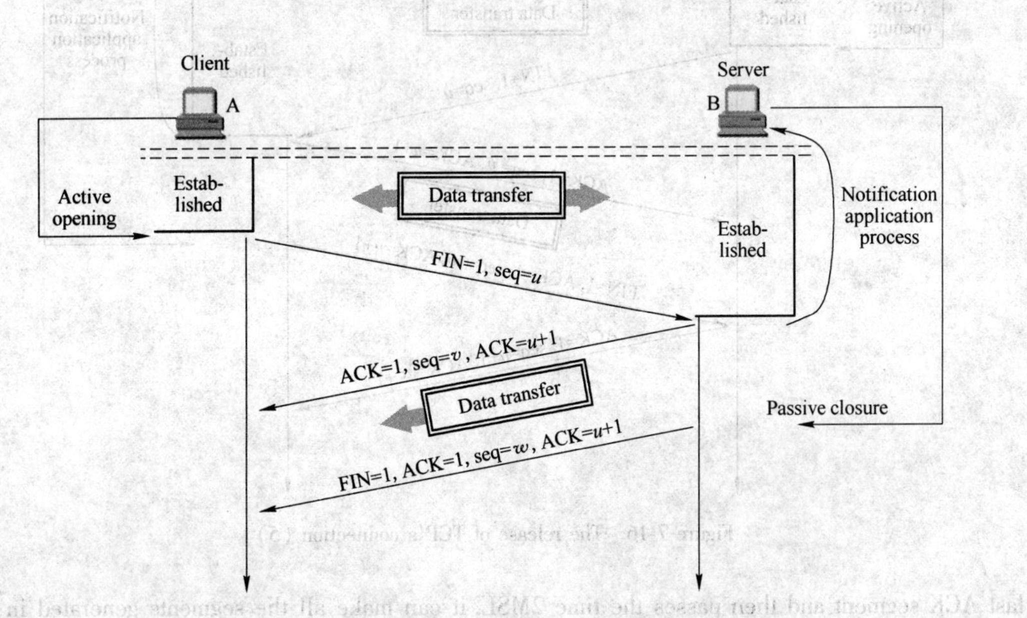

Figure 7-14 The release of TCP's connection (3)

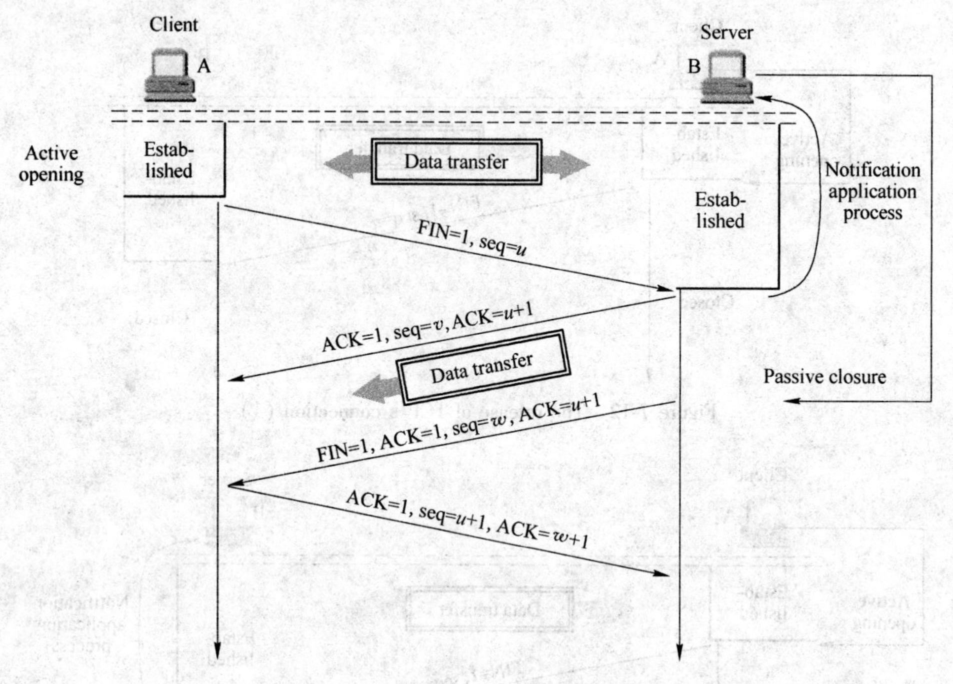

Figure 7-15 The release of TCP's connection (4)

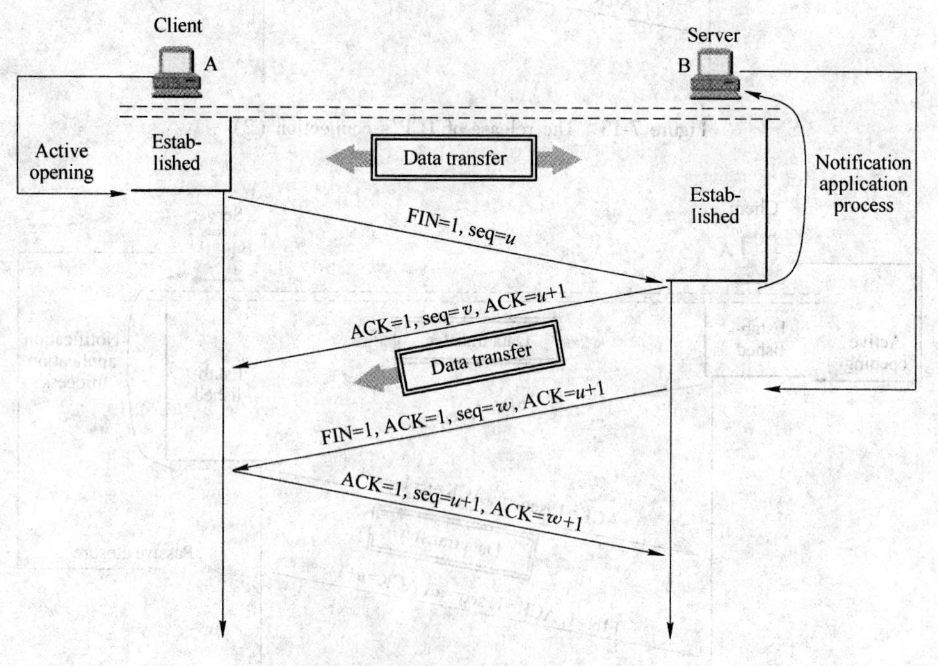

Figure 7-16 The release of TCP's connection (5)

last ACK segment and then passes the time 2MSL, it can make all the segments generated in the duration of the connection disappear from the network.

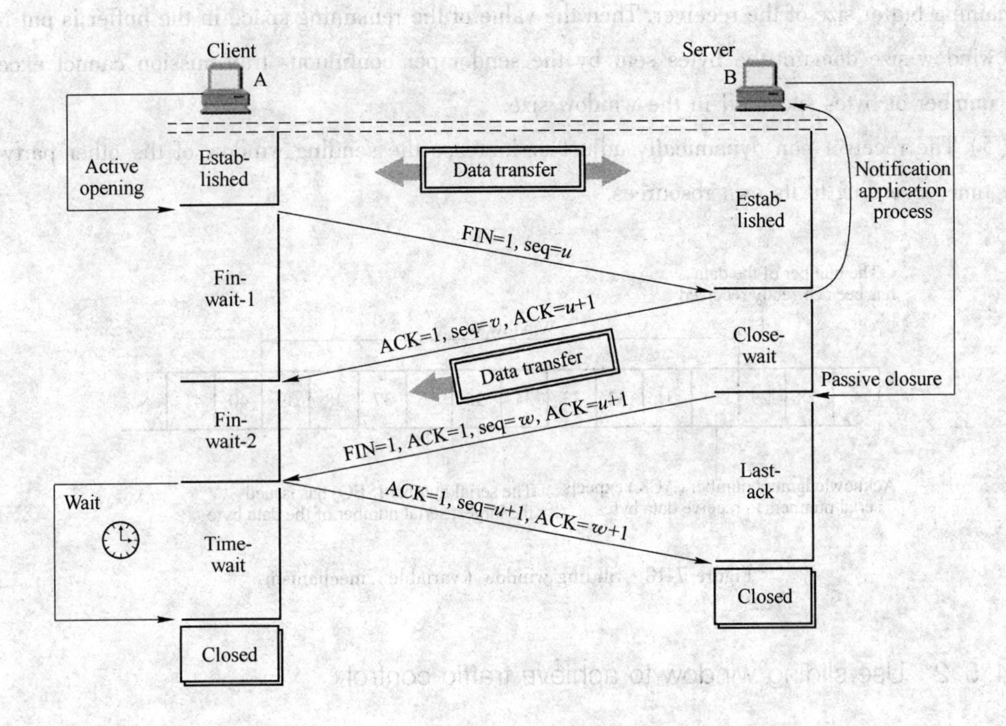

Figure 7-17 TCP connections must be delayed to truly release

TCP creates a connection using a three-way handshake process. During the connection creation process, many parameters are initialized; for example, the sequence number is initialized to ensure the robustness of sequential transmission and connection.

7.4.5 Traffic control

Traffic control limits the sender's sending rate from being too fast, so that the receiver can accept it. The sliding window mechanism makes it easy to control the flow of the sender on the TCP connection. The window unit of TCP is byte, not a segment, and the sender's window cannot exceed the value of the receiving window given by the receiver. TCP is a serial number for each byte of the data transmitted. The TCP sequence number field is 32bits, ensuring that the serial number is reusable.

7.4.5.1 Sliding window (variable) mechanism (as shown in Figure 7-18)

(1) TCP is a serial number for each byte of the data transmitted.

(2) TCP serial number field is 32bits, ensuring that the serial number can be reused.

(3) When a connection is established, a receive buffer is allocated for each connection at each end.

(4) When the data arrives, the receiver sends acknowledgement information, which includes the

remaining buffer size of the receiver. Then the value of the remaining space in the buffer is put into the window size domain, the bytes sent by the sender per continuous transmission cannot exceed the number of bytes specified in the window size.

(5) The receiver can dynamically adjust or increase the sending window of the other party at any time according to its own resources.

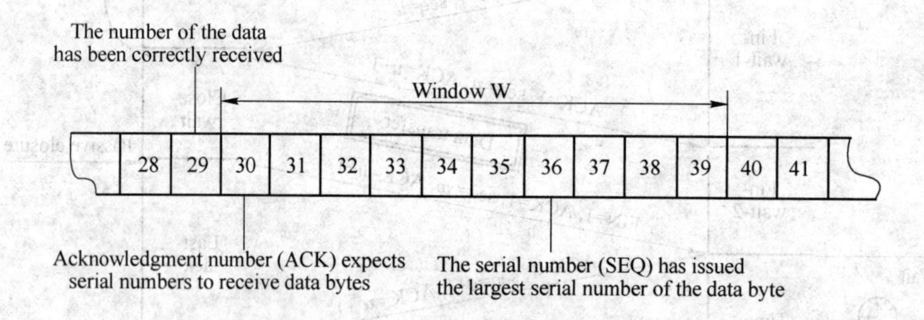

Figure 7-18 Sliding window (variable) mechanism

7.4.5.2 Use sliding window to achieve traffic control

(1) In general, it is always desirable to transfer data faster. However, if the sender sends the data too fast, the receiver may not have time to receive it, which will result in data loss. Traffic control is to prevent the sender's sending rate from being too fast, so that the receiver can receive it. Meanwhile, it ensures that the network is free of congestion. Traffic control can be easily implemented on a TCP connection using the sliding window mechanism.

(2) TCP has a duration timer for each connection. As long as the party connected to the TCP receives the zero window notification from the other party, the duration timer is started. If the time set by the duration timer expires, a zero window detection segment (only 1 byte of data is carried) is sent. The party gives the current window value when confirming this probe segment. If the window is still zero, then the party that receives this segment resets the duration timer. If the window is not zero, the deadlock can be broken.

7.4.5.3 Transmission efficiency must be considered

Different mechanisms can be used to control the timing of sending TCP segments.

(1) TCP maintains a variable equal to the maximum segment length MSS. As long as the data stored in the cache reaches the MSS byte, it is assembled into a TCP segment and sent out.

(2) The application process of the sender indicates that a message segment is required to be sent, that is, a push operation supported by TCP.

(3) The sender's timer expires, and the currently existing cached data is loaded into the segment (but the length cannot exceed MSS).

7.4.6 Congestion control

At some point, if the demand for a resource in the network exceeds the available part of the resource, the performance of the network will change. This is called congestion. Network congestion is often caused by many factors. Simply increasing the speed of the node processor or expanding the storage space of the node cache does not solve the congestion problem. The internet recommendation standard RFC 2581 defines four algorithms for congestion control, namely slow-start, congestion avoidance fast restrangsmit and fast recovery.

The rate of sending segment should be determined not only according to the receiving capacity of the receiver, but also considering the whole situation to avoid network congestion. This is determined by the two state quantities of the receiving window and the congestion window. The receiver window is the latest window value promised by the receiver according to the current size of the receiver buffer, and it is the traffic control from the receiver. The congestion window is a window value set by the sender according to its estimated degree of network congestion and it comes from traffic control of the sending side. Therefore, in addition to setting the receive notification window, the blocking window is set at the sending side:

Sending window = min (receiver announcement window, blocking window)

The size of the congestion window depends on the degree of congestion of the network and is changing dynamically. In the stable working state where congestion does not occur, the notification window and the congestion window of the receiving end are consistent.

Traffic control is the control of point-to-point traffic. It is the rate of controlling the sender's sending data so that the receiver can accept the data. However, since congestion control prevents excessive data injection into the network, it can overload routers or links in the network. Congestion control operates under the premise that the network can bear the load of the existing network.

7.4.7 Maintenance clock and timeout retransmission mechanism

One of the most important mechanisms in TCP reliability is handling data timeouts and retransmissions. The TCP protocol requires that each transmitter send a message segment at the transmitter, and then a timer will be started to wait for confirmation information, and the receiving end returns a confirmation message after successfully receiving new data. If the data cannot be confirmed before the timer timeout, TCP considers that the data in the packet segment has been lost or damaged, and needs to reorganize and retransmit the data in the packet segment. Although the concept of timeout retransmission is very simple, the mechanism of TCP handling timeout retransmission in implementation is quite complex compared with other reliable protocols.

TCP reliable data transmission protocol requires the receiver to confirm after receiving the TCP segment. The TCP can only use the ordinal number in the confirmation field to indicate that all the bytes are correctly received before the serial number, and there is no other denial of reply or selective retransmission function. These shows the necessity of carrying out the timeout retransmission.

For each connection, The TCP maintains a variable round-trip time RTT. The best estimated value of the round trip time of the sending end to the receiving end. When a segment is sent, the sender initiates a retransmission timer, which is used to measure the round-trip time from sending TCP to receiving acknowledgement. Retransmission occurs in case, it exceeds the estimated time interval. When acknowledgement is received, TCP obtains the current round-trip time from the timer, and uses the formula to estimate the new RTT value.

Problems

7-1 The following options describe TCP and UDP correctly ( ).

A. TCP is connection-oriented, such as dialing to establish a connection before making a call

B. TCP supports one-to-one, one-to-many, many-to-one and many-to-many interactive communication

C. TCP is oriented to byte streams. In fact, TCP regards data as a series of unstructured byte streams

D. UDP is connectionless, i. e. there is no need to establish a connection before sending data

7-2 The socket function does not return until it completes its task, which we call it ( ).

A. Message mechanism B. Event mechanism

C. Congestion method D. Non-congestion method

7-3 True or False.

(1) In some cases, only UDP can be used for network communication. ( )

(2) Sockets is a set of API provided by the operating system for network communication. ( )

(3) Send a UDP message, and the receiver may receive two UDP messages. ( )

(4) Send two TCP data, and the receiver may receive all the data at once. ( )

(5) TCP is only suitable for point-to-point communication. ( )

(6) The number of port numbers for UDP is less than that for TCP. ( )

7-4 Write the relationship between the transport layer and the network layer.

7-5 What is the sliding window protocol and what does it do?

7-6 Write the basic design idea of the stop-and-wait protocol.

7-7 Write the basic design idea of the selective retransmission protocol.

7-8 Compare the characteristics of three network protocols: stop-and-wait, Go-Back-N, and selective retransmission.

7-9 What is the three-way handshake which is established continuously?

7-10 How does the transport layer achieve error detection and error correction?

7-11 What is the maximum number of bytes in the data portion of a TCP segment and why is it? If the byte length of the data to be transmitted by the user exceeds the maximum sequence number that can be marked in the sequence number field in the TCP segment, can you still use TCP to transmit data?

7-12 What is the role of the flag bits SYN, ACK, FIN in the TCP protocol?

7-13 The communication channel bandwidth is 1 Gbit/s, the end-to-end delay is 10 ms, and the TCP send window is 65535 bytes. Please calculate the maximum throughput that can be achieved and the channel utilization rate.

7-14 What is a socket and where is it used?

7-15 What are the characteristics of the UDP protocol?

7-16 How are port numbers allocated and used?

7-17 The channel data transmission rate of a satellite is 50 kbit/s. The round-trip propagation delay is 500ms, and the length of the frame is 1000 bits.

(1) If the stop-and-wait protocol is used, t is used to indicate the time interval from the transmission of one frame of data to the transmission of the next frame. What is the value of it?

(2) What is the line utilization rate when using the stop-and-wait protocol?

(3) If the continuous ARQ protocol is used, how many frames can the sender send before receiving a response from a frame?

7-18 Why is there congestion in the network? How can we tell if there is congestion?

7-19 What is the difference between ports and sockets?

8 PROFINET

Goal:

(1) Understand the industrial Ethernet concepts.

(2) Master the PROFINET equipments.

(3) Be able to design and configure PROFINET.

8.1 Industrial Ethernet

With the continuous development of information technology, information exchange technology covers all walks of life. In the field of automation, more and more enterprises need to establish an integrated automation network management and control platform from the factory field equipment layer to the control layer and management level, and establish an enterprise information system based on industrial control network technology.

Industrial Ethernet provides an Ethernet standard for data transmission for manufacturing control networks. The technology is based on industry standards and utilizes a switched Ethernet architecture with high network security, operability and effectiveness to best meet the needs of users and manufacturers. Industrial Ethernet attracts more and more manufacturers with its unique low cost, high efficiency, high scalability and high intelligence.

8.1.1 Comparison of industrial Ethernet and traditional Ethernet

Comparison of industrial Ethernet and traditional Ethernet is shown in Table 8-1.

Industrial Ethernet products must be designed and manufactured to meet the needs of industrial network applications. Industrial site requirements for industrial Ethernet products include:

(1) The high temperature, humidity, air pollution and corrosive gas in the industrial production site environment require industrial grade products to have climatic environment adaptability, corrosion resistance, and dustproof and waterproof properties.

Table 8-1 Comparison of industrial Ethernet and traditional Ethernet

Class	Traditional Ethernet	Industrial Ethernet
Application	Ordinary office	Industrial conditions with harsh working conditions and high anti-interference requirements
Topology	Bus, ring, tree, etc.	Bus, ring, tree, etc; And easy to combine and convert various structures; simple installation; maximum flexibility and modularity; high expansion capability
Availability	General practical requirements to allow network failure time in seconds or minutes	Extremely high practicality; allowing network failure time <300ms to avoid production pauses
Network monitoring and maintenance	Network monitoring must be done by dedicated personnel using special tools	Network monitoring becomes part of the factory monitoring; network modules can be monitored by HMI software (such as WinCC), and faulty modules are easy to replace

(2) The presence of dust, flammable, explosive and toxic gases at industrial production sites requires explosion-proof measures to ensure safe production.

(3) The vibration and electromagnetic interference of the industrial production site are large, and the industrial control network must have mechanical environment adaptability (such as vibration resistance and impact resistance), electromagnetic environment adaptability or electromagnetic compatibility (EMC).

(4) The power supply of industrial network devices is usually based on the low-voltage DC power supply standard in the cabinet. In most industrial environments, the power required in the control cabinet is 24V DC.

(5) It is installed by standard rails and is easy to install. It is suitable for industrial environment installation. Industrial network devices should be easily installed in industrial field control cabinets and easily replaced.

8. 1. 2 Technical characteristics of industrial Ethernet

8. 1. 2. 1 Certainty and real-time

Since the Ethernet MAC layer protocol is CSMA/CD, this protocol makes collisions on the network, especially when the network load is too large. For an industrial network, if there are a large number of conflicts, it is necessary to retransmit the data multiple times, so that the uncertainty of the communication between the networks is greatly increased. In an industrial control network, this uncertainty from one place to another will inevitably lead to a reduction in system control performance.

In order to improve network congestion when the Ethernet load is heavy, an Ethernet switch can be used. It uses an effective collision domain partitioning technique that will share the local city network. Switching between the various conflict domains is a measure used to reduce conflicts and error transmission caused by the CSMA/CD mechanism. This can avoid conflicts and improve the certainty of the system.

In industrial control systems, real-time can be defined as the measurability of the system's response time to an event. That is, after an event occurs, the system must respond within a time frame that can be accurately foreseen. However, the real-time requirements for the delivery of data in the industry are very strict, and often the data is updated in tens of milliseconds. Also, due to the CSMA/CD mechanism of Ethernet, when a collision occurs, the data has to be retransmitted, and it can be tried up to 16 times. Obviously, this mechanism for conflict resolution comes at the cost of time. Moreover, once the line is dropped, even in just a few seconds, it may cause the entire production to stop or even cause equipment and personal safety accidents.

Real-time improvements in industrial Ethernet can be implemented from the device node side and the switch side that transmits data. The high-performance chip is used to design the embedded controller at the device node. Before the data transmission, the WBTPE (waiting priority Ethernet before transmission) algorithm is applied to make the important information priority control channel acquisition, realizing the regionalization of industrial data acquisition and closed-loop control, and avoiding overloading the network. In the process of industrial Ethernet data transmission, use IEEE802. 1P&Q-compliant switches to segment the collision domains and use its priority policy to improve the real-time exchange of emergency data again.

8. 1. 2. 2 Stability and reliability

Since Ethernet was an original design, it was not based on industrial network applications. When it is applied to industrial sites, facing harsh working conditions, severe line interference, etc. , these will inevitably lead to reduced reliability. In a production environment, industrial networks must have good reliability, recoverability, and maintainability. That is to say, if any component in a network system fails, it will not cause the crash of the application, operating system, or even the network system.

In order to solve this problem, we propose solutions from two aspects.

8. 1. 2. 3 Hardware device solutions

In terms of combating harsh environments, industrial Ethernet components and interfaces all meet industrial requirements, and are corrosion-resistant, dust-proof, and waterproof. For example, RJ-45 or DB9 with ruggedized RJ-45 interface is used instead of the general commercial RJ-45 interface; Industrial Ethernet equipment can operate over a wider temperature range: −40 to +85℃; electromagnetic compatibility standards comply with EN50081-2, the industrial grade EMC standard; MTBF can last at least 10 years.

Compared to ordinary commercial Ethernet, the hardware configuration of the industrial Ethernet control network has the following changes:

(1) Use high speed switches instead of hubs and bridges. In Ethernet, broadcast messages sent from any node are sent to devices on all hub ports. It does not isolate the broadcast area and can cause severe congestion. The switch internally establishes and maintains a mapping table of MAC addresses and ports, and configures virtual LAN functions (VLANs) to divide several ports into one group. Thus, packets sent from a certain port of the group only can be sent to other ports in this group, effectively isolating the port conflict area. In addition, the switch can also temporarily protect the data for the short-term competition for the same output port. In short, the use of switches greatly reduces the occurrence of conflicts, reduces network load, and effectively prevents network crashes caused by excessive load, improving network efficiency and improving network system reliability.

(2) Use Gigabit Ethernet. Gigabit Ethernet has 10 times the line speed of traditional 100M Ethernet, which can fully meet the requirements of high bandwidth application. 1000Mbit/s Ethernet has higher data collision recovery capability than 100Mbit/s Ethernet. The data collision back-off waiting time of the former is one tenth of the latter. When the network is heavily loaded and data collisions often occur, the performance of Gigabit Ethernet is significantly better than that of 100M Ethernet. In the case of less serious data conflicts, Gigabit Ethernet can provide more bandwidth and support more network load.

(3) Use a dual redundant network card. Each node in the network uses dual NICs for failover redundancy. Each node is connected to two switches (one of which acts as a hot backup for the other), and the switches are connected by cascading lines. In the dual NIC, the main NIC monitors whether the network connection is normal. The dedicated driver software running on it monitors the working status of the NIC. If the link fails due to a connection line, switch or NIC failure, the monitoring software will automatically transfer the MAC address and all connections from the primary NIC to the alternate NIC, and the information of the alternate NIC is broadcasted and the link is re-established through the standby switch to continue the session. Dual redundant network cards greatly increase the reliability of the entire control network.

8.1.2.4 Communication protocol solution

(1) Real-time Ethernet media access control protocol (RT-CSMA/CD). It is an improved Ethernet MAC protocol for industrial control, which is based on the three-time conditions that must be met according to the real-time nature of the communication subnet. The RT-CSMA/CD utilizes channel collisions to notify non-real-time nodes or low-priority real-time nodes to stop transmission, leaving the communication channel to high-priority real-time nodes. The protocol is a deterministic real-time communication protocol compatible with the common Ethernet MAC protocol standard, so that standard Ethernet nodes can be directly connected to the system for communication without any modification, which improves the real-time performance of industrial Ether-

net. With reliability, compatibility is guaranteed.

(2) Quality of service (QoS). IP QoS refers to the quality of service of IP, that is, the performance of IP data flows through the network. Its purpose is to provide end-to-end service quality assurance to users and improve system reliability. Using QoS technology in industrial Ethernet, it is possible to identify and prioritize data with higher priority from the control layer to improve real-time performance.

(3) IEEE 1588 network measurement and control system accurate clock synchronization protocol. It defines a precision time protocol (PTP) for sub-microseconds synchronization of clocks in sensors, actuators, and other end devices in industrial Ethernet or other distributed bus systems that use multicast technology. Its main principle is to periodically correct the clocks of all nodes in the network through a synchronization signal, so that the most accurate clock is synchronized with other clocks, so that the entire control system based on industrial Ethernet can be accurately synchronized.

(4) Industrial Ethernet application layer protocol. Traditional industrial fieldbus support vendors are committed to providing their own industrial Ethernet solutions in an effort to achieve the reliability and real-time requirements of industrial control networks, while at the same time hoping to achieve interoperability.

8. 1. 2. 5 Network maintainability and information security

On the one hand, compared with the complexity of the configuration and maintenance work of commercial Ethernet devices, industrial Ethernet devices must meet the convenience needs of users. Therefore, most industrial Ethernet devices currently support the Web mode, or the OPC method directly reflect the. information of the industrial Ethernet device is directly reflected on the HMI. In this way, engineers can easily maintain the network.

On the other hand, on the current world stage, various competitions are fierce. For many enterprises, especially those with leading technology, the production process that is embodied in their technology is often the fundamental interest of the company. The process flow and even the operating parameters of some key production processes may become the target of theft by the opponent, so it is stolen in the data transmission of industrial Ethernet.

Open interconnection is the advantage of industrial Ethernet. Remote monitoring, control, debugging, diagnosis, etc. greatly weaken the distribution and flexibility of control, breaking the limitations of time and space, but for these applications the legality of authorization and Examineability must be guaranted.

At present, the switch security technologies adopted by the information layer network mainly include the following:

Flow control technology limits the abnormal flow through the port to a certain range. Access control list (ACL) technology, which controls access and output control of network resources to ensure that network devices are not illegally accessed or used as attack springboards. The security socket layer (SSL) encrypts all HTP traffic, allowing access to the browser-based management GU on the switch. 802.1x and RADIUS network logins control port-based access for authentication and accountability. Source port filtering only allows designated ports to communicate with each other. Secure Shell (SSHv1/SSHv2) encrypts all data and ensures secure CLI remote access on the IP network. Secure FIP implements secure file transfer with the switch to avoid unwanted file downloads or unauthorized switch configuration file copying. However, there are still many practical problems in applying these security functions. For example, the traffic control function of the switch can only limit the rate of various types of traffic passing through the port, and limit the abnormal traffic of broadcast and multicast to a certain range. It is important to differentiate which are normal traffic and which are abnormal traffic. At the same time, how to set a suitable threshold is also difficult. Some switches have ACLs, but it is still useless if the ASIC supports fewer ACLs. A general switch cannot perform special processing on illegal ARP (the source destination MAC address is a broadcast address). Whether there are route fraud, spanning tree fraud and 802.1x DoS attacks, and DoS attacks on the switch network management system are all potential threats to the switch.

At the control layer, IE switches can learn from these security technologies on the one hand, but it must also be realized that IE switches are mainly used for fast forwarding of data packets, emphasizing forwarding performance to improve real-time performance. The application of these security technologies will face great difficulties in real-time and cost. At present, the application and design of industrial Ethernet is mainly based on engineering practice and experience. The network mainly includes control systems and operation stations, optimized system workstations, advanced control workstations. For data transmission between devices such as the database server, the network load is stable, and has a certain periodicity. However, with the need for system integration and expansion, the application of IT technology in automation system components, and the popularity of B/S monitoring methods, it is necessary to study the usability under network security factors, such as industrial Ethernet under burst traffic. Other salient issues include buffer capacity issues of network switches and the impact of changing from full-duplex switching to sharing on existing network performance. Therefore, on the other hand, industrial Ethernet must start with its own architecture and deal with it.

8.2 PROFINET

Launched by PROFIBUS International (PI), PROFINET is the next generation of automation bus standards based on industrial Ethernet technology. As a strategic technological innovation, PROFINET provides a complete network solution for the field of automation communications, covering topics such as real-time Ethernet, motion control, distributed automation fail-safe and network security, and is currently a hot topic in automation. As a cross-vendor technology, it is fully compatible with industrial Ethernet and existing fieldbus (eg. PROFIBUS) technologies to protect existing investments.

PROFINET is a complete solution for different needs, including 8 main modules, namely real-time communication, distributed field devices, motion control, distributed automation, network installation, IT-standard and information security, fail-safe and process automation.

8.2.1 PROFINET

8.2.1.1 PROFINET real-time communication

Depending on the response time, PROFINET supports the following three communication methods:

(1) TCP/IP standard communication: PROFINET is based on industrial Ethernet technology and uses TCP/IP and IT standards. TCP/IP is the de facto standard in the field of communication protocols. Although its response time is on the order of 100ms, the response time is sufficient for factory control level applications.

(2) Real-time (RT) communication: For data exchange between sensors and actuator devices, the system has stricter response time requirements, so PROFINET provides an optimized real-time communication channel based on Ethernet layer 2, through which the processing time of data is reduced in the communication stack. The typical response time of PROFINET real-time (RT) communication is 5~10ms.

(3) Isochronous real-time (IRT) communication: In field-level communication, the most demanding real-time communication is motion control. PROFINET's isochronous real-time (IRT) technology can meet the high-speed communication requirements of motion control. Under 100 nodes, the response time is less than 1ms and the jitter error is less than 1μs to ensure a timely and deterministic response.

8.2.1.2 PROFINET motion control

Control of the servo motion control system is easily achieved with PROFINET's isochronous real-time (IRT) function.

In PROFINET isochronous real-time communication, each communication cycle is divided into two different parts. One is the determined part of the loop, called the real-time channel; the other is the standard channel through which standard TCP/IP data is transmitted.

In the real-time channel, a time window of fixed-loop intervals is reserved for real-time data, and real-time data is always inserted in a fixed order. Therefore, real-time data is transmitted at a fixed interval, and the remaining time in the cycle is used to transmit standard TCP/IP data. Two different types of data can be transferred simultaneously on PROFINET without interfering with each other. Reliable control of the servo motion system is ensured by an independent real-time data channel.

8.2.1.3 PROFINET network installation

PROFINET supports star, bus and ring topologies. In order to reduce wiring costs and ensure high availability and flexibility, PROFINET offers a wide range of tools to help users easily install PROFINET. Specially designed industrial cables and durable connectors meet EMC and temperature requirements. They are also standardized within the PROFINET framework to ensure compatibility between different manufacturers' equipment.

8.2.1.4 PROFINET IT standards and network security

An important feature of PROFINET is the ability to simultaneously deliver real-time data and standard TCP/IP data. In the public channel that passes TCP/IP data, various proven IT technologies can be used (such as http, HTML, SNMP, DHCP, XML, etc.). When using PROFINET, we can use these IT standard services to enhance the management and maintenance of the entire network, which saves losts in commissioning and maintenance.

PROFINET realizes vertical communication integration from the field level to the management level. On the one hand, it facilitates the management to obtain field-level data. On the other hand, the data security problems originally existing in the management layer also extend to the field level. In order to ensure the safety of field-level control data, PROFINET offers a unique safety mechanism that protects the automation control system and minimizes the safety risks of the automation communication network by using dedicated safety modules.

8.2.1.5 PROFINET fail-safe

In the field of process automation, fail-safe is a very important concept. The so-called fail-safe means that when the system fails or a fatal error occurs, the system can be restored to a safe state (i.e., "zero" state). Here, safety has two meanings. On the one hand, it means the safety of the operator. On the other hand, it refers to the security of the entire system. In the field of process automation, a system failure or fatal error is likely to destroy the entire system or cause it to explode or destroy. The fail-safe mechanism is used to ensure that the system can automatically return to a safe state after a fault, without causing damage to the operator and the process control system.

PROFINET integrates the PROFT safe regulations to achieve the fail-safe SIL3 rating specified in IEC61508, which guarantees the safety of the entire system.

8.2.1.6 PROFINET and process automation

PROFINET can be used not only in factory automation but also in process automation applications.

With proxy server technology, PROFINET seamlessly integrates fieldbus PROFIBUS and other bus standards. Today, PROFIBUS is the world's only fieldbus standard that covers everything from factory automation to process automation. PROFINET with integrated PROFIBUS fieldbus solution is the perfect experience for applications in the field of process automation.

As an important part of the international standard EC61158, PROFINET is a completely open protocol. Members of PROFIBUS International have launched a large number of devices with PROFINET interface at the Hannover Fair in 2004, which promoted and popularized PROFINET technology. As time goes by, PROFINET will bring greater benefits and convenience to you and your automation control system as a new generation of industrial communication network standards for the future.

8.2.2 PROFINET IO

Within the totally integrated automation (TIA) framework, PROFINET IO is a logical further development of the following aspects: PROFIBUS DP, existing fieldbus, industrial Ethernet.

PROFINET IO has combined new concepts in common user operations and Ethernet technology based on 20 years of successful experience of PROFIBUS DP. This ensures the integration of PROFIBUS DP into the PROFINET domain. As an Ethernet-based automation technology standard for PROFIBUS/PROFINET international, PROFINET IO defines a cross-vendor communication platform, an automation system and engineering model.

The goals of PROFINET include:

(1) Industrial networking based on industrial Ethernet (open Ethernet standard);

(2) Compatibility of industrial Ethernet and standard Ethernet components;

(3) High stability with industrial Ethernet devices. Industrial Ethernet equipment is suitable for industrial environments (temperature, immunity, etc.);

(4) Use IT standards such as TCP/IP, http;

(5) Real-time function;

(6) Seamless integration with other fieldbus systems.

PROFINET communication is carried out via industrial Ethernet. The following transfer types are supported:

(1) Acyclic transmission of engineering and diagnostic data and interruption;

(2) Cyclic transmission of user data.

PROFINET IO communication takes place in real time.

The PROFINET devices of the SIMATIC product line have one or more PROFINET interfaces (Ethernet controllers/interfaces). The PROFINET interface has one or more ports (physical connection options). If the PROFINET interface has multiple ports, the device has an integrated switch. For PROFINET devices with two ports on one interface, the system can be configured as a linear or ring topology. A PROFINET device with three or more ports in an interface is suitable for setting up a tree topology. The naming properties and rules of the PROFINET interface and their representation are described below.

Each PROFINET device in the network is uniquely identified by its PROFINET interface. For this purpose, each PROFINET interface has:

(1) One MAC address (factory default);

(2) An IP address;

(3) PROFINET device name.

8. 3 SCALANCE X

8. 3. 1 Future-oriented switch

The Siemens totally integrated automation (TIA) concept has been successfully implemented countless times around the world, enabling a fully integrated solution through shared tools and standardized mechanisms. One of the most important foundational tasks is the development of industrial communication for SIMATICNET. SCALANCE is a milestone, a new active network component for building integrated networks. These active network components work perfectly together to create a network that is integrated, flexible, secure, and high performance in harsh industrial environments.

Industrial Ethernet switching SCALANCE X is an active network component that supports different network topologies: bus, star or ring fiber and electrical networks. These active network components can transfer data to a specified destination address.

The SCALANCE X product line is a new generation of SIMATIC NET industrial Ethernet switches. It consists of different modular product lines that are also suitable for PROFINET applications and work in conjunction with related automation tasks. Industrial Ethernet is a high-performance local area network that complies with IEEE802. 3 (Ethernet) and 802. 11 (wireless local area network) standards. With industrial Ethernet, users can build high-performance, wide-range communication networks. With the SCALANCE X switch, users can use the open industrial Ethernet standard PROFINET for real-time communication up to the field level.

The SCALANCE X industrial Ethernet switch has the following advantages:

(1) Robust, innovative and space-saving housing design, easy integration into SMAC solutions (standard 35 mm DIN rail, S7-300 DIN rail or direct wall mounting).

(2) Quick standard part design, as well as the PROFINET industrial Ethernet connection plug FastConnect RJ45 180 plug to remove stress and torque.

(3) Siemens HSR (high-speed redundancy technology), for SCALANCE X-200, SCALANCE X-300 or SCALANCE X-400. When the ring network structure consists of up to 50 switches, the network reconstruction time will be less than 0. 3s.

(4) Siemens stand-by technology for redundancy between ring networks of SCALANCE X-300/400 or SCALANCE X-200 IRT.

SCALANCE X is a new series of SIMATIC NET products with the following product lines.

8. 3. 1. 1 SCALANCE X-005 entry level switch and XB economy switch

The X-005 unmanaged switch with five RJ45 ports and diagnostics is available on small machine islands. Figure 8-1 SCALANCE X-055 and XB economical switch shows the switch of this model.

Figure 8-1 SCALANCE X-055 and XB economical switch

XB economy switch includes multiple port types and provides more interfaces. Expand the scope of X-005 mitigation and reduce the investment cost. Gigabit Ethernet links are also available.

8. 3. 1. 2 SCALANCE X-100 unmanaged switch

The SCALANCE X-100 unmanaged switch has electrical and fiber optic ports, and redundant power supplies and signal contacts are available for device-level applications. As shown in Figure 8-2 SCALANCE X-100 unmanaged switch.

Figure 8-2 SCALANCE X-100 unmanaged switch

8. 3. 1. 3 SCALANCE X-200 managed switch

SCALANCE managed switches are used for the device layer of plant-wide networking applications. With the STEP7 tool, the configuration and remote diagnostics functions can be integrated in the SIMATIC STEP7. This increases the level of plant availability. Devices with a high degree of protection do not need to be installed in the control cabinet. The XF series products are identical in function to the X-200 counterpart models, but with a more simplified structural design, saving the installation space of the control cabinet. This is shown in Figure 8-3 SCALANCE X-200 unmanaged switch.

Figure 8-3 SCALANCE X-200 unmanaged switch

For plant networks with enhanced real-time requirements and maximum availability requirements, the corresponding isochronous switches can be used. The SCALANCE X-200 IRT series switches are shown in Figure 8-4 SCALANCE X-200 series switch.

Figure 8-4 SCALANCE X-200 series switch

8.3.1.4 SCALANCE X-300 enhanced managed switch

The main application areas of the SCALANCE X-300 enhanced managed switches are the connection between the high-performance plant network and the enterprise network. The SCALANCE X-300 (excluding Layer 3 routing functions) enhanced network management product family combines the firmware functionality of the SCALANCE X-400 series and the compact structure of the SCALANCE X-200 series. Compared to SCALANCE X-200 switches, SCALANCE X-300 switches offer better management and better firmware. EEC products are specifically used in the power industry, and the switches have a more comprehensive power protocol certification, as shown in Figure 8-5 SCALANCE X-300 series switch and EEC production.

Figure 8-5 SCALANCE X-300 series switch and EEC production

The SCALANCE XR300 is rack-mounted and features a more flexible modular mounting structure. Compared with the traditional SCALANCE series, the XR300 series offers more ports and a more flexible combination, as shown in Figure 8-6.

Figure 8-6 The port of XR-300

8. 3. 1. 5 Modular SCALANCE X-400 switch

Modular SCALANCE X-400 switches are used in high-performance plant networks (eg factory networks with high-speed redundancy). Based on a modular structure, the switch can be precisely adjusted for the corresponding task. Thanks to support standards (eg VLAN, IGMP, RSTP), automation networks can be seamlessly integrated into existing office networks. The Modular SCALANCE X-400 switches us Layer3′s routing function to communicate between subnets of different IPs, as shown in Figure 8-7 SCALANCE X-400 series switch.

Figure 8-7 SCALANCE X-400 series switch

8. 3. 1. 6 SCALANCE X-500 core switch

Suitable for industrial-grade core networks that can be integrated into existing enterprise networks. From the control layer to the management layer, the SCALANCE X-500 core switches are well-used to provide high system availability while providing rich diagnostics and high-speed communication. The scalability of the base unit and the optional Layer 3 switching feature allow the network to be modified and extended at any time for specific applications. The switch builds a ring topology up to 10Gbit/s, regardless of whether it is an electrical or fiber link. The fault communication resumes the nominal time of 50ms, as shown in Figure 8-8 SCALANCE X-500 series switch.

Figure 8-8 SCALANCE X-500 series switch

8. 3. 2 SCALANCE X technical features

8. 3. 2. 1 SCALANCE X-005 entry level switch and XB economical switch

Industrial Ethernet X-005 entry-level switch and XB economical switch are not managed switch.

This product is inexpensive and can be constructed with small star or linear structures with switching functions in the machine or equipment island. The SCALANCE X-005′s metal body (IP30) is rugged and compact, suitable for installation in control cabinets, standard mounting rails, the S7-300, or directly on the wall. The SCALANCE XB series offers a more economical plastic housing as well as a wider selection of interface types.

The SCALANCE X-005 and XB series products have the following features:

(1) Device diagnostics via LED (power, link status, data communication);

(2) RJ45 port with fixed protection, designed for industrial Ethernet connection plugs that meet the requirements of PROFINET standard, IE FC RJ45 socket, with stress and torque relief;

(3) Integrated Auto-Crossing function, can use non-crossing cable;

(4) Automatic detection and negotiation of data transmission rate.

8.3.2.2 SCALANCE X-100 unmanaged switch

The SCALANCE X-100 series are unmanaged switches with up to 24 electrical ports and are ideal for on-site diagnostics of bus and star network configurations (10/100Mbit/s), Ethernet and related equipment applications. The rugged metal housing is ideal for industrial field environments and saves space when installed in a control cabinet. SCALANGE X-100 switches can be used both in electrical and electrical/optical forms, and when they are added to the network, fiber and electrical conversions can be achieved.

The SCALANCE X-100 has the following features:

(1) Device diagnosis through LED (power supply, link status, data communication) and signal contacts;

(2) Power supply redundancy;

(3) RJ45 port with fixed protection, specially designed for industrial Ethernet connection conforming to the requirements of PROFINET standard, IE FC RJ45 socket, with the function of removing stress and torque;

(4) Integrated Auto-Crossing function for non-crossover cables;

(5) Automatic detection and negotiation of data transmission rate.

8.3.2.3 SCALANCE X-200 managed switch

The SCALANCE X-200 series managed switches are ideal for Ethernet in bus and star configurations (10/100Mbit/s). The SCALANCE X-200 has a compact housing that saves space when installed in the control cabinet and is powerful enough to easily expand the network.

When combined with the SCALANCE X-400/X-200 IRT or OSM/ESM as a redundancy manager, the SCALANCE X-200 offers ring redundancy, so it provides efficient network reliability.

With the C-PLUG card, the device can be quickly replaced even without a programming device; it ensures that the configuration and data applications in the C-PLUG are transferred to another SCALANCE X without much prior knowledge. The SCALANCE X-200 switch series can be integrated into STEP7 to meet the PROFINET standard process and system diagnostics. The SCALANCE X-200 text changer is available in both electrical and electrical/optical forms. When the device is added to the network, electrical and fiber conversion is possible.

The SCALANCE X-200 has the following features:

(1) Device diagnosis through LED (power supply, link status, data communication) and signal contacts;

(2) Power redundancy;

(3) RJ45 port with fixed protection, specially designed for industrial Ethernet connection plugs complying with PROFINET standard, IE FC RJ45 socket, with the function of removing stress and torque;

(4) Compact structure;

(5) Integrated auto-crossing function, and it can use non-crossing and cable;

(6) Automatic detection and negotiation of data transmission rate;

(7) Remote diagnosis via signaling contacts, SNMP, web browser and PROFINET IO monitoring functions;

(8) Supports PROFINET RT communication, and integrates configuration and diagnostic functions in STEP7;

(9) Implementing the management of the network switch through SNMP;

(10) Automatic E-mail sending function;

(11) The C-PLUG exchange medium can be used to quickly replace the equipment;

(12) It supports redundant ring network, and can be used as redundancy manager.

8.3.2.4 SCALANCE X-200 IRT managed switch

The SCALANCE X-200 IRT switch can be used to configure isochronous real-time synchronization networks, and a network can withstand a large amount of real-time data communication and standard data communication (TCP/IP), thus avoiding the need for dual infrastructure.

High-speed redundancy for fast recovery response times (less than 0.3s) with 50 switches.

With the SCALANCE X-200 IRT switch, you can set up ring redundancy and make ring redundancy with other SCALANCE X. The SCALANCE N-200 IRT switch is available in both electrical and electrical/fiber versions.

The SCALANCE X-200 IRT has the following features:

(1) Device diagnostics via IED (power, link status, data communication) and signal contacts;

(2) Power redundancy;

(3) RJ45 port with fixed protection, designed for industrial Ethernet connection plugs complying with PROFINET standards, IE FC RJ45 socket, with stress and torque relief;

(4) Compact structure;

(5) Integrated auto-crossing function: it can use non-crossing and cable;

(6) Automatic detection and negotiation of data transmission rate;

(7) Remote diagnostics via signaling contacts, SNMP, web browser and PROFINET IO monitoring;

(8) Supports for PROFINET IRT communication and integrates configuration and diagnostics in STEP7;

(9) Implementing the management of the network switch through SNMP;

(10) Automatic E-mail sending function;

(11) The C-PLUG exchange medium can be used to quickly replace the equipment;

(12) It supports redundant ring network, and can be used as redundancy manager;

(13) Supports ring network redundancy.

8.3.2.5 SCALANCE X-300 bare management industrial Ethernet switch

The Gigabit Ethernet switches of the SCALANCE X-300 switch series are suitable for building high-performance Ethernet in bus, ring and star configurations (10/100/1000Mbit/s). Flexible to build fiber and/or electrical networks, SCALANCE X-300 switches produce efficient network reliability, support redundant network structures, integrate redundant controllers, and support redundant power supplies.

With the help of the C-PLUG card, the device can be quickly replaced even without the aid of a programmer. It can guarantee that C-PLUG can be used to transfer original configuration and data applications to another switch without professional operation knowledge. Gigabit Ethernet connection between SCALANCE X-300 switches can be established.

SCALANCE X-400 switches can be ideally suited, for example to handle systems such as the PCS7 control system. The following combinations of network topologies and topologies exist.

(1) Gigabit Ethernet ring network with high-speed medium redundancy; in order to prevent transmission link or switch failure, up to 50 SCALANCE X-300 switches are connected end-to-end in a ring network, and multi-mode fiber can be used to form a line with a total length up to 150km, or a single-mode fiber can be used to form a line of 1300km, and a long-distance fiber can reach a total length of 3500km. When a transmission path in a ring network or a SCALANCE X-300 switch fails, the transmission path can be reconstructed within 0.3s;

(2) Star structure with SCALANCE X-300 switches: each SCALANCE X-300 switch can be

connected to up to 10 nodes or subnets via electrical or optical ports;

(3) For the data terminal connected to the twisted pair, the speed of the network segment is 10/100/1000Mbit/s.

The SCALANCE X-300/XR300 has the following features:

(1) Device diagnostics via IED (power, link status, data communication) and signal contacts;

(2) Power redundancy;

(3) RJ45 port with fixed protection, designed for industrial Ethernet connection plugs complying with PROFINET standards, IE FC RJ45 socket, with stress and torque relief;

(4) Integrated auto-crossing function for non-crossover cables;

(5) Automatic detection and negotiation of data transmission rate;

(6) Integrated;

(7) Remote diagnosis via signal contacts, SNMP, web browser and PROFINET IO diagnostics;

(8) Support for PROFINET RT communication, configuration and diagnostic functions that can be integrated in SIMATIC STEP7;

(9) Manage the switch through SNMP;

(10) Automatic E-mail sending function;

(11) C-PLUG exchange medium for simple equipment replacement;

(12) Support for multicast and broadcast restrictions;

(13) Integrated with Enterprise Security Policy (Enterprise Security Policies) through support for VLANs;

(14) IGMP Snooping and IGMP Query support multicast filtering and restriction;

(15) Compatible with OSM, ESM and ELS SCALANCE X products to protect existing investments;

(16) Supports redundant ring network, can be used as redundancy manager;

(17) Supports ring network redundancy.

8.3.2.6 SCALANCE X-400 modular industrial Ethernet switch

The modular switch of the SCALANCE X-400 is an industrial Ethernet switch suitable for high-performance networks with a bus-type, star-ring structure (10/100/1000Mbit/s).

Flexible construction of electrical or fiber-optic industrial Ethernet: Network topologies, port types and quantities can be easily adapted to different network structures. Its benefits include high network reliability, support for redundant ring network architecture, integrated redundancy manager, redundant power supply and the ability to exchange and expand media modules during operation.

With the plug-in C-PLUG card, the device can be replaced without the help of a programmer. Configuration data or application data can be safely stored in the C-PLUG. Even without the expertise, these settings can be easily transferred the C-PLUG card to the replaced SCALANCE X-400.

(1) High-speed Ethernet and Gigabit Ethernet ring with high-speed media redundancy; To prevent transmission links or switch failures, up to 50 SCALANCE X-400 switches can be connected into a ring network, using a multimode fiber to form a line with a length. Alternatively of up to 150km a single-mode fiber can be used to form a line with a total length of 1300km, and using a long-distance fiber can reach 3500km. When a transmission path in the ring network or a SCALANCE X-400 service fails, the transmission path can be reconstructed within 0. 38.

(2) The SCALANCE X-400 supports redundant connections to the ring network structure and can be connected to the corporate network using fast spanning tree redundancy.

(3) With the star structure of the SCALANCE X-400 switches, each SCALANCE X-400 switch can be connected to up to 26 nodes or subnets via electrical or optical ports.

(4) A data terminal connected to a twisted pair. The network segment speed is 10/100/1000Mbit/s.

They have the following features:

(1) Device diagnostics via IED (power, link status, data communication) and signal contacts;

(2) Power redundancy;

(3) RJ45 port with fixed protection, specially designed for industrial Ethernet connection plugs complying with PROFINET standard, IE FC RJ45 socket, with the function of removing stress and torque;

(4) Integrated auto-crossing function for non-crossover cables;

(5) Automatic detection and negotiation of data transmission rate;

(6) Remote monitoring via SNMP, web browser and PROFINET IO monitoring via signal contacts;

(7) Support for PROFINET RT communication, integration of configuration and diagnostic functions in SIMATIC STEP7;

(8) Manages the switch through SNMP;

(9) Automatic E-mail sending function;

(10) C-PLUG card for simple device replacement;

(11) Support for ancestral broadcast and broadcast restrictions;

(12) Through the support of VLAN, it can be integrated into enterprise security polices ESP;

(13) Protocols IGMP Snooping and IGMP Query support multicast filtering and restriction;

(14) Compatible with SCALANCE X products with OSM, ESM and ELS to protect investment;

(15) Layer 3 routing function (SCALANCE X414-3E);

(16) Supports redundant ring network and can be used as redundancy manager;

(17) Supports ring network redundancy.

8.4 VLAN

8.4.1 Introduction

A VLAN is a logical network segment that divides a physical network into multiple logical workgroups as shown in Figure 8-9. This technology divides a LAN into multiple logical LANs—VLANs. Each VLAN is a broadcast domain. The communication between the devices in a VLAN is the same as in a LAN. Broadcast packets are restricted to one VLAN. Devices belonging to different VLANs cannot directly access each other, and communication between them depends on routing. The special advantage of VLANs is to reduce network load for nodes and other VLAN segments.

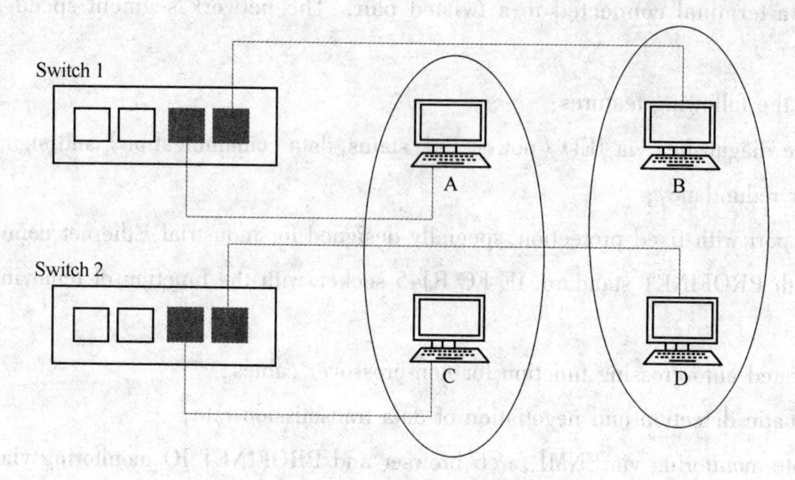

Figure 8-9 VLAN introduction

VLANs are defined in the IEEE 802. 1Q standard. These include VLAN based port, VLAN based MAC address, VLAN based network layer and VLAN based IP address. VLANs are usually marked with a VLAN ID and a VLAN name. According to the IEEE 802. 1Q standard, the VLAN ID is represented by 12 bits and can support 4096 VLANs. 1 ~ 1005 is the standard range, and 1025 ~ 4096 is the extension range. At present, a considerable number of switches only support the standard range 1 ~ 1005, and the VLAN ID that can be used for Ethernet is 1 ~ 1000, and 1002 ~ 1005 is the VLAN ID used by FDDI and Token Ring. The VLAN name is represented by 32 characters and can be letters and numbers. When creating a VLAN, you need to give a name. If no name is given, the system automatically creates the default VLAN name.

VLAN technical characteristics include:

(1) The VLAN works in the second layer of the OSI reference model (data link layer).

(2) Each VLAN is an independent logical network segment, an independent broadcast domain.

The broadcast information of a VLAN is sent only to members of the same VLAN and is not sent to other VLAN members.

(3) Each VLAN has a unique subnet number. VLANs cannot communicate directly and must be completed through Layer 3 routing.

VLANs are defined by using software to define VLAN members in the entire network. Currently, there are three commonly used methods for VLAN division.

8.4.1.1 Port-based VLAN

Port-based VLANs define VLAN members by switch port. Each switch port belongs to a VLAN. It is statically assigned by the network administrator to the VLAN's ports, which maintain the specified VLAN settings until the administrator changes it. Therefore, this method of dividing VLANs is also called a static VLAN.

The method of port-based VLAN division is very simple, effective, secure, and easy to monitor and manage. It is a commonly used VLAN division method.

8.4.1.2 Based on MAC address

VLANs based on MAC addresses are defined by the MAC address of each device connected to the switch device. Since it can divide VLANs by end users, it is often referred to as a user-based VLAN partitioning method.

This method of partitioning often requires a VLAN configuration server that holds the VLAN management database. When the workstation initially connects to a switch port that does not specify a VLAN, the switch checks the MAC address entry in the VLAN management database based on the end user's MAC address and dynamically sets the connection port and corresponding VLAN settings. Compared with port-based VLAN division, this method can be said to be a dynamic VLAN division. In a dynamically divided VLAN, the switch port can automatically set the VLAN.

The advantage of dynamic VLAN partitioning is that it reduces the management work in the wiring closet when new users or users move, but the network administrator maintains a correct VLAN management database. In addition, when a VLAN is used to divide a VLAN, a switch port may belong to multiple VLANs. A port must receive broadcast information of multiple VLANs, which may cause port congestion.

8.4.1.3 Based on layer 3 protocol type or address

This method allows you to define VLAN members according to the network layer protocol type (TCP/IP, IPX, DECNET, etc.) or you can define VLAN members by network address (logical address), for example the IP address of the TCP/IP protocol or the subnet. It is also a way of dynamic VLAN partitioning, except when the switch dynamically sets the VLAN, where it is based on the protocol type or logical address of the datagram instead of the MAC address. The method of dividing the amount of VLAN based on the layer 3 protocol type or address has the advantage of facilitating the formation of an application-based VLAN.

8.4.2 Port-based VLAN configuration

In order to understand the function and principle of VLAN, we will introduce how to configure port-based VLAN with switch. The topology diagram we used is shown in following Figure 8-10.

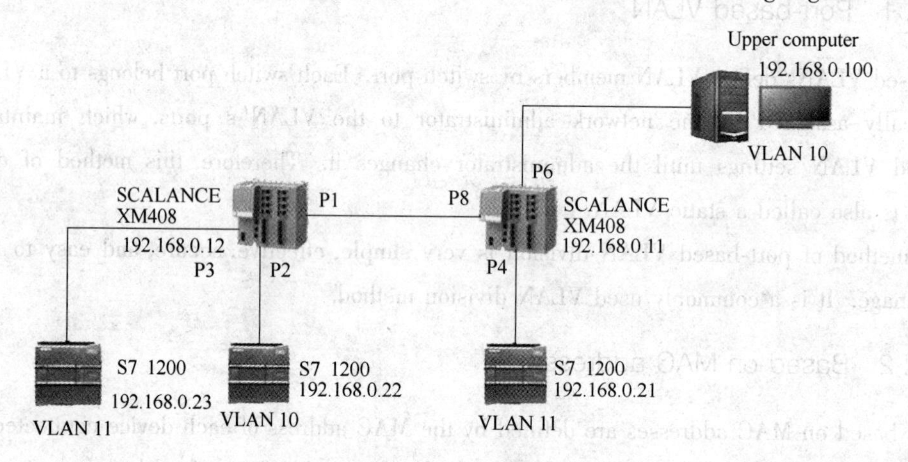

Figure 8-10 VLAN configuration topology

(1) Using industrial Ethernet cables, connect the modules according to the network structure logical topology and its port number.

We assume, except for P4, P6, and P8, which need to be assigned VLANs, other ports belong to VLAN 1 by default. First, connect the host computer to the P5 port of the XM408. Because the subsequent configuration divides the P6 port into VLAN 10, you can no longer access the XM408′s web interface through the P6 port. To access the web interface, you must pass the default VLAN Port 1.

(2) Configure the IP address of the host computer, switch and S7 1200 PLC with PST.

(3) Configure the VLAN for the SCALANCE XM408.

1) Enter the IP address 192.168.0.11 of the SCALANCE XM408 in the browser. After entering the username and password on the login screen that opens, enter the configuration interface. On the left of the configuration page, select the VLAN item under layer 2 and enter the VLAN configuration interface as shown in Figure 8-11.

2) Enter 10 in the "VLAN ID" edit box and click the "Create" button to create VLAN 10. In the row with ID=10, double-click under the "P1.6" title bar and select "u" in the pop-up list. Click the "Set Values" button as shown in Figure 8-11.

3) In the same way, enter 11 in the "VLAN ID" edit box to create VLAN 11. In the line with

VLAN ID = 11, double-click under the "P1.4" title bar and select "u" in the pop-up list. Click the "Set Values" button as shown in Figure 8-11.

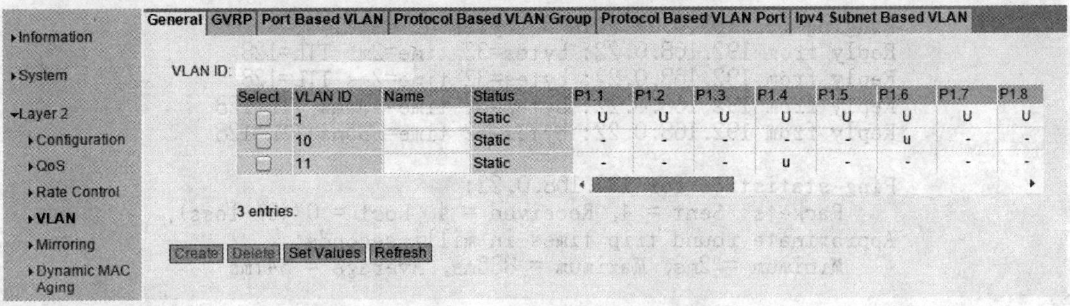

Figure 8-11 VLAN configuration interface

4) In the VLAN configuration interface, click the "Port Based VLAN" tab on the upper page, and find the "P1.6" row in the table on the page. In the "Port VID" column of the row, select VLAN 10, which would assign the P1.6 port of XM408 to VLAN 10. Similarly, set port P1.4 to VLAN 11. Finally click on the "Set Values" button as shown in Figure 8-12.

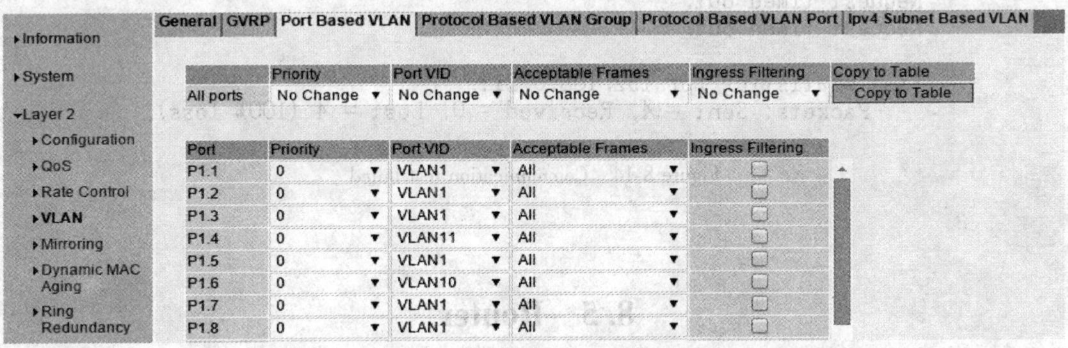

Figure 8-12 VLAN configuration interface

5) The port is configured in trunk mode. Generally, the port interconnected by the switch is configured as a trunk, allowing data from multiple VLANs to pass through the port. Click the General tab of the VLAN configuration interface. In the row with VLAN ID = 10, select M in the P1.8 list. Similarly, in the row with VLAN ID = 11, select "M" in the "P1.8" column. Finally click on the "Set Values" button as shown in Figure 8-11.

(4) Configure the VLAN for the SCALANCE XM408 (right) in a similar way to the XM408 (left).

(5) Communication test. The ping command is used to test the network. It can determine the connectivity of the network. The ping command sends a data packet to the destination host and waits for the response from the destination host. Based on the result, the path reliability, link delay time and whether the destination host can be reached can be known. So it determines whether the network is connected.

1) The host computer pings the IP address of the PLC in VLAN 10 and checks the communication result, as shown in Figure 8-13 .

```
Pinging 192.168.0.22 with 32 bytes of data:
Reply from 192.168.0.22: bytes=32 time=2ms TTL=128
Reply from 192.168.0.22: bytes=32 time=2ms TTL=128
Reply from 192.168.0.22: bytes=32 time=499ms TTL=128
Reply from 192.168.0.22: bytes=32 time=885ms TTL=128

Ping statistics for 192.168.0.22:
    Packets: Sent = 4, Received = 4, Lost = 0 (0% loss),
Approximate round trip times in milli-seconds:
    Minimum = 2ms, Maximum = 885ms, Average = 347ms
```

Figure 8-13 Communication test succeeded

2) The host computer pings the IP address of the PLC in VLAN 11 and checks the communication result, as shown in Figure 8-14.

```
Pinging 192.168.0.21 with 32 bytes of data:
Request timed out.
Request timed out.
Request timed out.
Request timed out.

Ping statistics for 192.168.0.21:
    Packets: Sent = 4, Received = 0, Lost = 4 (100% loss),
```

Figure 8-14 Communication test failed

8.5 Router

8.5.1 Introduction

A large industrial enterprise control system includes management, control and equipment layers. In order to ensure the convenience and security of different system management and control as well as the stability of the overall network operation, VLAN technology is usually used for virtual network division. For example, in the industrial control system, each production workshop and command dispatch center are divided into VLANs. By configuring routes to solve communication problems between different VLANs, each production workshop and the command and dispatch center can be interconnected.

Routers interconnect multiple different networks or segments to exchange data between different logical subnets. They can understand OSI layer 3 packets and protocols and can calculate the best path. Routing refers to the process by which a router receives a packet, analyzes the destination address information, selects a packet path, and then forwards the packet to the next router in the selected network. The packet arrives at the next device called a hop.

A gateway is essentially an entry point for a network to other networks. When the host finds that the destination address of the data packet is not in the local network (by determining the destination network number by the destination address and the gateway mask), the data packet is forwarded to the gateway for processing. Routing is the choice of different network paths for packets from source address to destination address.

A static route is a fixed routing table set in the router. Static routes do not change unless the network administrator intervenes. Because static routes cannot reflect changes in the network, they are generally used in networks with small network sizes and fixed topologies. The network administrator must know the topology connection of the router and specify the routing path manually. When the network topology changes, the NMS needs to manually modify the routing path. The advantages of static routing are simple, efficient, and reliable.

Routers can learn routing information from other routers through process called dynamic routing. Dynamic routing updates information from other routers. Dynamic routing requires the processing overhead of the router, but the management cost is low after the initial setup. (The static route has the highest priority among all routes. When a dynamic route conflicts with a static route, the static route takes precedence.)

8.5.2 Routing configuration

8.5.2.1 Local routing configuration

In order to understand the characteristics of local routes, we will introduce how to configure routes. The topology diagram we used is shown in Figure 8-15.

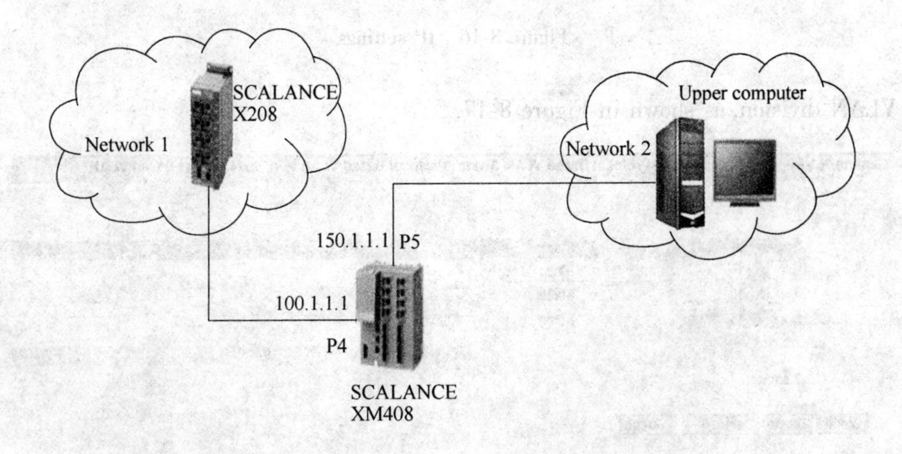

Figure 8-15 Routing configuration topology

(1) Configure the IP address and gateway of the host computer, XM408 switch and X208 switch. Configure the IP address of the host computer to be 150. 1. 0. 2 and the gateway to 150. 1. 1. 1 (the gateway must be set), as shown in Figure 8-16. Use the PST tool to configure the IP address of the SCALANCE XM408 to 192. 168. 0. 11 and the X208 to 100. 1. 0. 2. The gateway is set to 100. 1. 1. 1 (set the gateway through the Ind. Ethernet interface "Use router" function).

Internet 协议版本 4 (TCP/IPv4) Properties

General

You can get IP settings assigned automatically if your network supports this capability. Otherwise, you need to ask your network administrator for the appropriate IP settings.

Obtain an IP address automatically

Use the following IP address:

IP address: 150 . 1 . 0 . 2

Subnet mask: 255 . 255 . 0 . 0

Default gateway: 150 . 1 . 1 . 1

Obtain DNS server address automatically

Use the following DNS server addresses:

Preferred DNS server: . . .

Alternate DNS server: . . .

Validate settings upon exit Advanced...

Figure 8-16 IP settings

(2) VLAN division, as shown in Figure 8-17.

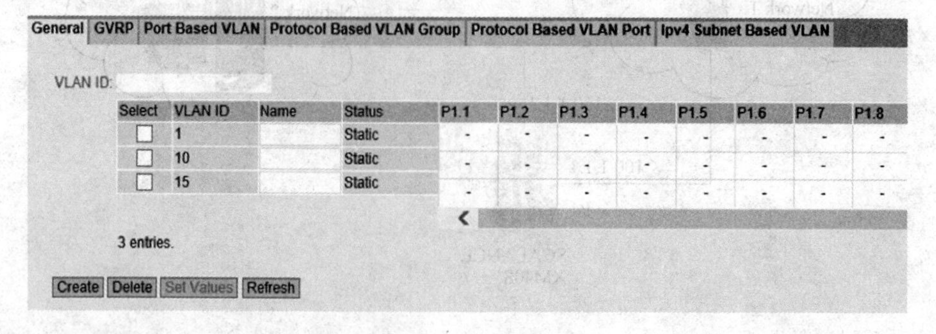

Figure 8-17 VLAN division

(3) Start routing.

In the directory tree "layer 3" → "configuration", check the check box in front of the "routing" item and click the "set values" button, as shown in Figure 8-18.

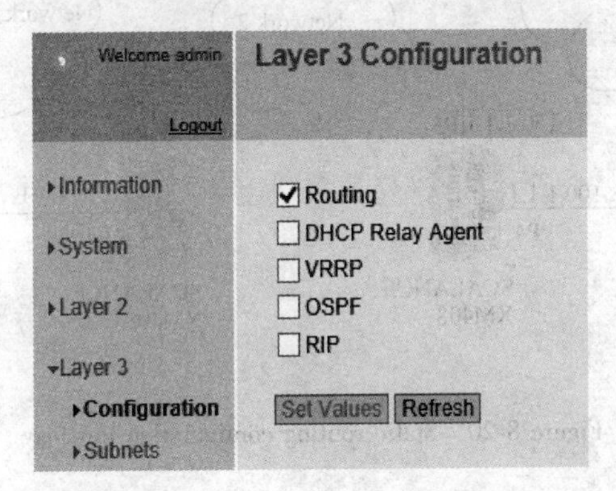

Figure 8-18 Start routing

(4) Assign gateway IP to two VLANs. Assign a gateway IP to VLAN 100 and VLAN 150 separately. In the directory tree "layer 3" → "subnets", select the "overview" tab. In this interface, select VLAN 100 in the "Interface" drop-down list, then click the "create" button, which will add "VLAN 100 entry", then select "VLAN 100" under "interface" in the table to enter the "configuration" tab page. On this page, enter the IP address as 100. 1. 1. 1, enter the subnet Mask as 255. 255. 0. 0, and finally click the "set values" button, as shown in Figure 8-19.

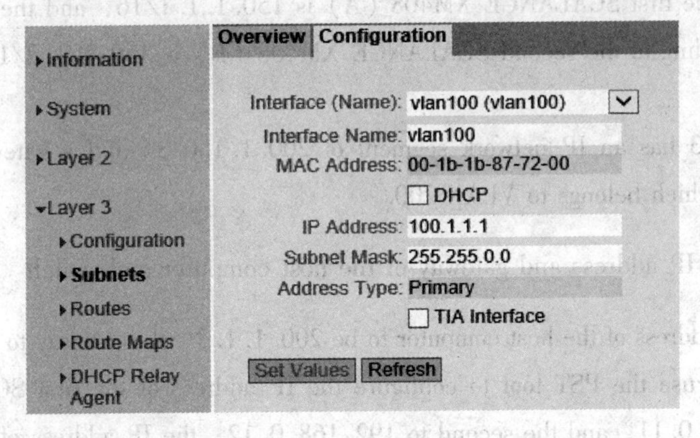

Figure 8-19 Assign gateway IP to two VLANs

(5) Communication test. Communication test by the ping command of the host computer.

8. 5. 2. 2 Static routing configuration

In order to understand the IP address planning of the network and understand the principle of static routing, we introduce how to configure static routes. The topology diagram we used is shown in Figure 8-20.

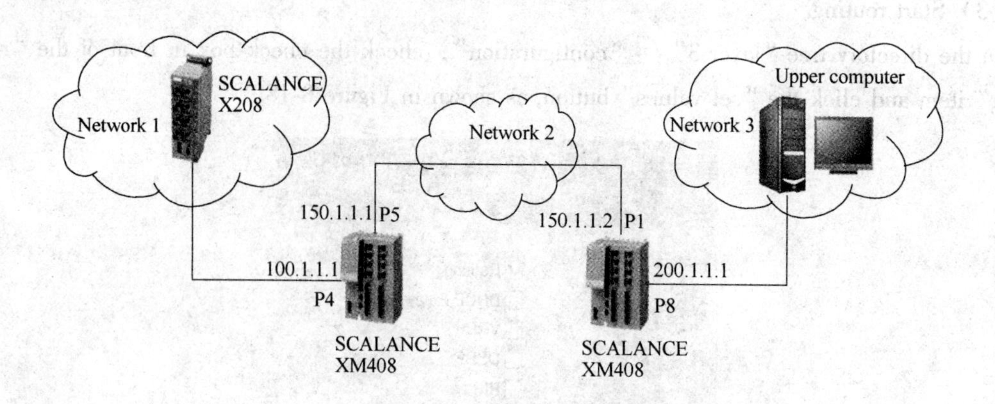

Figure 8-20 Static routing configuration topology

A Network planning

(1) The IP address of the first SCALANCE XM408 (left) is 192. 168. 0. 11;

(2) The IP address of the second SCALANCE XM408 (right) is 192. 168. 0. 12;

(3) Network 1 has an IP network segment of 100. 1. 0. 0/16 and a gateway IP address of 100. 1. 1. 1/16, which belongs to VLAN 100;

(4) The IP network segment of Network 2 is 150. 1. 0. 0/16, the IP address of the gateway corresponding to the first SCALANCE XM408 (A) is 150. 1. 1. 1/16, and the IP address of the gateway corresponding to the second SCALANCE XM408 (B) is 150. 1. 1. 2/16, which belongs to VLAN 150;

(5) Network 3 has an IP network segment of 200. 1. 1. 0/24 and a gateway IP address of 200. 1. 1. 1/24, which belongs to VLAN 200.

B Configure the IP address and gateway of the host computer and switch

Configure the IP address of the host computer to be 200. 1. 1. 3, the gateway to 200. 1. 1. 1 (must set the gateway); use the PST tool to configure the IP address of the first SCALANCE XM408 (left) to 192. 168. 0. 11, and the second to 192. 168. 0. 12; the IP address of the X208 is configured to 100. 1. 0. 41, and the gateway is set to 100. 1. 1. 1 (PST Set Gateway).

C Configuring the SCALANCE XM408 (left)

(1) Add a VLAN ID to the corresponding port (the port connected to other three layers switches (or routers) cannot be set to M, i. e. trunk mode), as shown in Figure 8-21.

(2) Enable routing function, as shown in Figure 8-22.

(3) Assign gateway IP to VLAN 100 and VLAN 150, as shown in Figure 8-23 and Figure 8-24.

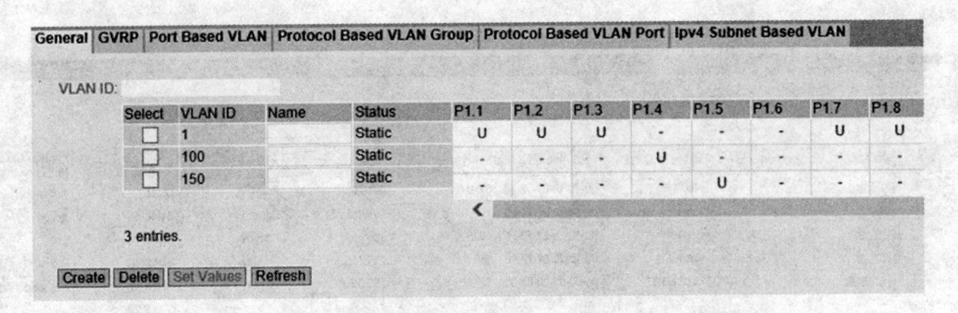

Figure 8-21 VLAN settings

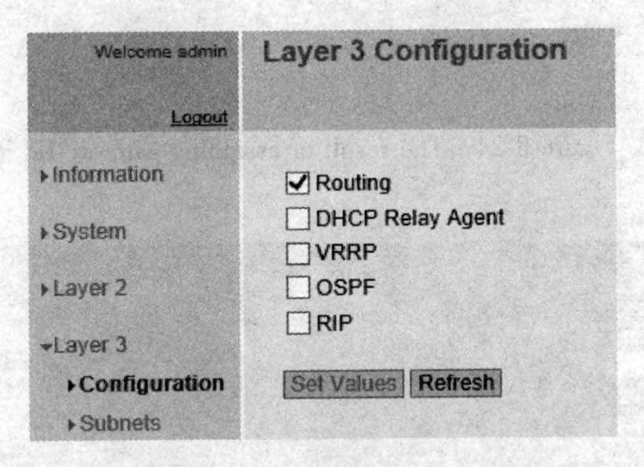

Figure 8-22 Enable routing function

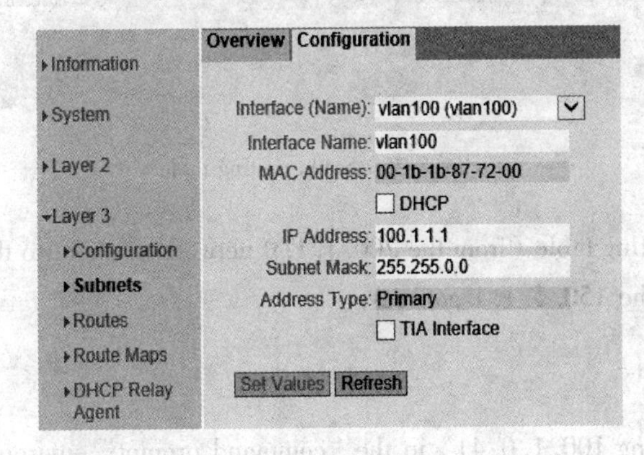

Figure 8-23 Assign gateway IP

(4) Add a static routing table.

In the directory tree "layer 3" → "routes", fill in the destination network as 200. 1. 1. 0, subnet mask as 255. 255. 255. 0, gateway as 150. 1. 1. 2, and then click the "create" button, as shown in Figure 8-25. From the 100. 1. 0. 0 network segment to the 200. 1. 1. 0 network segment, the 150. 1. 1. 2 gateway is required.

D Configure the 2nd SCALANCE XM408 (right)

(1) Add a VLAN ID to the corresponding port.

(2) Enable routing function.

(3) Assign the gateway IP to the corresponding VLAN.

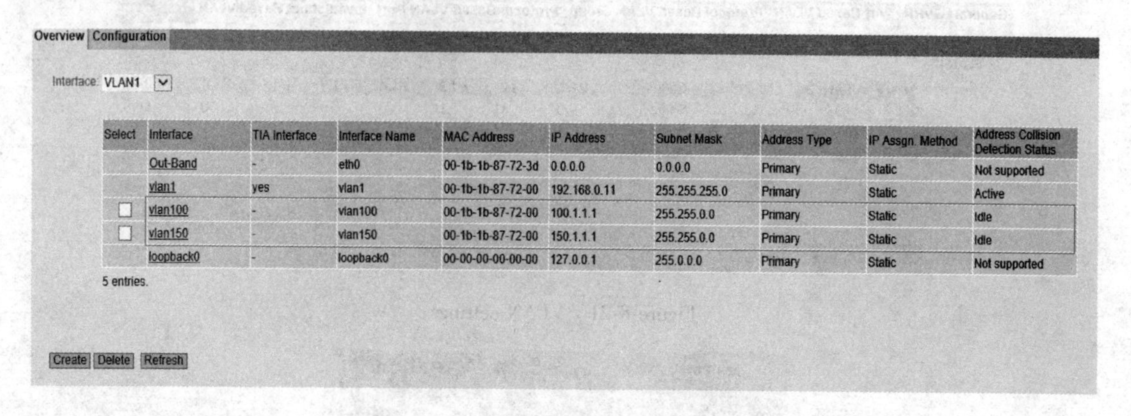

Select	Interface	TIA Interface	Interface Name	MAC Address	IP Address	Subnet Mask	Address Type	IP Assgn. Method	Address Collision Detection Status
	Out-Band	-	eth0	00-1b-1b-87-72-3d	0.0.0.0	0.0.0.0	Primary	Static	Not supported
	vlan1	yes	vlan1	00-1b-1b-87-72-00	192.168.0.11	255.255.255.0	Primary	Static	Active
☐	vlan100	-	vlan100	00-1b-1b-87-72-00	100.1.1.1	255.255.0.0	Primary	Static	Idle
☐	vlan150	-	vlan150	00-1b-1b-87-72-00	150.1.1.1	255.255.0.0	Primary	Static	Idle
	loopback0	-	loopback0	00-00-00-00-00-00	127.0.0.1	255.0.0.0	Primary	Static	Not supported

Figure 8-24　The result of assigning gateway IP

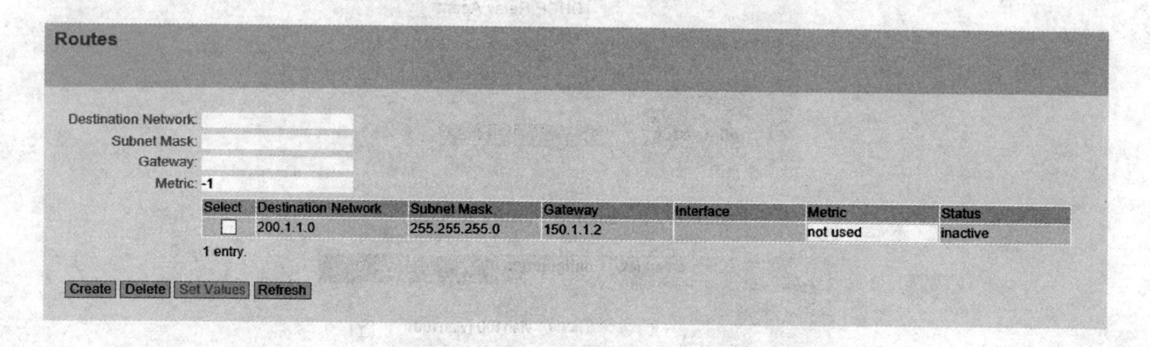

Select	Destination Network	Subnet Mask	Gateway	Interface	Metric	Status
☐	200.1.1.0	255.255.255.0	150.1.1.2		not used	inactive

Figure 8-25　Static routing table

(4) Add a static routing table (from the 200. 1. 1. 0 network segment to the 100. 1. 0. 0 network segment to go through the 150. 1. 1. 1 gateway.).

E　Communication test

Enter the command "ping 100. 1. 0. 41" in the "command prompt" environment of the host computer to view the communication results, as shown in Figure 8-26.

```
Pinging 100.1.0.41 with 32 bytes of data:
Reply from 100.1.0.41: bytes=32 time<1ms TTL=128
Reply from 100.1.0.41: bytes=32 time<1ms TTL=128
Reply from 100.1.0.41: bytes=32 time<1ms TTL=128
Reply from 100.1.0.41: bytes=32 time<1ms TTL=128

Ping statistics for 100.1.0.41:
    Packets: Sent = 4, Received = 4, Lost = 0 (0% loss),
Approximate round trip times in milli-seconds:
    Minimum = 0ms, Maximum = 0ms, Average = 0ms
```

Figure 8-26　Communication test

8.6 Redundancy

8.6.1 Introduction

Network redundancy is a guaranteed strategy for industrial networks. As a fast-reaction backup system, the purpose of network redundancy is to mitigate the risk of unplanned outages and to ensure continuous production through immediate response, thereby reducing the impact of any failure on critical data streams.

Industrial networks have high requirements for availability, and ring redundancy is an important means to improve network availability. The ring industrial Ethernet technology is developed based on Ethernet. It inherits the advantages of fast Ethernet and low cost, and provides a redundant link for data transmission on the network, improving the availability of the network.

Each switch is connected in sequence through a redundant ring port, which constitutes a ring network structure, as shown in Figure 8-27. One of the switches acts as a redundancy manager (RM) and manages the redundant ring network. In a ring network, only one switch can be configured as a redundancy manager. The redundancy manager (RM) monitors the network link status by sending a monitoring frame. When the network is normal, one of the redundant ring ports of the RM is in a logically disconnected state, so that the entire network maintains a linear type in the logical structure, avoiding broadcast storms (when there is a loop in the network, each frame will be broadcast repeatedly in the network, causing a broadcast storm). The redundancy manager monitors the network status. When the cable on the network is disconnected or when the switch fails, it is restored to another logical line structure by connecting to an alternate path. If the fault is eliminated, the network logic structure will restore the original line structure.

8.6.2 Ring network redundancy configuration

In order to understand the role and principle of ring redundancy, we will introduce how to configure ring redundancy. The topology diagram we used is shown in following Figure 8-28.

Assume that the three switches are not connected into a ring before the configuration.

8.6.2.1 Configure the SCALANCE XM408-8C

(1) Set the IP address of the host computer to 192.168.0.100. The IP address of the SCALANCE X-400 is configured via the PST to be 192.168.0.11.

(2) Log in to the XM408 web configuration interface and select "ring redundancy" under "layer 2". Enable the "ring redundancy" function; select "HRP manager" in the "ring re-

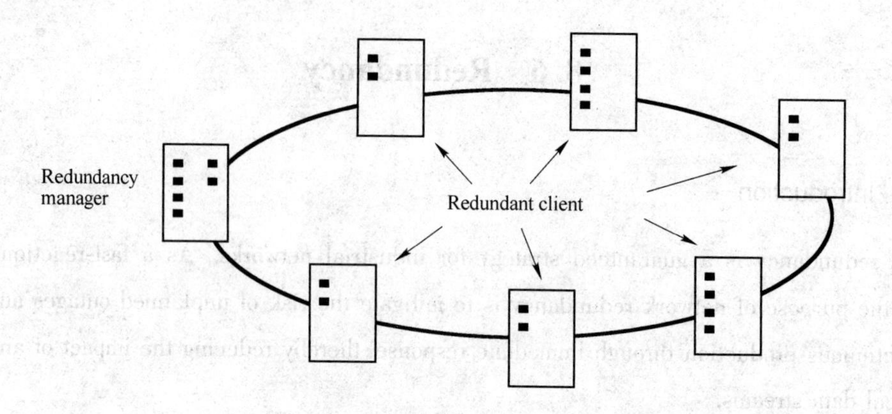

Figure 8-27 Ring network structure

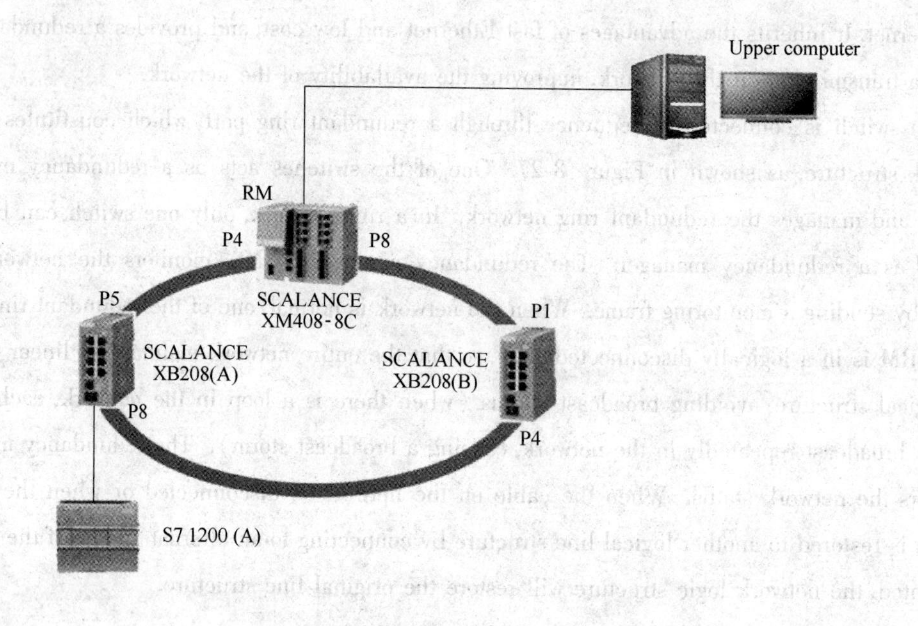

Figure 8-28 Ring network redundancy configuration topology

dundancy mode" drop-down list; then configure the "ring ports" used in the redundant ring, P1. 4 and P1. 8 (where the left port P1. 4 is the alternate port, and the right port P1. 8 is the normal port), as shown in Figure 8-29.

(3) Configuring the switch for the time. When the check button in front of "ring redundancy" is checked, a prompt will pop up, as shown in Figure 8-30. This is because the switch defaults to selecting the spanning tree at the factory. At this point, you need to enter the spanning tree interface under layer 2. In this screen, remove "√" in the check box in front of "spanning tree" and click the "set values" button, as shown in Figure 8-31.

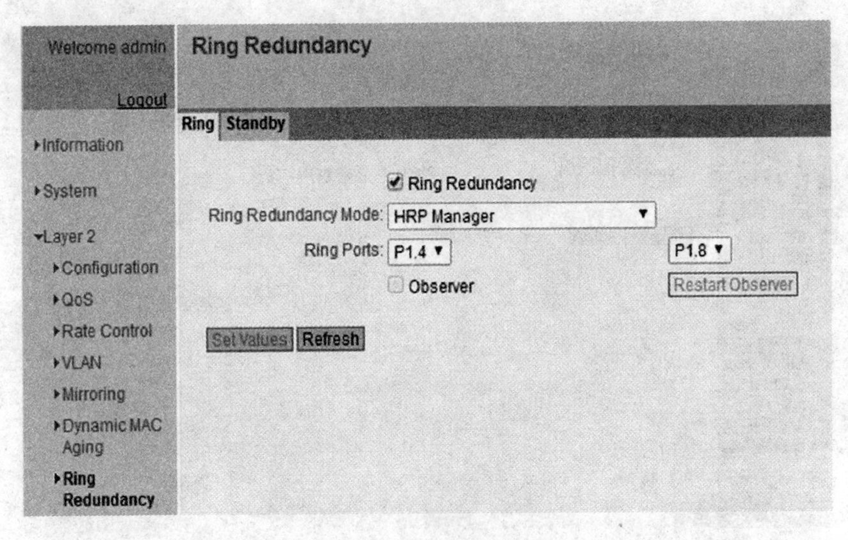

Figure 8-29 Ring network redundancy configuration

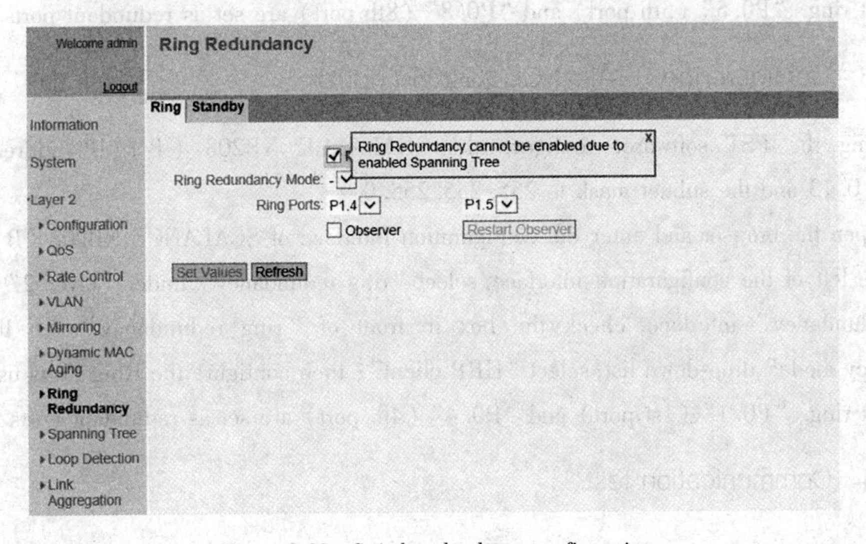

Figure 8-30 Switch redundancy configuration

8.6.2.2 Configure the SCALANCE XB208 (left)

(1) Using the PST software, configure the SCALANCE XB208 (A) IP address to be 192.168.0.12 and the subnet mask to 255.255.255.0.

(2) Open the browser and enter the network configuration interface of SCALANCE XB208. In the list on the left of the network configuration interface, select "ring redundancy" under "layer 2". In the "ring redundancy" interface, check the box in front of "ring redundancy"; in the "ring redundancy mode" drop-down list, select "HRP client"; then configure the ring ports used in the

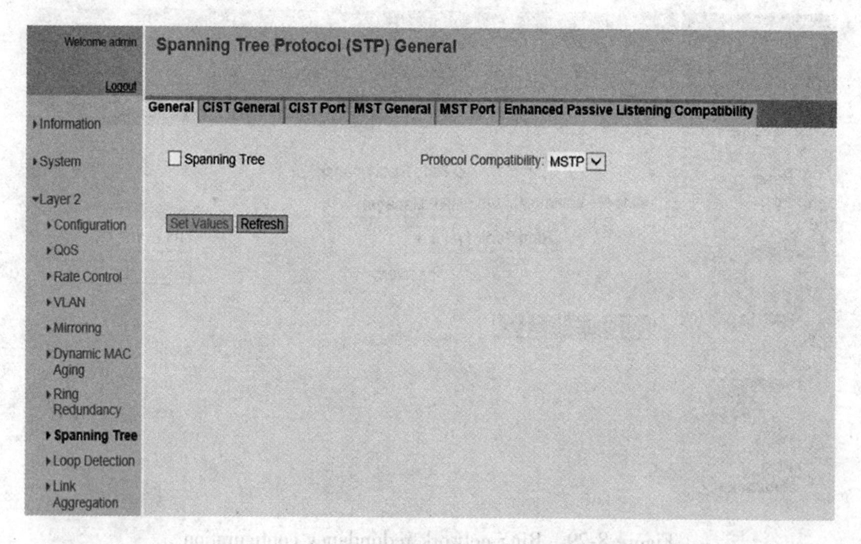

Figure 8-31 Cancel the spanning tree function

redundant ring. "P0. 5" (5th port) and "P0. 8" (8th port) are set as redundant ports.

8.6.2.3 Configure the SCALANCE XB208 (right)

(1) Using the PST software, configure the SCALANCE XB208 (B) IP address to be 192. 168. 0. 13 and the subnet mask to 255. 255. 255. 0.

(2) Open the browser and enter the configuration interface of SCALANCE XB208 (B). In the list on the left of the configuration interface, select "ring redundancy" under "layer 2". In the "ring redundancy" interface, check the box in front of "ring redundancy"; in the "ring redundancy mode" drop-down list, select "HRP client"; then configure the Ring Ports used in the redundant ring. "P0. 1" (1st port) and "P0. 4" (4th port) are set as redundant ports.

8.6.2.4 Communication test

A Normal communication

During normal communication, the XM408 is the HRP manager, the RM indicator of the SCALANCE XM408 is always on, the port number 8 is flashing fast, and the port number 4 is flashing slowly, indicating that the device is normal. Use the ping command to test if the communication is normal.

B Fault communication

Unplug the network cable plugged into the P8 port of the SCALANCE XM408 to simulate a fault or damage to the communication line between the XM408 and XB208 (right) switches in the ring network as shown in Figure 8-32. At this time, the 4 ports of the XM408-8C immediately start flashing, indicating that the port is "activated" and the RM indicator starts flashing, indicating that the network structure has changed and there is already a place in the network. Use the ping command to test if the communication is normal.

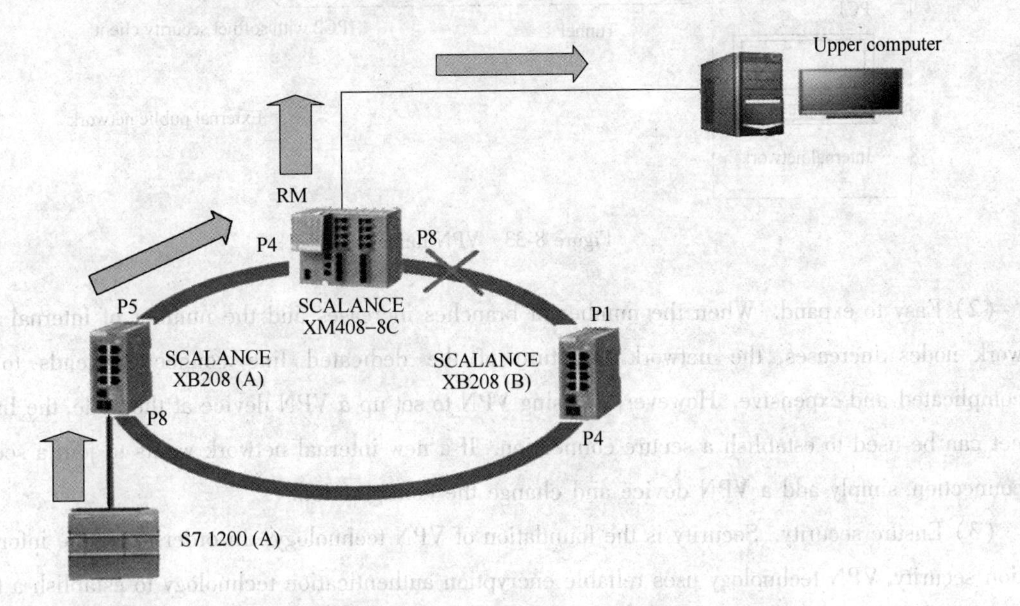

Figure 8-32 Communication test

8.7 Security module

8.7.1 VPN

With the development of the electronic information industry, enterprises have higher requirements on the flexibility, security and economy of their own networks, which makes the virtual private network technology (VPN) develop rapidly. VPN is not an entity, but rather a concept and is a general term for building a "virtual private network" connection technology through public communication facilities. Its function is to establish a private network on the public network for encrypted communication. The VPN gateway realizes remote access by encrypting the data packet and converting the destination address of the data packet, as shown in Figure 8-33. It can be realized by various methods such as server, hardware, software, etc., and has broad application prospects.

Advantages of VPN:

(1) Reduces costs. VPN uses the existing Internet to set up a virtual private network, which can realize data security transmission without using dedicated lines, and provides lower cost than other communication methods (such as dedicated lines and long-distance calls).

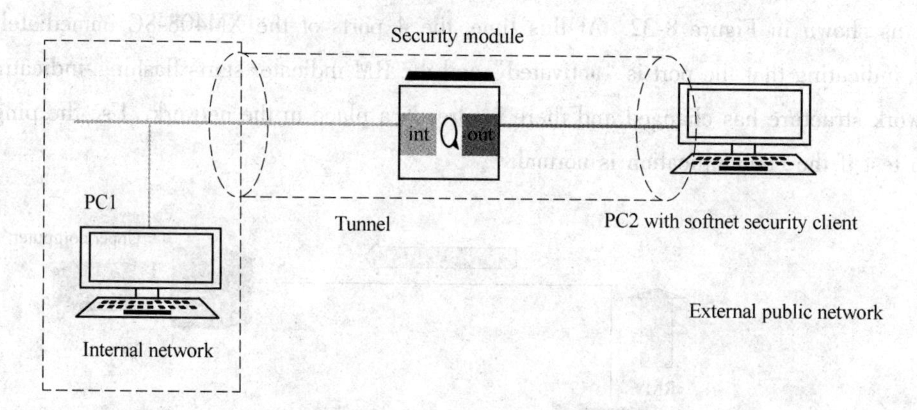

Figure 8-33 VPN tunnel

(2) Easy to expand. When the number of branches increases and the number of internal network nodes increases, the network structure of the dedicated line connection tends to be complicated and expensive. However, by using VPN to set up a VPN device at the node, the Internet can be used to establish a secure connection. If a new internal network wants to join a secure connection, simply add a VPN device and change the configuration.

(3) Ensure security. Security is the foundation of VPN technology. In order to ensure information security, VPN technology uses reliable encryption authentication technology to establish a tunnel on the internal network to prevent information from being leaked, falsified and copied.

According to different classification criteria, VPNs can be classified according to several criteria:

(1) Classified by VPN protocol:

There are three types of VPN tunneling protocols: PPTP, L2TP, and IPSec. PPTP and L2TP work in the second layer of the OSI model, also known as the layer 2 tunneling protocol. IPSec is the layer 3 tunneling protocol.

(2) Classified by VPN application:

Access VPN: The client-to-gateway uses the public network as the backbone network to transmit VPN data traffic between devices.

Intranet VPN: A gateway-to-gateway that connects resources from the same company through the company's network architecture.

Extranet VPN: An extranet is formed with a partner enterprise network to connect one company to another company's resources.

(3) Classfied by the type of equipment used:

Network equipment providers have developed different VPN network devices for different customer needs, mainly switches, routers and firewalls.

Router-style VPN: Router-based VPN deployment is easier, as long as the VPN service is added to the router.

Switched VPN: mainly used to connect VPN networks with fewer users.

Firewall VPN: Firewall VPN is the most common VPN implementation, and many vendors offer this type of configuration.

(4) According to the implementation principle:

Overlapping VPN: This VPN requires users to establish VPN links between end nodes, including GRE, L2TP, and IPSec.

Peer-to-Peer VPN: The network operator completes the establishment of the VPN tunnel on the backbone network, mainly including MPLS and VPN technologies.

VPN implementation methods include:

(1) VPN server: in a large LAN, VPN is implemented by setting up a VPN server in the network center.

(2) Software VPN: VPN can be implemented through dedicated software.

(3) Hardware VPN: VPN can be implemented through dedicated hardware.

(4) Integrated VPN: Some hardware devices, such as routers and firewalls, contain VPN functions, but hardware devices that generally have VPN functions are usually more expensive than those without this function.

At present, VPN mainly adopts four technologies to ensure security. These four technologies are tunneling, encryption and decryption, key management technology, and user and device authentication technology (Authentication).

(1) Tunnel technology: Tunnel technology is the basic technology of VPN. It is similar to point-to-point connection technology. It establishes a data channel (tunnel) in the public network, and allows data packets to be transmitted through this tunnel. The tunnel is formed by a tunneling protocol and is divided into second and third layer tunneling protocols.

The second layer tunneling protocol first encapsulates various network protocols into PPP, and then wraps the entire data into the tunneling protocol. The data packet formed by this two-layer encapsulation method is transmitted by the second layer protocol. The second layer tunneling protocol includes L2F, PPTP, L2TP, etc. The L2TP protocol is currently the standard of the IETF, and is formed by the IETF combining PPTP and L2F.

The third layer tunneling protocol directly loads various network protocols into the tunneling protocol, and the formed data packets are transmitted by the third layer protocol. The third layer tunneling protocol includes VTP, IPSec, etc. IPSec (IP Security) consists of a set of RFC documents that define a system to provide security protocol selection, security algorithms, and services such as keys used by the service to provide security at the IP layer.

(2) Encryption and decryption technology: Encryption and decryption technology is a more mature technology in data communication, and VPN can directly use the existing technology.

(3) Key management technology: The main task of key management technology is how to securely pass keys over public data networks without them being stolen. The current key management technology is divided into SKIP and ISAKMP/OAKLEY. SKIP mainly uses Diffie-

Hellman's algorithm to transmit keys on the network; in ISAKMP, both sides have two keys for public and private use.

(4) User and device authentication technology: User and device authentication technologies are most commonly used in user name and password or card authentication.

VPN application:

(1) Access VPN: The client-to-gateway uses the public network as the backbone network to transmit VPN data traffic between devices.

With today's "mobile office" and the growing demand for telecommunications, providing remote access to corporate networks for corporate employees is a must for new business forms, which has led to the emergence of remote access VPNs.

Remote access VPN (as shown in Figure 8-34) enters the local ISP through dial-up, ISDN, digital subscriber line (xDSL), etc.

Enter the Internet and connect to the enterprise's VPN gateway to establish a secure "tunnel" between the user and the VPN. Secure access to the remote intranet through the tunnel saves both communication costs and security.

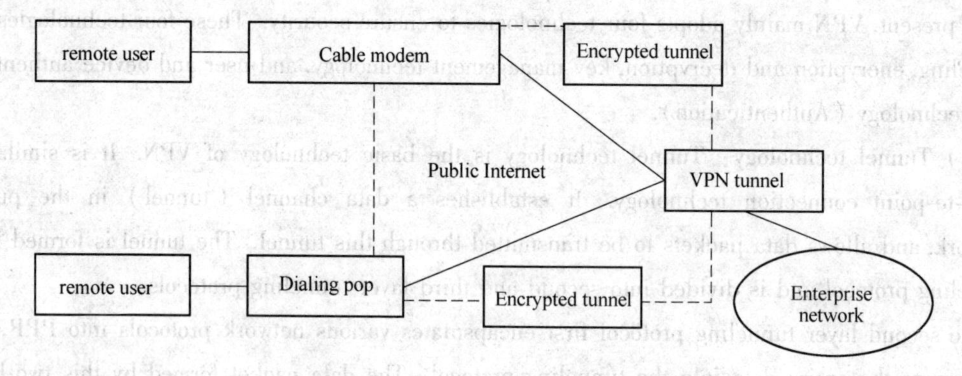

Figure 8-34 Remote access VPN architecture

(2) Intranet VPN: A gateway-to-gateway configuration that connects resources from the same company through the company's network architecture.

Site-to-site intranet VPN is the main method to solve the security of intranet architecture and connection security and transmission security. As shown in Figure 8-35, a typical network architecture for a site-to-site intranet VPN is presented. A VPN gateway is located on the boundary between the private enterprise network and the shared public Internet. It is used to encrypt data communications (Because the key used for encryption is known only to the VPN gateway device, no third party can destroy the data). It transfers data to other VPN gateways over the Internet. After receiving the encrypted data from the Internet, the destination VPN gateway decrypts the data and sends it to the private enterprise network.

The VPN gateway device at one site in Figure 8-35 is capable of handling communications to other sites and from other sites. No dedicated physical lines are required between sites, so creating and deleting a VPN tunnel requires simply changing the configuration of the device, making the VPN very flexible and dynamic.

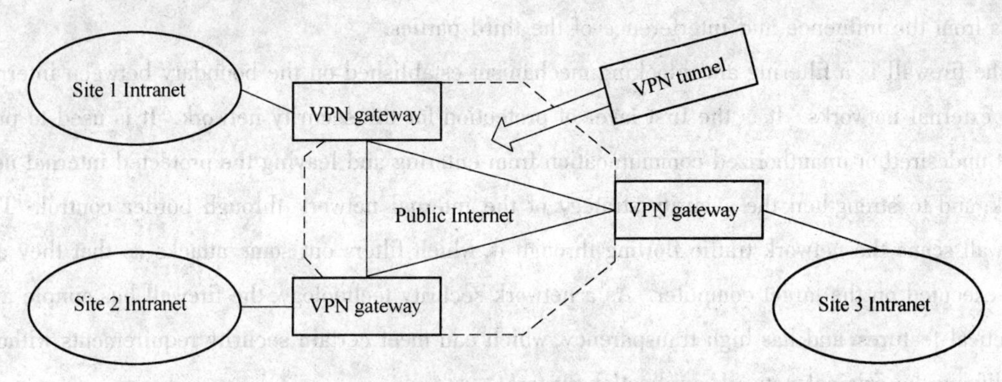

Figure 8-35 Site-to-site Intranet VPN architecture

(3) Extranet VPN: An extranet is formed with a partner enterprise network to connect one company to another company's resources, as shown in Figure 8-36.

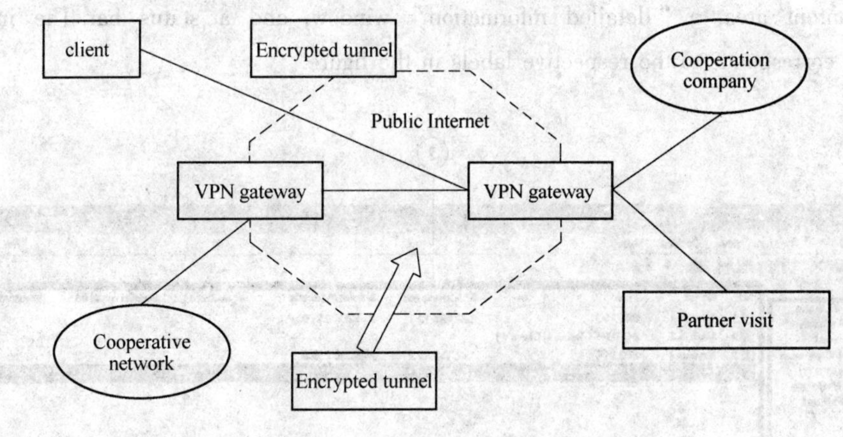

Figure 8-36 External network VPN architecture

A VPN tunnel is established between these gateways and a VPN gateway located within a business partner network over the Internet. In another case, a VPN client on a single computer host allows the customer to access the extranet from a designated remote access computer. Tunnels use specialized access rules and filters that allow only certain commercial applications to communicate through secure tunnels. The same VPN gateway can be used to establish a secure tunnel from multiple business partners.

8.7.2 Firewall

Network security is an important factor in industrial information security. When the internal office network Intranet is connected to the production control network ControlNet, in order to ensure that on-site production control is not affected and to protect its information security, certain technical means are needed to isolate the factory production network and the factory office network to restrict the access of the office network to the control network, while protecting the network and workstations from the influence and interference of the third parties.

The firewall is a filtering and blocking mechanism established on the boundary between internal and external networks. It is the first layer of protection for the security network. It is used to prevent undesired or unauthorized communication from entering and leaving the protected internal network, and to strengthen the security strategy of the internal network through border control. The firewall scans the network traffic flowing through it, which filters out some attacks so that they are not executed on the target computer. As a network security technology, the firewall has simple and practical features, and has high transparency, which can meet certain security requirements without modifying the original network application system.

8.7.3 SCT

As shown in Figure 8-37, the user interface is mainly composed of four parts, namely a navigation panel, a content area, a "detailed information" window, and a status bar. The inter face sequentially corresponds to the respective labels in the figure.

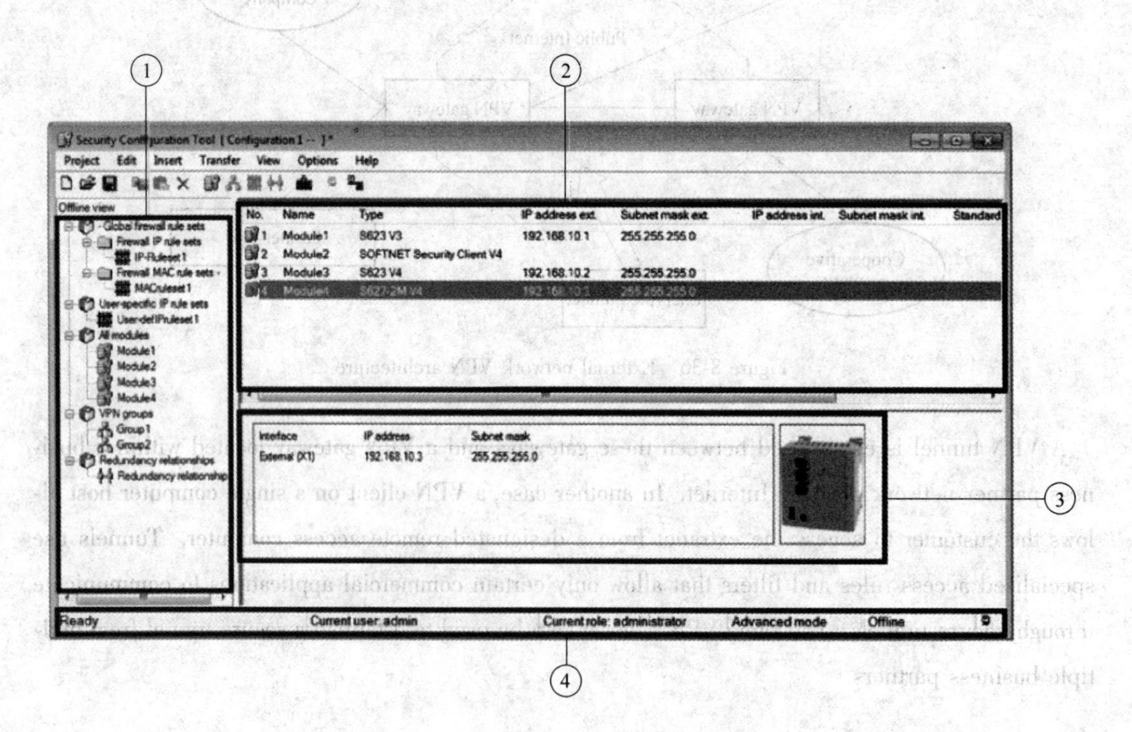

Figure 8-37 User interface structure in advanced mode

①Navigation panel:

- Global firewall rule set-this object contains the configured global firewall rule set.

Other folders:

——Firewall IP rule set.

——Firewall MAC rule set.

- User-specific IP rule sets.
- All modules-include all configuration modules and SOFTNET configurations in the project.
- VPN group-contains all generated VPN groups.
- Redundancy relationship-contains all generated redundancy relationships in the project.

②Content area:

When you select an object in the navigation panel, you will see detailed information about the object in the content area. In addition, for some security modules, you can view and adjust some of the interface configuration in this area. If these security modules provide the corresponding configuration options, double-click on these security modules to open the properties dialog where you can further enter the parameters.

③ "Detailed information" window:

The detailed information window contains additional information about the selected object and allows you to configure VPN properties for specific connections in the context of the VPN group. Use the view menu to hide and show the details window.

④Status bar:

The status bar displays the status of the run and the current status message. These include:

- Current user and user type.
- Operator View-Standard Mode/Advanced Mode.
- Mode-Online/Offline.

8.7.4 SCT firewall configuration

Taking "firewall in configuration standard mode" as an example as shown in Figure 8-38, we will introduce how to configure SCT firewall.

With this configuration, the transmission of IP traffic can only be initiated from the internal network; only the response from the external network is allowed.

As shown in Figure 8-39, the process flow for configuring the firewall in the standard mode is configured.

(1) Set up SCALANCE and network (connect power, plug in network cable).

(2) IP configuration for the PC.

(3) Create project and security modules using SCT.

Start the security configuration software, select the menu command "project>new...", the dialog box shown in Figure 8-40 pops up, and then enter the user name and password to create a new user.

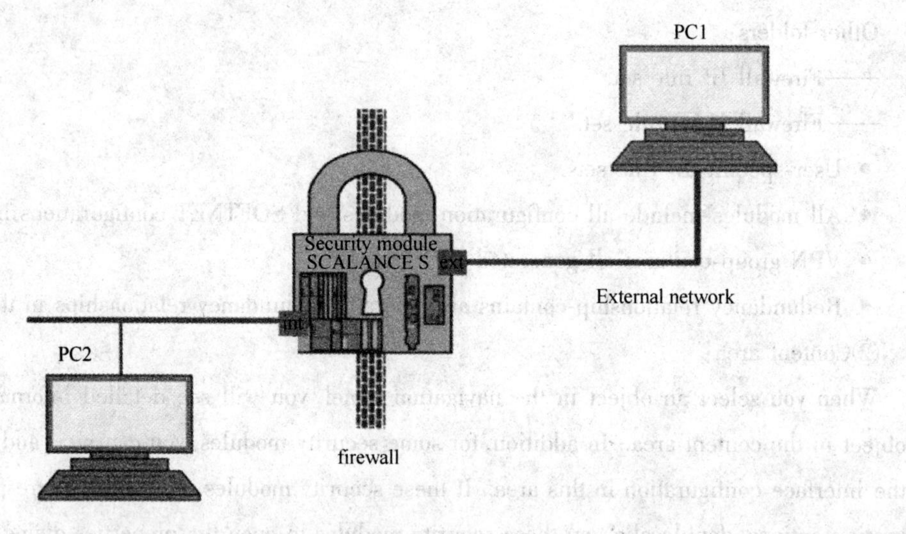

Figure 8-38 Firewall in standard mode

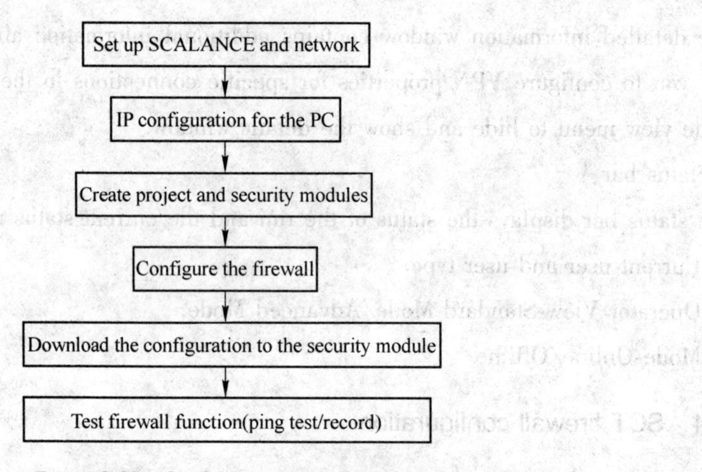

Figure 8-39 Configuration step flow chart

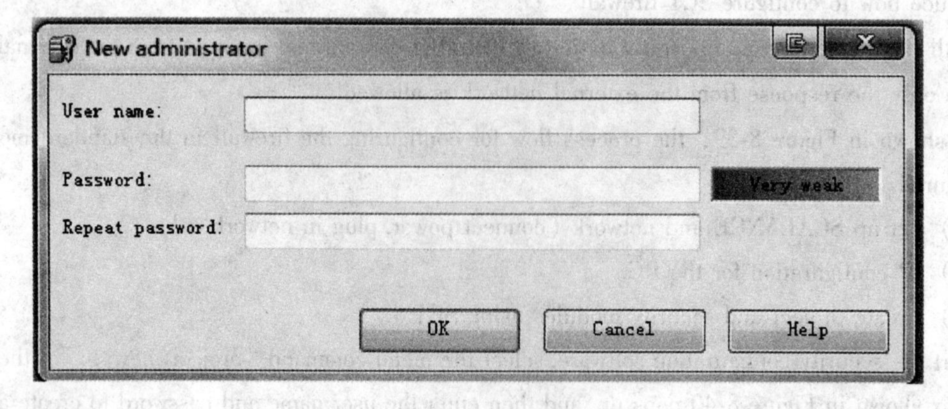

Figure 8-40 User creation interface

After clicking the "OK" button, the "selection of a module or software configuration" dialog is opened (as shown in Figure 8-41) and configured according to the device information. In the "configuration" area, enter the MAC address, external IP address and external subnet mask, then click "OK" to confirm the dialog box, and the interface shown in Figure 8-42 pops up.

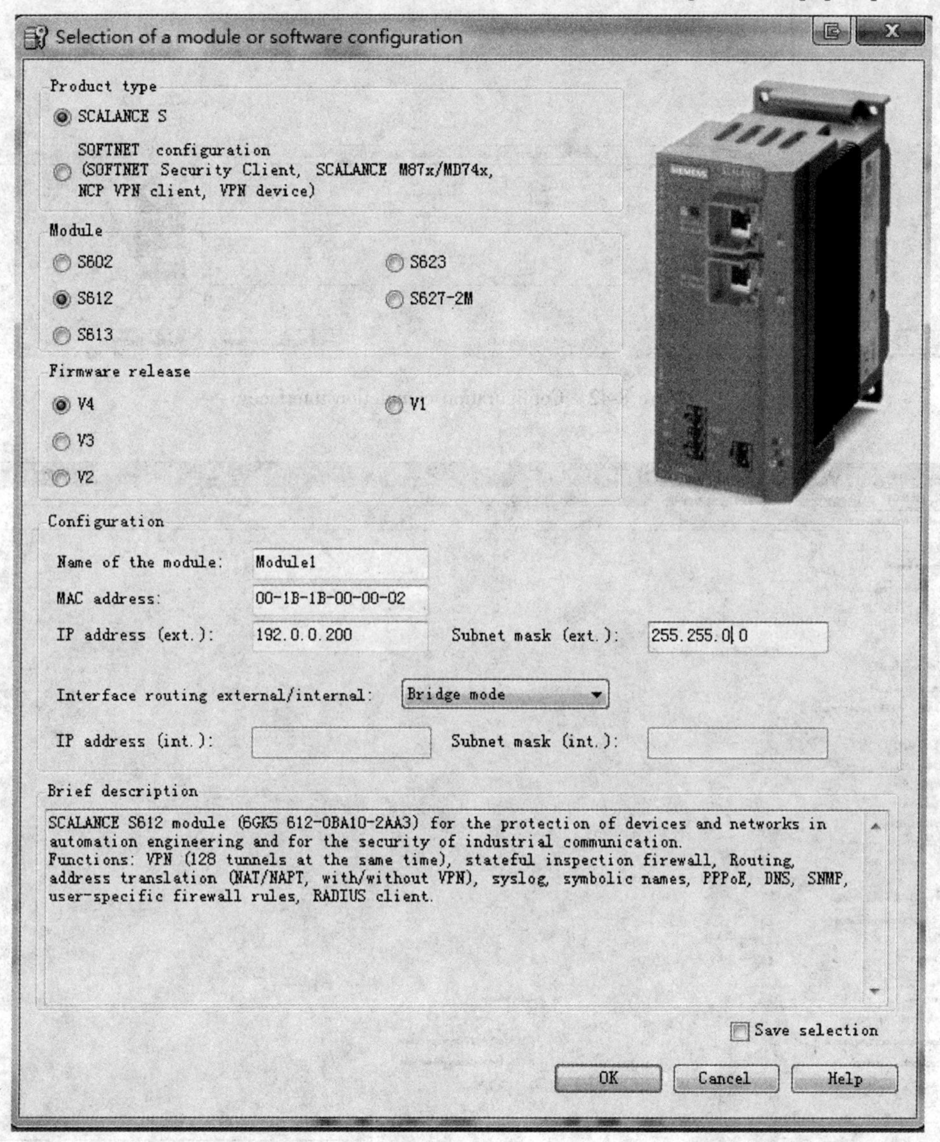

Figure 8-41 Security module configuration interface

(4) Configure the firewall.

First select the security module in the content area, then select the menu command "Edit > Properties..." and select the "Firewall" tab in the displayed dialog box to enable the settings shown Figure 8-43; select the "logging" option to log data traffic. Finally, click "OK" to close the dialog and save the project with the menu command "Project > Save".

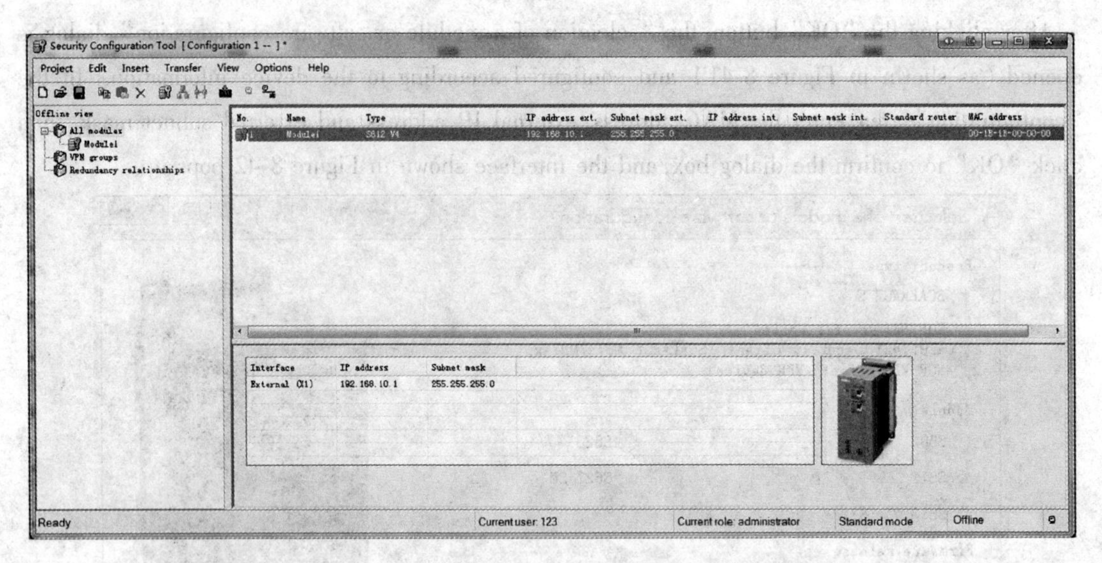

Figure 8-42 Configuration completion interface

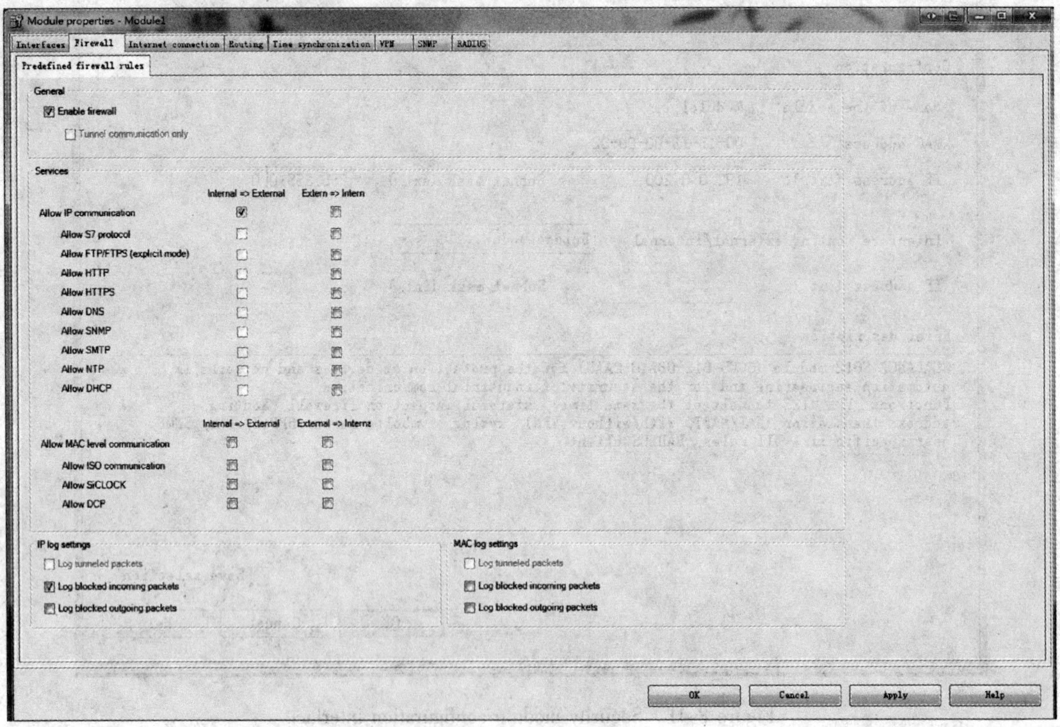

Figure 8-43 Firewall configuration interface

(5) Download the configuration to the security module.

Select the security module in the content area, then select the menu command "Transfer>To module(s)..." and finally click the "Start" button to start the download, as shown in Figure 8-44.

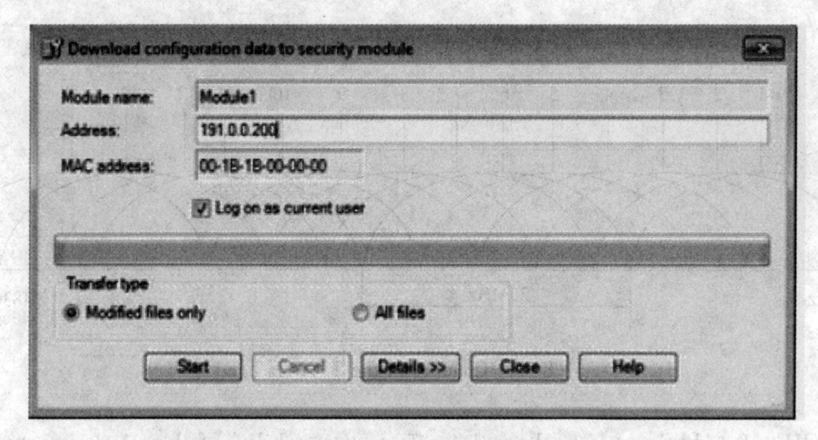

Figure 8-44 Download the configuration to the security module

8.8 Wireless

8.8.1 Introduction

Industrial wireless communication technology is another hot core technology in the field of industrial control after fieldbus. It is a revolutionary technology to reduce the cost of industrial measurement and control systems and improve the application range of industrial measurement and control systems. It is also a new direction for industrial automation development in the next few years.

Compared with traditional wired transmission communication, data transmission in the form of wireless transmission is less affected by the environment, and reliable information transmission can be realized even in areas where traditional technology is difficult or impossible to achieve. It also opens up new resources with efficient automation solutions.

8.8.1.1 Wireless

(1) Considering the redundancy of wireless signals, the coverage of each AP should have overlapping areas, in the event of failure of individual APs and other devices, in order to achieve real-time seamless information transmission in the entire line;

(2) Wireless communication band. In wireless communication, 2.4GHz and 5GHz are common frequency bands, of which 2.4GHz is the most commonly used frequency band, including standard protocols such as 802.11 a/b/g/n. The channel frequency distribution in the 2.4GHz band is shown in Figure 8-45.

The 2.4GHz band has better penetration and a longer coverage distance, but there are more devices at 2.4GHz, which are likely to cause signal interference. The transmission rate is higher in the 5GHz band, but its signal attenuation is more severe, resulting in a coverage area that is not as

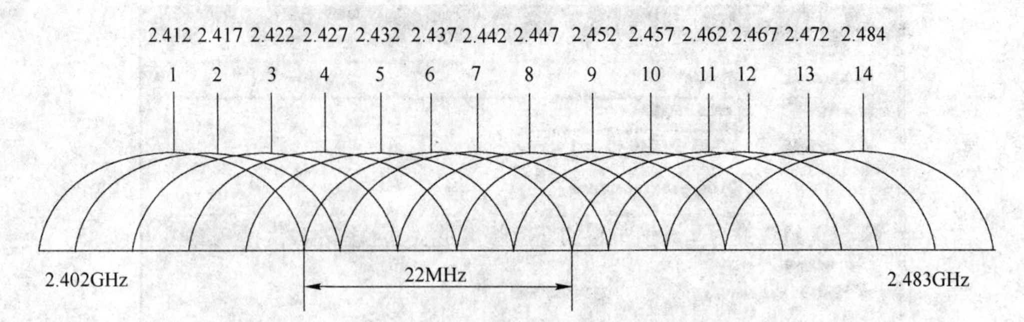

Figure 8-45 Channel in the 2. 4GHz band

wide as 2. 4GHz. In addition, channel overlap affects the stability of the wireless network. There are only three groups of channels that do not interfere with each other in the 2. 4GHz band, namely 1, 6, 11; 2, 7, 12, and 3, 8, 13. The 1-11 channel is an international standard. The 12 and 13 channels are used in China, so some devices do not support 12 and 13 channels, which means that 1, 6, and 11 channels in the 2. 4GHz band can be selected.

(3) Seamless switching. When the wireless client disconnects from the previous connected AP during the mobile process, the data cannot be transmitted while accessing another AP, and this time is called roaming time. Civil or commercial communication does not require roaming switching time, but the roaming time in the process cannot be too large; otherwise it will be prone to security incidents.

Using the fast roaming technology of Siemens wireless devices, i · e. iPCF seamless roaming, can reduce information interaction time and ensure safe production.

(4) Packet loss rate: refers to the ratio of the number of lost packets in the test to the data group sent.

$$\text{Packet loss rate} = \frac{\text{Input packet} - \text{output packet}}{\text{Input packet}} \times 100\%$$

In the process of wireless network propagation, if obstacles are encountered, or interfered with other signals, and the transmission distance is too far, the signal strength will be weakened and the packet loss rate will increase. Generally, when the signal strength of the wireless coverage area is high, the packet loss rate satisfies 1% or less, and reliable data transmission can be realized.

8. 8. 1. 2 SCALANCE W

SIMATIC NET's IWLAN product portfolio combines reliability, ruggedness and security for wireless communication solutions.

(1) High efficiency:

1) Next-generation IWLAN access points, client modules and IWLAN controllers.

2) High data transfer rate for powerful wireless communication.

3) MIMO technology increases bandwidth and range, while also improving the reliability of data transmission.

(2) Reliability:

1) Periodic, predictable data communication (determinism) .

2) Mobile device quickly transfers from one wireless cell to another (fast roaming) .

3) Monitor wireless connections, wireless redundancy and automatic switching channels.

(3) Durability:

1) Anti-dust, waterproof construction.

2) Metal case, impact resistance, seismic protection, mechanical stability.

3) Anti-condensation.

(4) Security:

1) Standard mechanism for identifying users (authentication) .

2) Data encryption (TKIP, RADIUS) .

3) WPA2 and encryption technology for high security and preventing illegal access (AES) .

4) Suitable for areas with explosion hazards and fail-safe communication (safety) .

8. 8. 2 Wireless access point and client configuration

(1) After configuring the IP address using the PST, enter the W774/W734 configuration interface through a web browser, as shown in Figure 8-46.

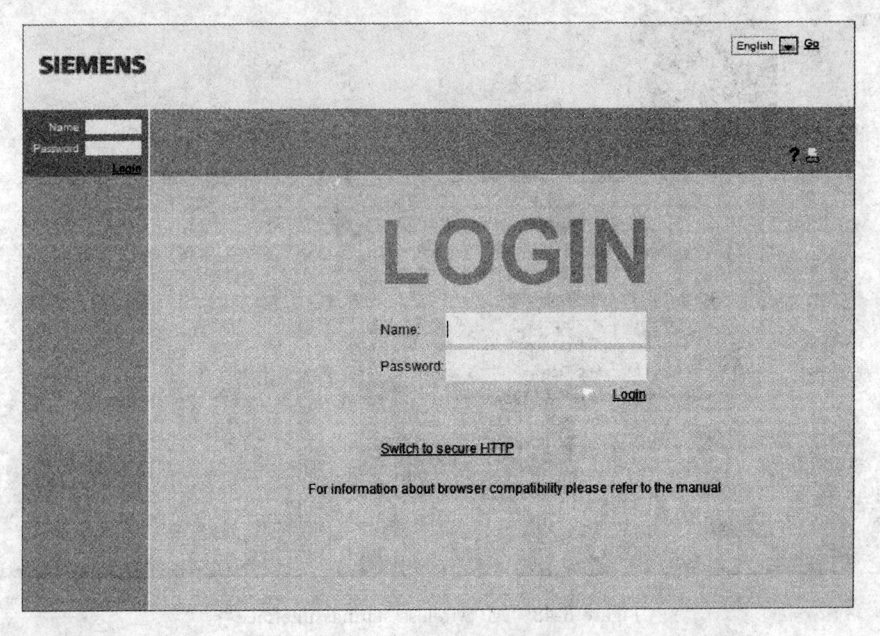

Figure 8-46 Web browser login interface

(2) Enter the wizard interface for the first time, select "Device Type" to enter the main interface (the device type is divided into client and access point), as shown in Figure 8-47.

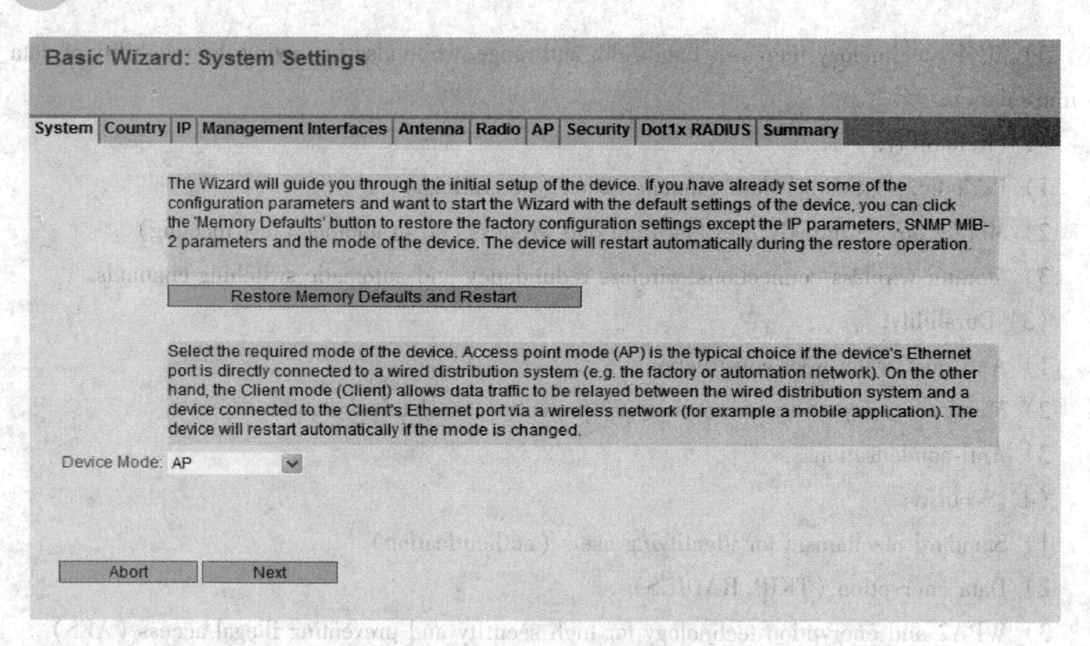

Figure 8-47 Wizard interface

(3) Click on the "WLAN" sub-item "Interfaces" on the left side of the configuration interface for wireless configuration, as shown in Figure 8-48.

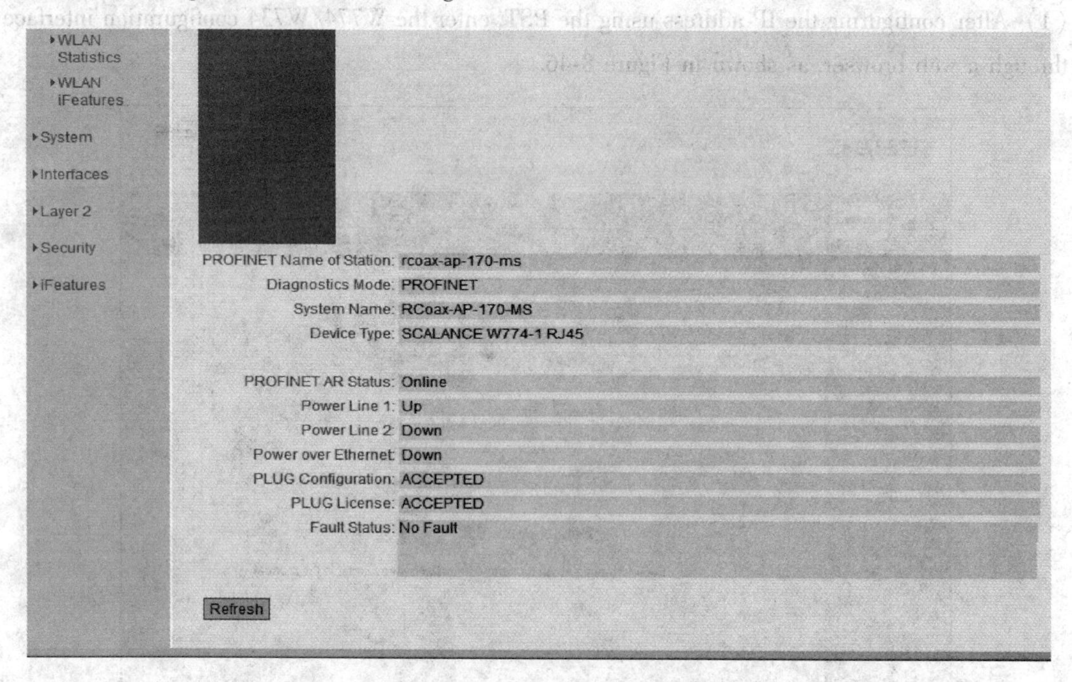

Figure 8-48 AP wireless settings interface

(4) Under the "Basic" tab, select "China as the (Country Code), select the check box under the "Enabled" title bar in the table, and select the "max. Tx Power" value as 17dBm for "Tx Power. Check" is displayed as "Allowed", other configurations remain unchanged, and finally

click the "Set Values" button, as shown in Figure 8-49.

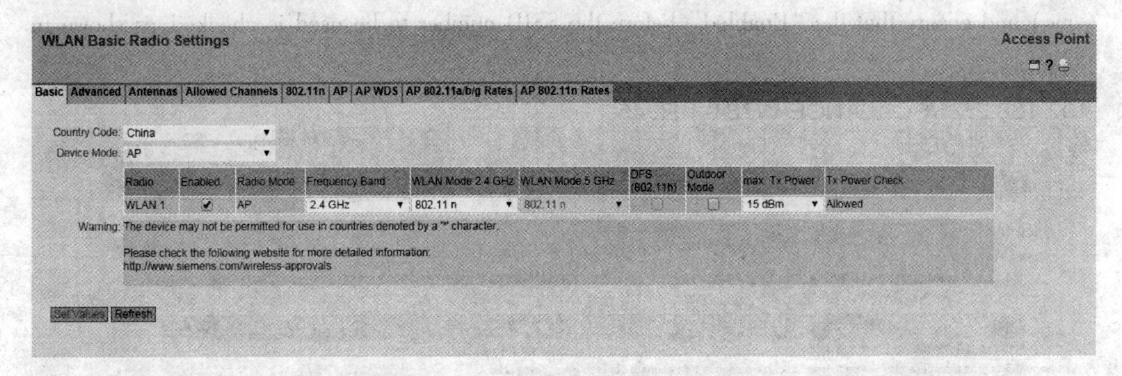

Figure 8-49 Basic configuration interface

(5) Set the access point antenna type: Under the "Antennas" tab, select "Antenna Type" as "ANT795-4MA", the other configurations remain unchanged, and finally click the "Set Values" button, as shown in Figure 8-50.

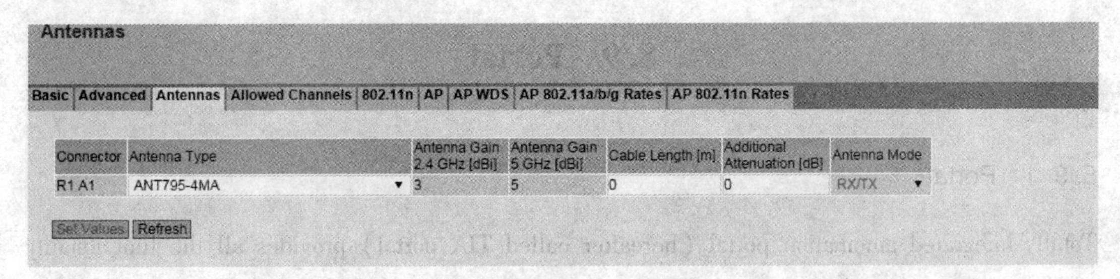

Figure 8-50 Access point antenna type setting interface

(6) On the "Allowed Channels" tab, list the channels that can be selected when the selected frequency bandwidth is 2.4G or 5G. You can keep the default settings; that is, all channels are checked (the device adaptively selects channels according to the wireless environment), or after selecting "Use Allowed Channels only", select a specific channel, as shown in Figure 8-51.

Basic | Advanced | Antennas | Allowed Channels | 802.11n | AP | AP WDS | AP 802.11a/b/g Rates | AP 802.11n Rates

Radio	Use Allowed Channels only
WLAN 1	☐

Frequency Band: 2.4 GHz

☑ Select / Deselect all

Radio	Radio Mode	1	2	3	4	5	6	7	8	9	10	11	12	13
WLAN 1	AP	☑	☑	☑	☑	☑	☑	☑	☑	☑	☑	☑	☑	☑

Frequency Band: 5 GHz

☑ Select / Deselect all

Radio	Radio Mode	149	153	157	161	165
WLAN 1	AP	☐	☐	☐	☐	☐

Set Values | Refresh

Figure 8-51 Channel setting interface

(7) On the AP tab, you can change the SSID number (the AP and Client settings must be the same) and ensure that the "Enabled" before the SSID number to be used is checked, as shown in Figure 8-52.

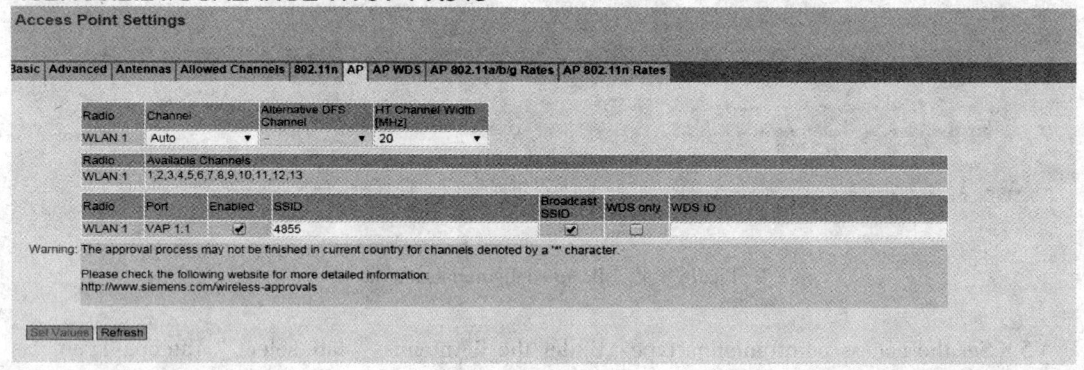

Figure 8-52 SSID modification interface

8.9 Portal

8.9.1 Portal

Totally integrated automation portal (hereafter called TIA portal) provides all the functionality needed to automate tasks in a single cross-software platform, as shown in Figure 8-53.

Figure 8-53 Integrated environment

As the first shared working environment for integrated engineering, the TIA portal offers a wide range of SIMATIC systems in a single framework. Therefore, for the first time, TIA portal supports reliable and convenient cross-system collaboration. All required software packages, from hardware configuration and programming to process visualization, are integrated into a comprehensive engineering framework.

When using the TIA portal, the following features provide efficient support during the implementation of an automation solution.

(1) Integrated engineering with unified operating concept. Process automation and process visualization go hand in hand.

(2) Consistent centralized data management with powerful editors and common symbols. Once created, it is available in all editors.

Changes and corrections will be automatically applied and updated throughout the project.

(3) Complete library concept. Ready-to-use instructions and existing parts of the project can be used over and over again.

(4) Multiple programming languages. Automated tasks can be implemented in five different programming languages.

8.9.2 Portal creation project

In the following steps, a new project will be created. All data generated during the creation of the automation solution is saved in the project file. The data will be stored as an object. In a project, objects are arranged in a tree structure (project hierarchy). The project hierarchy is based on devices and stations and their configuration data and procedures.

The following hardware and software devices are required to create the project:

(1) Hardware:

——CPU 1511-1 PN for installation and wiring in the hardware section of the Getting Started.

——Ethernet connection to the PG/PC.

(2) Software:

The following packages must be installed on the PG/PC and can be executed:

——SIMATIC STEP 7 Professional V13.

——SIMATIC WinCC Advanced V13 or SIMATIC WinCC Professional V13.

Next, we will illastrate how to create a new project.

8.9.2.1 Create a new project

(1) Click "Create new project".

(2) Enter a project name, as shown in Figure 8-54.

(3) Click "Create" to create a new project. All data, such as hardware configuration data, CPU programming and visualization data in the HMI, is saved in the project, as shown in Figure 8-55.

8.9.2.2 Create S7-1500 CPU

In the following steps, an unspecified CPU will be created. Unspecified CPUs are placeholders for specific CPUs in the hardware catalog that will be defined later.

(1) Open the device & network portal.

(2) Insert a new device.

(3) Enter "Color _ Mixing _ CPU" as the CPU name, as shown in Figure 8-56.

(4) Open the "SIMATIC S7-1500" folder.

(5) Select the CPU that has not been specified.

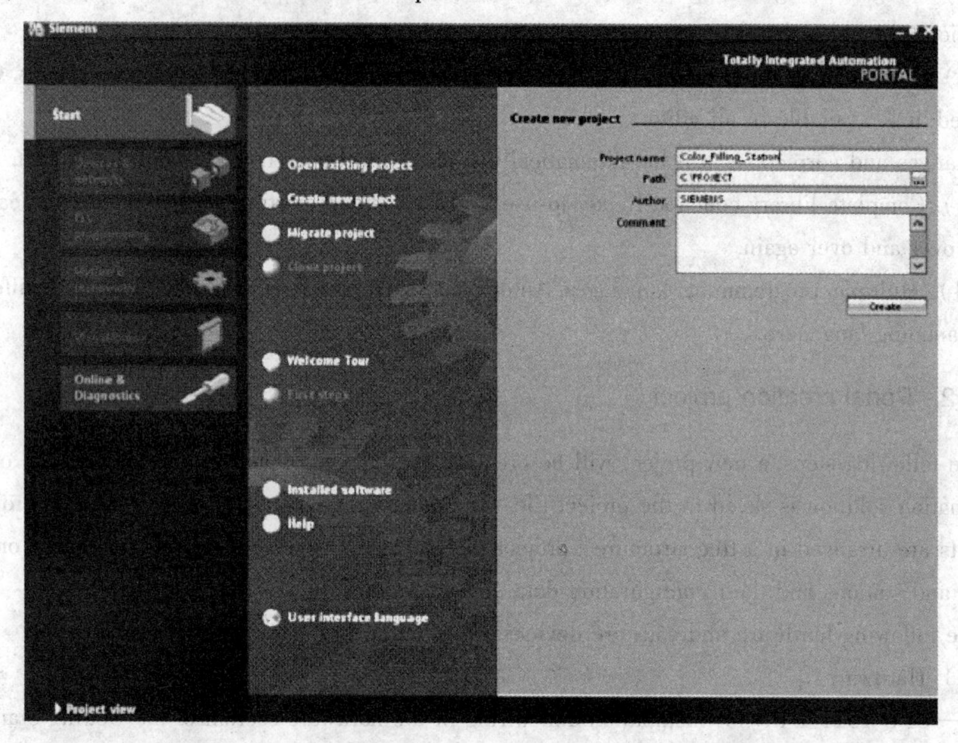

Figure 8-54 Create a new project

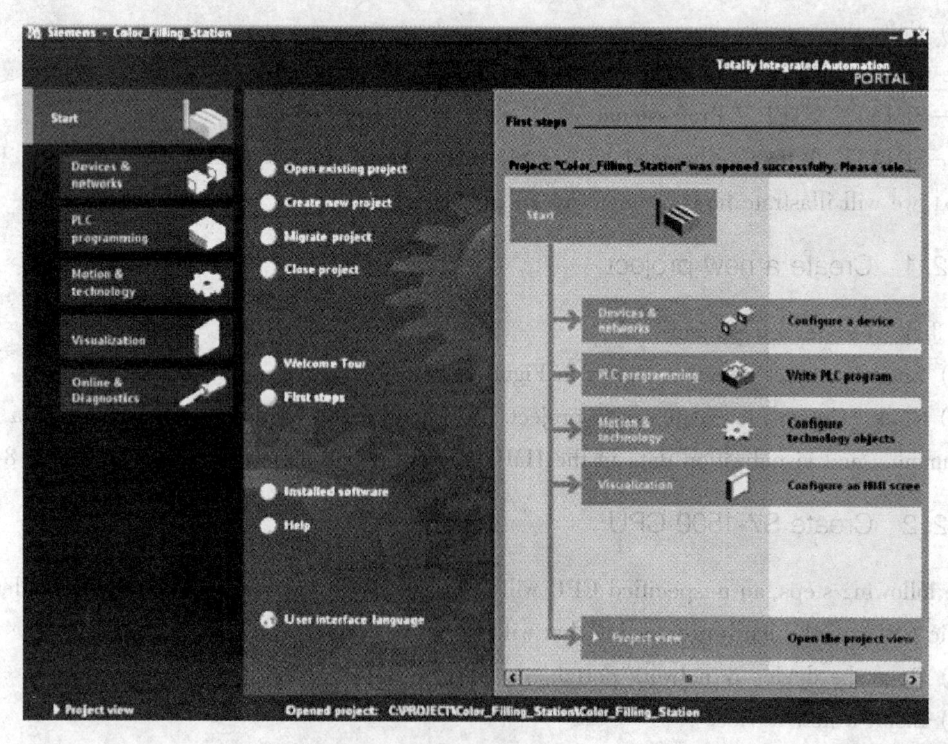

Figure 8-55 Project view

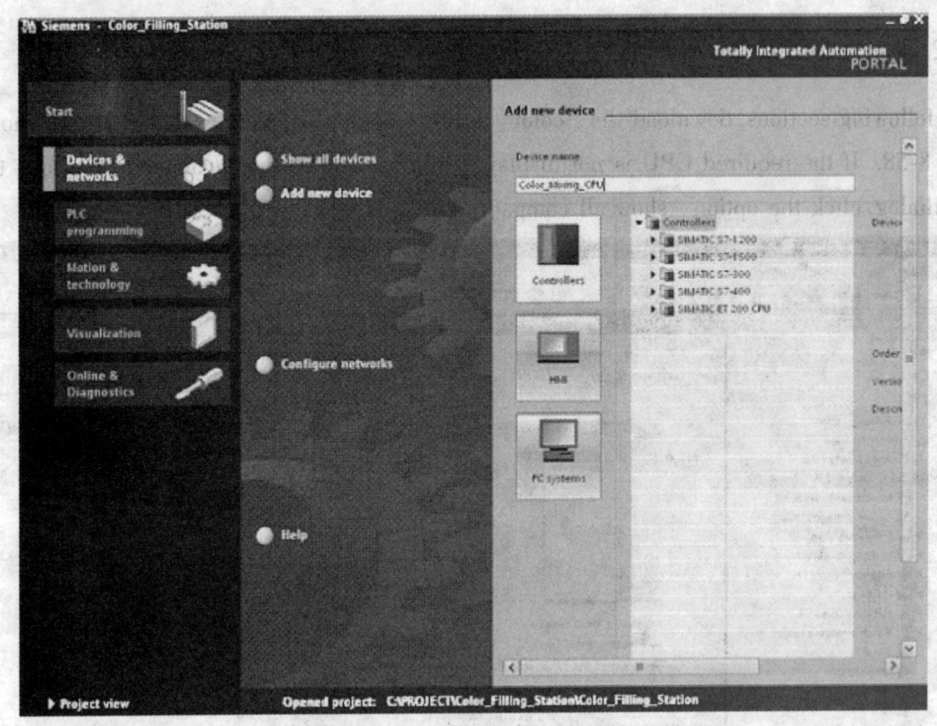

Figure 8-56　Equipment and network

(6) Double-click to create the CPU. An unspecified CPU was created in the project file. Here you can create user program content for the CPU, as shown in Figure 8-57.

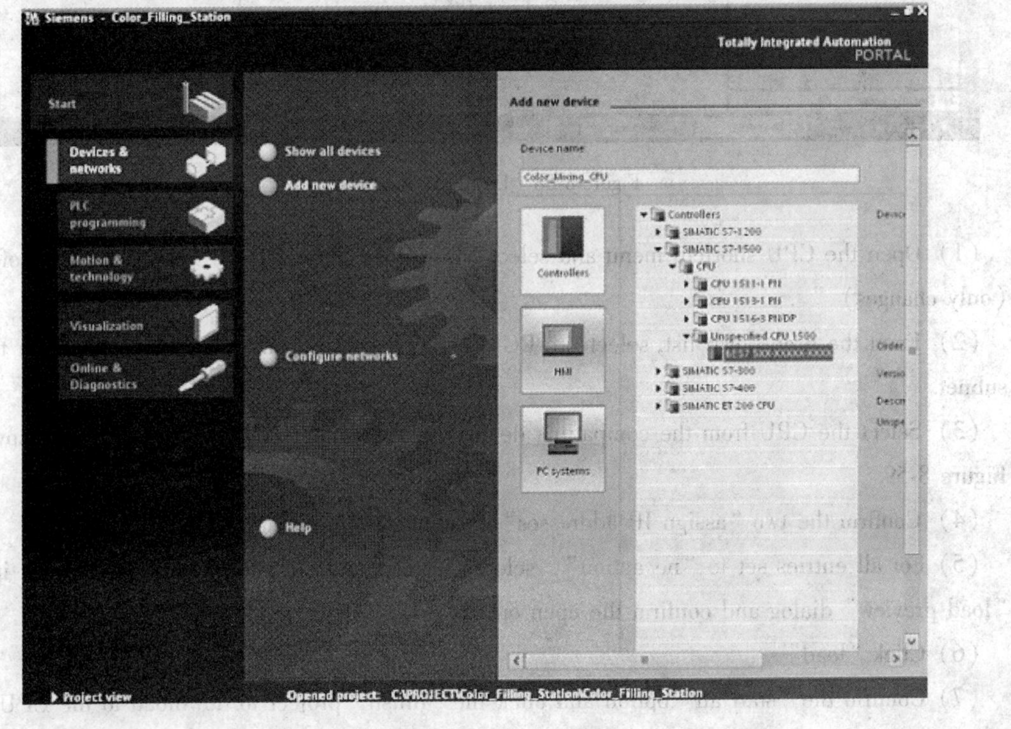

Figure 8-57　Create CPU

8. 9. 2. 3 Load the project into the CPU

In the following sections, download the "color _ filling _ station" project to the CPU, as shown in Figure 8-58. If the required CPU is not displayed after setting in the "extended download to device" dialog, click the option "show all compatible devices".

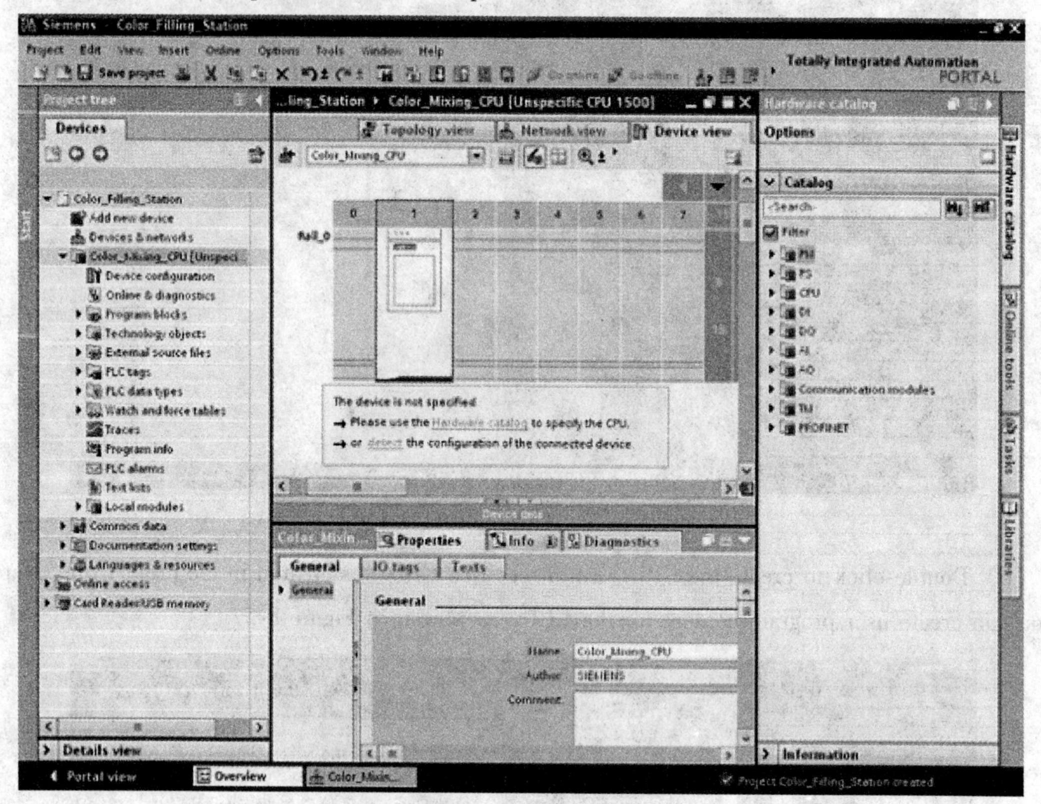

Figure 8-58 Download the project

(1) Open the CPU shortcut menu and select "Download to device" > "Hardware and software (only changes)".

(2) From the drop-down list, select the PG/PC interface type, interface and connection to the subnet.

(3) Select the CPU from the compatible devices in the subnet and click "Load", as shown in Figure 8-59.

(4) Confirm the two "assign IP addresses" by clicking "Yes" and "OK" Dialog.

(5) For all entries set to "no action", select the alternate entry in the drop-down list in the "load preview" dialog and confirm the open option.

(6) Click "load".

(7) Confirm the "start all" option and click the "finish" project to download to the CPU. The software interface is shown in Figure 8-60.

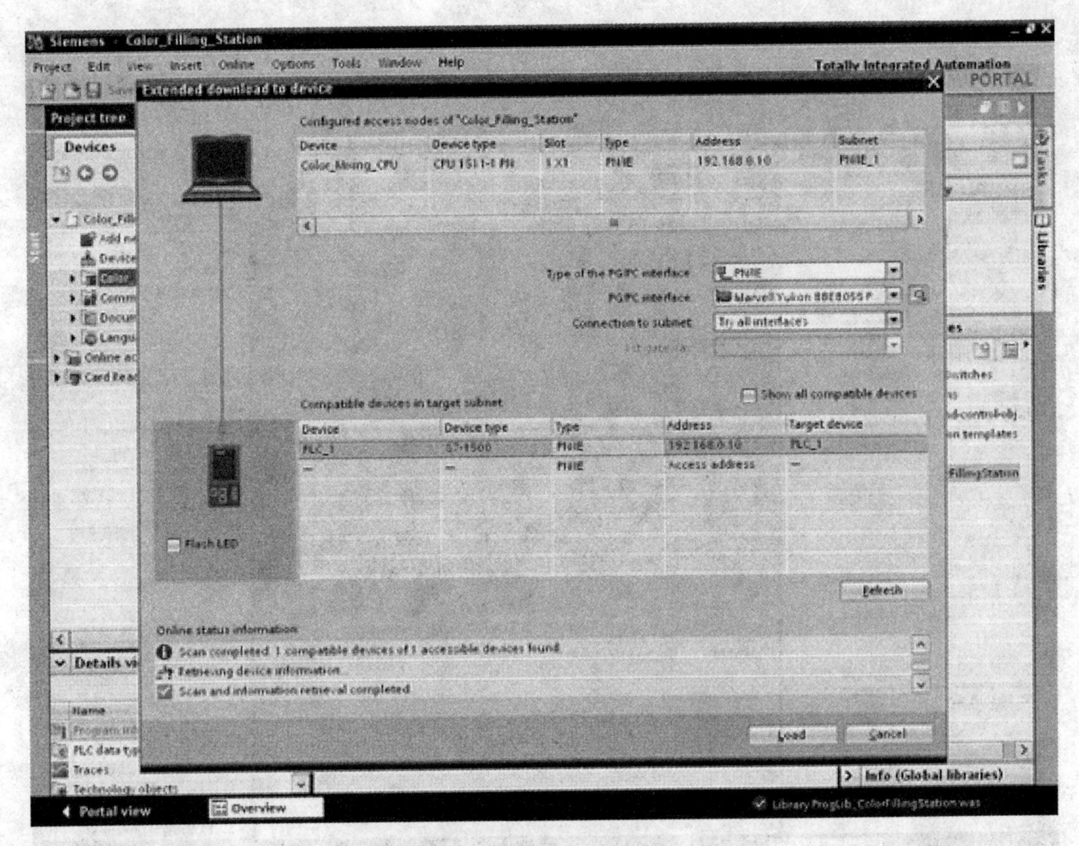

Figure 8-59 Specify IP address

8.9.3 Assign an IO device to an IO controller

The PROFINET IO system consists of a PROFINET IO controller and its assigned PROFINET IO devices. Once these devices are in place in the network or topology view, STEP 7 assigns default values to them . At this stage, we only consider the assignment of IO devices to the IO controller. Assume that you are already in the network view of STEP 7, a CPU (eg CPU 1516-3 PN/DP) has been placed and an IO device (eg IM 155-6 PN ST) has been placed. To assign an IO device to an IO controller, follow these steps:

(1) Move the mouse pointer over the interface of the IO device.

(2) Hold down the left mouse button.

(3) Move the mouse pointer. The pointer will now use the networking symbol to indicate the "networked" mode. At the same time, you can see that the lock symbol appears on the pointer. The lock symbol disappears when the pointer is moved to a valid target position.

(4) Now move the pointer over the interface of the IO controller. When this operation is performed, the left mouse button can be kept pressed all the time, or the left mouse button can be released.

(5) Now release the left mouse button or hold it down again (depending on the previous action).

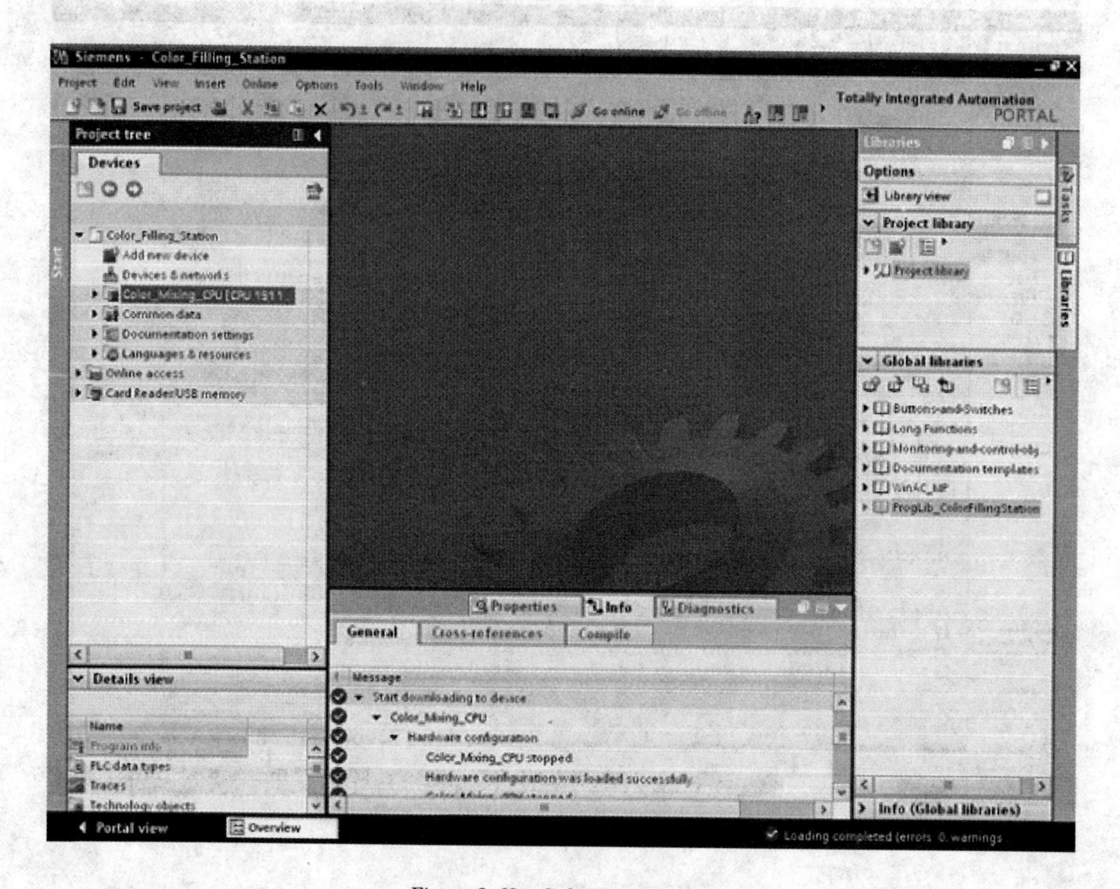

Figure 8-60 Software interface

The network view of assigning on IO controller is shown in Figure 8-61.

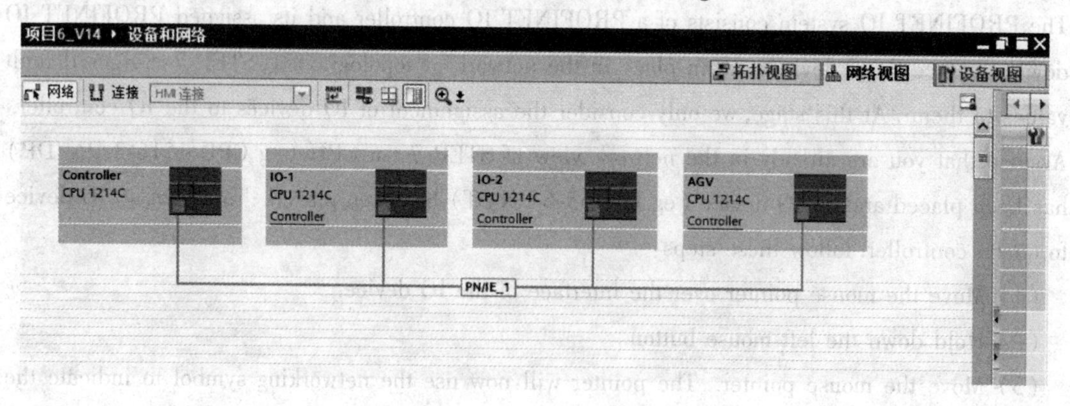

Figure 8-61 Assign an IO device to an IO controller in the network view

Now check the distribution:

An overview of the communication relationships can be found in the "IO communication" tab in the table area of the network view. The table is a context-sensitive table that can be selected in the graphics area:

(1) The interface selection shows the I/O communication of the corresponding interface.

(2) The interface selection shows all I/O communication of the CPU (including PROFIBUS).

(3) The station interface selection (shown above) shows the I/O communication for the entire station.

8. 9. 4 Intelligent IO device

The "smart device" (smart IO device, as shown in Figure 8-62) function of the CPU simplifies data exchange and CPU operation with the IO controller (eg. as an intelligent preprocessing unit for subprocesses). The smart device can be linked as an IO device to the "upper" IO controller.

The preprocessing process is done by a user program in the smart device. The processor values acquired in the centralized or distributed (PROFINET IO or PROFIBUS DP) I/O are preprocessed by the user program and supplied to the IO controller.

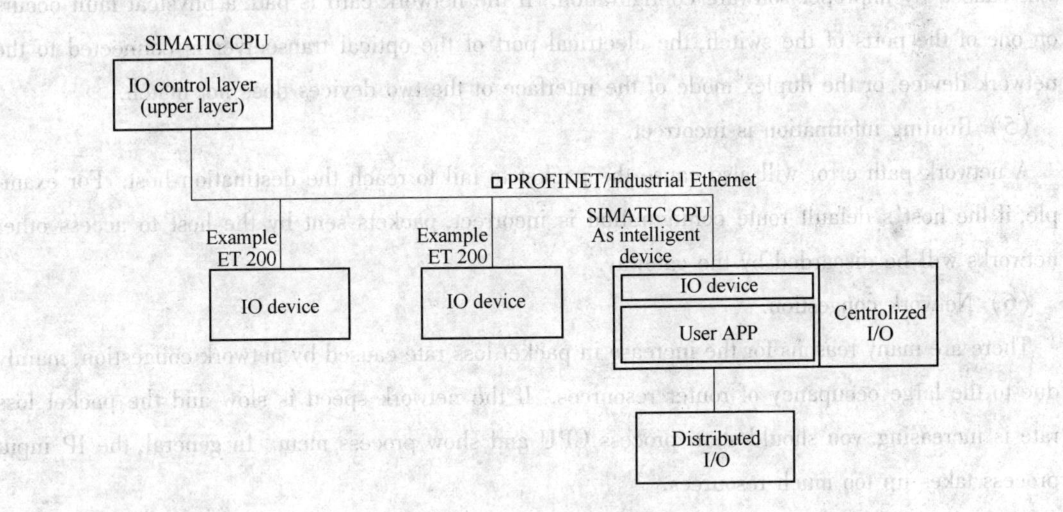

Figure 8-62 Smart device

8. 10 Network packet loss rate and its testing method

Packet loss rate refers to the ratio of the number of lost packets in the test to the data group sent.

Packet loss rate = (input message−output message) / input message × 100%

The packet loss rate is related to the packet length and the packet transmission frequency. It is also affected by many factors such as signal attenuation, network quality, and so on. Generally, when the Gigabit NIC has a traffic greater than 200 Mbps, the packet loss rate is less than 5/10000; when the traffic is greater than 60 Mbps, the packet loss rate is less than 1 in 10000.

8.10.1 Reasons for network packet loss

(1) Network problem itself.

(2) Physical line failure.

There are many packet loss phenomena caused by physical lines, such as fiber connection problems, jumper misalignment with device interfaces, twisted pair and RJ-45 connectors. In addition, the communication line is subject to random data or data error caused by burst noise, and the interference of the radio frequency signal and the attenuation of the signal may cause loss of the data packet. The quality of the line can be checked with the aid of a network tester.

(3) Virus attacks.

(4) Equipment failure.

A device fault mainly refers to a fault in the hardware of the device, and does not include packet loss caused by improper software configuration. If the network card is bad, a physical fault occurs on one of the ports of the switch, the electrical port of the optical transceiver is connected to the network device, or the duplex mode of the interface of the two devices does not match.

(5) Routing information is incorrect.

A network path error will also cause the packet to fail to reach the destination host. For example, if the host's default route configuration is incorrect, packets sent by the host to access other networks will be discarded by the gateway.

(6) Network congestion.

There are many reasons for the increase in packet loss rate caused by network congestion, mainly due to the large occupancy of router resources. If the network speed is slow and the packet loss rate is increasing, you should show process CPU and show process mem. In general, the IP input process takes up too much resources.

There are still many situations in the application that cause network congestion, such as a large amount of UDP traffic, which can be solved by solving the spoof attack. Another occasion of network congestion is when a large number of multicast streams and broadcast packets traverse the router; similarly the network may be conqested if the router is configured with IP NAT and there are many DNS packets traversing the router. After the above situation causes the network to be congested, the two communicating parties take traffic control and discard the packets that cannot be transmitted.

(7) A port in the network forms a bottleneck.

(8) System resources are insufficient.

8.10.2 Ping scan tool Fping

Fping is one of the most tested ping scan tools in the UNIX environment. Most of the early ping scanning tools need to wait for the previous detected host to return a certain response message before continuing to detect the existence of the next host, but Fping can send a large number of ping requests in parallel in a round-robin manner. Therefore, using the Fping tool to scan multiple IP addresses is much faster than ping.

Fping is a ping-like program that responds to requests via the ICMP protocol to detect the presence of a host. The difference between Fping and ping is that it can specify the number of hosts to ping on the command line, or specify the host list file to be pinged. Different from the ping to wait for a host connection timeout or send back feedback information, Fping sends a data packet to the next host after sending a data packet to the host, so that multiple hosts can ping at the same time. If a host pings, the host will be marked and removed from the waiting list. If it is not pinged, the host cannot be reached, and the host remains in the waiting list, waiting for subsequent operations.

Advantages of Fping:

(1) Can ping multiple hosts at once;

(2) Can ping from the host list file;

(3) The results are clear, easy to handle scripts;

(4) Fast.

How to use Fping:

(1) Put Fping .exe in the root directory of the C drive.

(2) Cmd into the dos system implementation (not double-click to run).

1) Use the "CD C: /" command to enter the root directory of the C drive.

2) Enter Fping, you can see the various command parameters provided by the Fping program, as shown in Figure 8-63.

```
C:\windows\system32>CD C:/

C:\>FPing

Fast pinger version 3.00
(c) Wouter Dhondt (http://www.kwakkelflap.com)

Usage:
FPing <host(-list)> [-t time] [-w timeout] [-c] [-n count] [-s data_size]
        [-S size1/size2] [-R min/max] [-d ping_data] [-h TTL] [-v TOS]
        [-r routes] [-f] [-j] [-g host1/host2] [-H filename] [-a] [-A]
        [-p(x)] [-i] [-b(-)] [-T] [-D] [-1] [-o] [-L filename]

Options:
        -t : time between 2 pings in ms up to 1000000
        -w : timeout in ms to wait for each reply
        -c : continuous ping (higher priority than -n)
             to see statistics and continue - type Control-Break;
             to stop - type Control-C.
        -n : number of pings to send to each host
        -s : amount of data in bytes up to 65500
        -S : size sweep: ping with size1, size1 + 1, ..., size2 bytes
        -R : random length between min and max (disabled when using -S)
        -d : ping with specified data
        -h : number of hops (TTL: 1 to 128) + print hops
        -v : Type Of Service (0 to 255) (IPv4-only)
        -r : record route (1 to 9 routes) (IPv4-only)
        -f : set Don't Fragment flag in packet (IPv4-only)
        -j : print jitter with each reply (only when pinging one host)
        -g : ping IP range from host1 to host2 (IPv4-only)
        -H : get hosts from filename (comma delimited, filename with full path)
        -a : resolve addresses to hostnames
        -A : print addresses with each reply
        -p : use a thread pool to ping multiple hosts (enables ICMP dll)
             x is optional and allows you to choose the number of threads
             e.g. -p uses a thread for every host
                  -p5 uses a pool of 5 threads/core
        -i : use ICMP dll instead of raw socket (disables -r)
```

Figure 8-63 View Fping program command parameters

(3) Example:

If you want to monitor the packet loss online, type the following command:

Fping 192. 168. 0. 2 (target host IP address) -s 1428 -w 100 -t 10 -n 10.

——s: Specify the packet size, up to 65500 Byte.

——w: Rreturn timeout, ms.

——t: Ping interval, ms.

——n: Total number of ping packets.

The sample results are shown in Figure 8-64, including packet loss (Figure 8-65) and packet loss statistics (Figure 8-66).

```
C:\>Fping 192.168.0.2 -s 1428 -w 100  -t 10 -n 10

Fast pinger version 3.00
(c) Wouter Dhondt (http://www.kwakkelflap.com)

socket() - An attempt was made to access a socket in a way forbidden by its access permissions.
switching to ICMP dll
Pinging 192.168.0.2 with 1428 bytes of data every 10 ms:

Reply[1] from 192.168.0.2: bytes=1428 time=7.1 ms TTL=128
Reply[2] from 192.168.0.2: bytes=1428 time=3.0 ms TTL=128
Reply[3] from 192.168.0.2: bytes=1428 time=3.1 ms TTL=128
Reply[4] from 192.168.0.2: bytes=1428 time=3.6 ms TTL=128
Reply[5] from 192.168.0.2: bytes=1428 time=9.1 ms TTL=128
Reply[6] from 192.168.0.2: bytes=1428 time=3.7 ms TTL=128
Reply[7] from 192.168.0.2: bytes=1428 time=3.8 ms TTL=128
Reply[8] from 192.168.0.2: bytes=1428 time=3.5 ms TTL=128
Reply[9] from 192.168.0.2: bytes=1428 time=3.5 ms TTL=128
Reply[10] from 192.168.0.2: bytes=1428 time=3.5 ms TTL=128

Ping statistics for 192.168.0.2:
        Packets: Sent = 10, Received = 10, Lost = 0 (0% loss)
Approximate round trip times in milli-seconds:
        Minimum = 3.0 ms, Maximum = 9.1 ms, Average = 4.4 ms
```

Figure 8-64 Sample test results

```
Reply[641] from 192.168.0.2: bytes=1428 time=1.2 ms TTL=128
Reply[642] from 192.168.0.2: bytes=1428 time=1.2 ms TTL=128
Reply[643] from 192.168.0.2: bytes=1428 time=1.3 ms TTL=128
Reply[644] from 192.168.0.2: bytes=1428 time=1.6 ms TTL=128
Reply[645] from 192.168.0.2: bytes=1428 time=1.3 ms TTL=128
Reply[646] from 192.168.0.2: bytes=1428 time=1.3 ms TTL=128
Reply[647] from 192.168.0.2: bytes=1428 time=1.5 ms TTL=128
Reply[648] from 192.168.0.2: bytes=1428 time=1.3 ms TTL=128
192.168.0.2: request timed out
Reply[650] from 192.168.0.2: bytes=1428 time=1.5 ms TTL=128
Reply[651] from 192.168.0.2: bytes=1428 time=1.4 ms TTL=128
Reply[652] from 192.168.0.2: bytes=1428 time=1.6 ms TTL=128
Reply[653] from 192.168.0.2: bytes=1428 time=1.3 ms TTL=128
Reply[654] from 192.168.0.2: bytes=1428 time=1.4 ms TTL=128
Reply[655] from 192.168.0.2: bytes=1428 time=1.3 ms TTL=128
Reply[656] from 192.168.0.2: bytes=1428 time=1.8 ms TTL=128
Reply[657] from 192.168.0.2: bytes=1428 time=1.4 ms TTL=128
Reply[658] from 192.168.0.2: bytes=1428 time=1.2 ms TTL=128
Reply[659] from 192.168.0.2: bytes=1428 time=1.1 ms TTL=128
Reply[660] from 192.168.0.2: bytes=1428 time=1.3 ms TTL=128
Reply[661] from 192.168.0.2: bytes=1428 time=1.6 ms TTL=128
```

Figure 8-65 Packet loss

```
Ping statistics for 192.168.0.2:
        Packets: Sent = 2000, Received = 1996, Lost = 4 (0% loss)
Approximate round trip times in milli-seconds:
        Minimum = 1.0 ms, Maximum = 4.3 ms, Average = 1.4 ms
```

Figure 8-66 Packet loss statistics

Problems

8-1 What are the advantages and disadvantages of port-based VLANs?

8-2 If the trunk mode is not set, can the host computer access the PLC in VLAN 10?

8-3 What are the characteristics of local routing?

8-4 If you do not configure the gateway of the PC, what can you observe when you execute the ping command? Why?

8-5 To further verify and view the role of the set static route, enter the routing command "tracert 100. 1. 0. 41" in the "command prompt" environment, please take a screenshot and analyze the results.

8-6 The HRP redundancy protocol is used in the test. What are the its advantages?

8-7 When the cable between the two XB208s is unplugged during the test, what is the change of the indicator light of the XM408?

References

[1] Xianhui Yang. Fieldbus Technology and its Application [M]. Tsinghua University Press, 1999.

[2] Xianhui Yang. Fieldbus Technology and its Application (the Second Edition) [M]. Tsinghua University Press, 2008.

[3] Xiren Xie. Computer Networks (the Seventh Edition) [M]. Publishing House of Electronics Industry, 2017.

[4] Fengying Wang, Zhen Cheng, Jinling Zhao. Computer Networks [M]. Tsinghua University Press, 2010.

[5] Taijun Li, Yuanguai Lin, Jin Zhang, et al. Computer Networks [M]. Tsinghua University Press, 2009.

[6] Andrew S. Tanenbaum. Computer Networks (the fourth Edition) [M]. Tsinghua University Press, 2004.

[7] Andrew S. Tanenbaum. Computer Networks (the fifth Edition) [M]. China Machine Press, 2016.

[8] Xianglin Wang. Computer Networks—Principle, Technology and Application (the Second Edition) [M]. China Machine Press, 2016.

[9] Deji Wang. SIEMENS Industrial Network Communication Technology [M]. China Machine Press, 2012.

[10] https: //www. siemens. com/cn/en/home. html.